The

DESCENT *of*
ARTIFICIAL INTELLIGENCE

The
DESCENT *of*
ARTIFICIAL INTELLIGENCE

A Deep History *of an* Idea 400 Years *in the* Making

Kevin Padraic Donnelly

University *of* Pittsburgh Press

Published by the University of Pittsburgh Press, Pittsburgh, Pa., 15260
Copyright © 2024, University of Pittsburgh Press
All rights reserved
Manufactured in the United States of America
Printed on acid-free paper
10 9 8 7 6 5 4 3 2 1

Cataloging-in-Publication data is available from the Library of Congress

ISBN 13: 978-0-8229-4796-7
ISBN 10: 0-8229-4796-X

COVER ART: Marcel Duchamp, *Nude Descending a Staircase (No. 2)*, 1912. Oil on canvas, 57 7/8 × 35 1/8 inches © Association Marcel Duchamp / ADAGP, Paris / Artists Rights Society (ARS), New York 2024.

COVER DESIGN: Alex Wolfe

For Jo and Lou

CONTENTS

PART III. SOCIAL SCIENCE IN AMERICA

ACKNOWLEDGMENTS

TO WRITE A BOOK ABOUT ARTIFICIAL INTELLIGENCE IN THE THIRD DECADE of the twenty-first century is to be mindful of one of the many great lines attributed to Yogi Berra: "You've got to be very careful if you don't know where you are going, because you might not get there." As discussed in the book, the modern features of artificial intelligence are marked—and, some might say, plagued—by an uncertainty over what, precisely, the machines are supposed to be doing. This uncertainty is not unique to AI, however, as it can be attributed to the overall problem of most of us people not knowing what, precisely, *we* are supposed to be doing. As I try to show, one particular way of trying to figure out why human beings think and act the way they do—the mathematical and quantitative social sciences—has left a considerable influence on the idea of artificial intelligence. I cannot be sure the book got there, but I do at least think I knew where it was going.

To write a book about artificial intelligence in the third decade of the twenty-first century is also to enter into a realm of polemics and panegyrics, especially now that AI is so tied to modern American technology companies, companies that have their own detractors and enthusiasts for reasons only tangentially related to thinking machines. In late 2022, as the final draft of this manuscript was nearing completion, the fanfare surrounding the release of new chatbots based on "generative" artificial intelligence only intensified such debates. To be clear, the book that follows is neither polemic nor panegyric, and it likely has little to say about the big philosophical questions AI can sometimes inspire, or the many legal, moral, and ethical questions that AI promises to provoke. Rather than write another book about either terrifying or incredible machines, I have chosen to write a history of how we have defined ourselves.

Last, to try to write a book about artificial intelligence in the third decade of the twenty-first century is to incur so many intellectual debts, in so many diverse fields, that the list of "thanks" below can account for only a fraction of the assistance I received. Though much of the strength of the book is due to the work of those below, the errors or mistakes are mine alone.

To start, I would like to thank the four anonymous reviewers at the University of Pittsburgh Press, two who responded to a very rough proposal and two who helped shape the final version of the book. As editor, Abby Collier did excellent work in difficult times to find readers and offer helpful suggestions to make the book much better than it otherwise would have been. Rebecca Makas yet again took on the project of reading the entire book near deadline, and her help can be found on almost every page.

The inspiration for the book occurred during a conference on Mary Poovey's *A History of the Modern Fact*, held at Penn State's Center for Humanities and Information in the fall of 2017. I would like to thank Eric Hayot, the center's director, for inviting me to talk about quantification and statistics; Dr. Poovey, for providing a model of how to write a long history of a complex subject that people might not realize has a history; and the other conference participants for the excellent talks and conversations.

Thanks are also due to the many fine commenters, fellow panelists, and respondents at a series of conferences where draft chapters of the book were first presented. This includes the European Symposium Series in Societal Challenges to Computational Social Science, held in Cologne in 2018; the Society for Utopian Studies Conference, held in East Lansing, Michigan, in 2019; the Social Science History Association, held in Chicago in 2019; and the *Scientiae* 2021 conference, scheduled for Amsterdam but (alas) held virtually.

While all of the people who organized and ran these conferences deserve thanks, special acknowledgments are due to the organizers of two conferences that helped inform the direction of the book, both held in 2019. First, thank you to Adrianna Link for organizing "Networks: The Creation and Circulation of Knowledge from Franklin to Facebook," held at the American Philosophical Society in Philadelphia. Thanks as well to the organizers of the Sixth Annual Conference of the History of Recent Social Sciences, held in Berlin: Jamie-Cohen Cole, Phillippe Fontaine, Jeff Pooley, and Susanne Schmidt. For a historian who imagined Durkheim and Veblen as "recent" social science, both the organizers and the group assembled were invaluable in helping to think about the final few chapters of the book.

For the ability to look into the archives of modern social scientists, I would like to thank the Hanna Holborn Gray Special Collections Research Center at the University of Chicago Library, which provided the Robert L. Platzman Memorial Fellowship to write and research for a month in Chicago. Also, thank you to Nanci Young and the Sophia Smith Collection at Smith College for a Travel to Collections Grant to spend several weeks in Northampton,

Massachusetts. Closer to home, a special thanks to the librarians, research assistants, and archivists at the Fisher Fine Arts Library and Van Pelt-Dietrich Library Center at the University of Pennsylvania, as well as the Charles Library at Temple University. Their help during the Covid pandemic was invaluable, as was the ability to use interlibrary loan, JSTOR, the Internet Archive, and (yes) Google Books.

I also have to thank Alvernia University. Just weeks before I sent a draft manuscript to the press, a *New York Times* article revealed that many companies, including seven of the ten largest employers in the world, track the daily work habits of their employees through monitoring software in their home and office computers. I feel fortunate that not only did the university not put a camera in my home, but they actually provided substantial time and resources to complete a project of this length and scope. Most helpful was a fellowship provided by Ray and Carol Neag that has supported a number of scholarly works in the Department of the Humanities, as well as a sabbatical supported by the university. As president of the university during this time, Tom Flynn deserves a special thanks for encouraging an atmosphere of scholarship and intellectual curiosity at every level, including allowing faculty to visit his house, drink beer, and discuss research three Fridays each semester.

At Alvernia, Jim Foster deserves a thank you for taking a chance on an ABD student many years ago. Additionally, many provosts, deans, and chairs have made accommodations to support this work, including Glynis Fitzgerald, Kevin Godfrey, Corey Harris, Elizabeth Matteo, and Shirley Williams. In Humanities Hall, thank you to Tom Bierowski, Tim Blessing, Scott Davidson, Josh Hayes, Bongrae Seok, Janae Sholtz, and Victoria Williams for their conversations. Carrie Fitzpatrick, Spence Stober, and Donna Yarri also provided important mentorship, as did the late Beth DeMeo, whose quick wit, smile, and always apposite Shakespeare quotes were the centerpiece of the Peanut Bar gang. Off campus, thank you to colleagues Darren Koch, Peter Rampson, Erin Way, and Mary Ellen Wells for our joyous commiserations. Further down Route 222, thank you to the friends at Franklin and Marshall College (and Brendee's), including Tim and Christine Wyman McCarty for always talking about everything interesting, and Nina Kollars for sending me books and thinking about technology in novel ways.

It seems almost too small a space here to acknowledge my parents—Mom, Dad, Joan, and Nancy—but if I attempted to put down in words what you have given me it would leave time for nothing else. Johanna and Lucinda also receive a special thanks, given that they were the only people allowed to regularly interrupt this work, as their smiles, smirks, and squawks were all welcome and encouraged at any hour. And R!, thank you for everything *not* related to your intellectual curiosity, keen eye for repetition, and willingness to follow every turn that attends the process of writing a book. These things I have already thanked you for many times. Thanks instead for the new things I learn about you every day. Hard as it seems to believe, it just keeps getting better.

ABBREVIATIONS

CMP	*Correspondance mathématique et physique*
DML	Jean le Rond d'Alembert, *Mélanges de littérature, d'histoire et de philosophie*
DD	Denis Diderot, *Œuvres philosophiques*
EEE	Earle Edward Eubank Papers
HM	La Mettrie, *L'homme machine*
JTP	John W. Tukey Papers
LLB	Luther Lee Bernard Papers
MAS	*Mind amongst the Spindles: A Miscellany, Wholly Composed by the Factory Girls*
OC	*Œuvres de Condorcet*
OCB	*Œuvres complètes de M. de Bonald*
OCH	Claude Adrien Helvétius, *Œuvres complètes d'Helvétius*
OCM	Joseph Marie de Maistre, *Œuvres complètes*
OCR	Jean-Jacques Rousseau, *Œuvres complètes*
OPD	René Descartes, *Œuvres philosophique*
OPH	Paul-Henri Thiry d'Holbach, *Œuvres philosophique*
WN	Adam Smith, *An Inquiry into the Nature and Causes of the Wealth of Nations*

The

DESCENT *of*
ARTIFICIAL INTELLIGENCE

INTRODUCTION

Turing's Paradox *and the* Failure *of the* Sciences of Man

> The works and customs of mankind do not seem to be very
> suitable material to which to apply scientific induction.
> —Alan Turing, "Computing Machinery and Intelligence" (1950)

> We may accomplish the AI dream by stripping humans
> of so much singular identity that people are reduced to
> mere agents.
> —Illah Reza Nourbakhsh, *Robot Futures* (2015)

THE ORIGINS OF ARTIFICIAL INTELLIGENCE (AI) CAN BE FOUND IN THE history of simple people, not smart machines. This claim requires some qualification. To say that AI is the result of simple—or, more precisely, simplified—people is not to deny the incredible increase in complexity in modern thinking machines, and it certainly is not to deny the intelligence of those responsible for the algorithms, computer programs, and data collection tools behind them. Machines can certainly do more things that appear to be thought-like than they previously have been able to do, and the creators of these machines were and are unquestionably intelligent people, at least by the standards set for intelligence in most of the world in the twenty-first century. Rather, to say that simple people are responsible for AI is to argue for the historical antecedents of roboticist Illah Nourbakhsh's provocative claim that the "AI dream" may occur through "stripping humans" of their "singular identity."[1] As will be seen, in both the abstract and the concrete sense, simple people have been central to the idea of artificial intelligence for centuries.

More specifically, this book will argue that the idea of artificial intelligence emerged out of reductive and simplistic methodologies pursued in the collective fields of knowledge generally known as the social sciences, or what were for centuries called the "sciences of man."[2] Going back four centuries to the idea of "mechanical philosophy"—which attempted to describe nature through machine analogies—the book covers a number of approaches that sought to study human thought and behavior using the assumptions, tools, and techniques of the natural sciences. This includes long-forgotten ideas like the "geometric spirit," "social physics," "the hedonistic calculus," and the "iron laws" of classical political economy, as well as more recent approaches like eugenics, statistical sociology, positivist economics, and behaviorism. Each, in their own way, attempted to reduce the wide diversity and seeming randomness of human thought and behavior into a model that could predict, often quantitatively, the actions of human subjects. In addition to presenting the abstract theories that simplified human behavior, the book also explains how the failure of such ideas to accurately predict human actions led to interventionist actions in the lives of real people. As the messy business of history frequently interfered with the attempts to simplify human behavior to lawlike consistency, social thinkers often found that such "laws" could only obtain through education, social reform, and political intervention. Finally, the book also covers the arguments of a number of *critics* of such approaches, including social thinkers, scientists, and mathematicians who worried that theories of human action driven exclusively by scientific methodologies had the potential to create, rather than reveal, a new type of human subject. In their view, reductive theories produced a very real kind of artificial intelligence, one that often seemed to originate in the practices of scientific work itself.

Remarkably, the successes and failures of the sciences of man and social sciences to understand human thought and action over nearly four hundred years have been almost completely absent in discussions of the history of artificial intelligence. In part, this oversight can be explained by Alan Turing's comment above, that science was unsuited to studying the "works and customs of mankind."[3] In what may seem paradoxical, Turing's skeptical stance on using "induction"—i.e., the process of deriving laws from observation and experiment—to study human action occurred in the same 1950 paper in which he first proposed his famous "Turing Test," a thought experiment that predicted smart machines would one day be able to fool human examiners by "imitating" a person. As a prophet of artificial intelligence, founder of computer science, and champion of the scientific method, it may seem odd that Turing doubted that science could understand people. How, after all, could scientists fashion a machine that imitated something that could not be understood through scientific methodology? If people's actions and behavior could not be understood through "scientific induction," then how could science build machines that imitated this action? If *social scientists* could not make reliable predictions about people based on previous actions, how could *computer scientists* build something that could imitate this unpredictability? The answer, explored in this book, was to simplify what it means to be a person.

Although human thought, behavior, and communication have been the goal for AI since Turing, little has been accomplished in the subsequent seventy years that might suggest computer scientists have looked to the history of the social sciences as a guide to understanding their ultimate goal. In fact, to look at the state of artificial intelligence today is to see that questions about the nature of human thought, behavior, and communication have remained unasked, unanswered, or ignored, in large part because AI history has been so narrowly focused on machines and their makers as to bracket out discussions of what kind of *person* AI is supposed to replicate. For example, in interviews with fifty leading AI experts published in 2018, the futurist and writer Martin Ford asked each subject to provide a year when "human-level AI might be achieved."[4] While the predictions ranged from a few decades to a century, neither Ford nor any of those interviewed defined "human level" or considered how such a standard could be determined scientifically. Though no scientific account was offered of what it actually means for a person to act and think, all fifty engineers and computer scientists were confident that AI would one day reach this undefined goal.

In perhaps the most vivid illustration of how often the "human level" is usually ignored in discussions of AI, in both Turing's legendary "imitation game" and its many subsequent iterations, there are few accounts of what kind of *people* might participate alongside the machine. For example, in 2002 the futurist Ray Kurzweil and businessman Mitch Kapor made a bet about the possibility that a machine might one day pass the Turing Test. Although the "rules" of the test ran close to two thousand words, the only mention of the human players was the stipulation that three human "foils" were needed, without any mention of, say, the age, race, gender, religious background, or personality of the people selected to participate.[5] Even putting a barrier like culture aside, it could be argued that machines would have varying levels of success at the Turing Test with eccentrics, artists, mystics, or children, and it is perhaps unsurprising that the first computer to "pass" the test did so through mimicking the random interruptions, non sequiturs, and "b*llshit" of a teenager.[6] In fact, one could imagine in all seriousness that the easiest way for a machine to pass the Turing Test would be to have either a very boring human competitor or a dumb interlocutor. Similarly, to give just one notable example of how AI success might depend upon the simplification of human activity—the Writers Guild of America strike from the summer of 2023—it has been argued that years of simplistic, "derivative," and "formulaic" script writing has made the machines' task that much easier.[7] Machine success, therefore, is not necessarily tied to a fixed goal but might be greatly helped by a lowering, or simplification, of the "human level."

The idea that new technologies might inhibit human agency and intellect is as old as warnings from Socrates and Plato that written language harmed human memory, and the perceived threats of various machines have been debated many times over.[8] Rather than adopt a simplistic technological determinism that blames machines themselves for a simplification of human thought and behavior, this book instead argues that the "human level" that AI researchers are trying to reach would be unimaginable without the long history of scientific attempts to understand human thought, action, and behavior. People have of course come up with many other methods for trying to understand one another throughout history outside of the social sciences—from divine explanation to art and literature to intuition—but these approaches are particularly unsuited as forerunners of today's AI, as they tend to make people's "singular identity" paramount. An AI system based on individual human beings as divine creations possessing willing souls, complex literary figures rent by deeply personal family histories, or completely absurd and random beings would not make it very far. Conversely, many things that AI does do well would seem completely absurd or pointless to a vast number of people in different times and places who have not relied on scientific explanations for human behavior. To a medieval European peasant, Māori oral historian, Tang dynasty chronicler, or Roman centurion, for example, there would be very little utility or intelligence in today's machines, or anything "human level" in AI success in playing games, scheduling meetings, summarizing legal texts, or producing middling undergraduate papers. What this book attempts to show, therefore, is that the multifaceted phenomenon we know today as AI has been made possible because of a significant effort over the past four hundred years to reduce the "works and customs of mankind" to a point where scientific methodologies could be used to understand them. For the harshest critics of reductive social science traced in this story, it was theories about simple people, rather than smart machines, which proved the stuff of nightmares.

The Quest

In shifting the focus of the story of artificial intelligence from technical developments in machines to scientific theories of people, this book significantly expands the chronological scope of AI history.[9] Most standard histories of the field begin approximately seventy to ninety years ago and are centered around a few crucial figures in computer science and mathematics.[10] In this narrative, the "quest" for AI begins in 1936 with the twenty-four-year-old Turing's remarkable "Entscheidungsproblem" paper, which required Turing to define the idea of a "computable number."[11] In the process of doing so, Turing also imagined a machine that could compute such numbers, which led to two groundbreaking papers that suggested such a hypothetical machine could therefore replicate the process of human computation and even thought.[12] Inspired by Turing and the growth of electronic computers, in 1955 four researchers requested $13,500 for a summer conference at Dartmouth to study "Artificial Intelligence," with the belief that any "aspect of learning or any other feature of intelligence can in principle be so precisely described that a machine can be made to simulate it." As they ambitiously declared, the Dartmouth group believed that a "significant advance" could be accomplished if a "carefully selected group of scientists work on it together for a summer."[13]

For most of early AI history, researchers were split between two approaches that Turing had suggested for building an intelligent machine. The first, supported by two of the Dartmouth Conference organizers, Marvin Minsky and John McCarthy, envisioned that machines could be programmed with sufficient rules to both process and produce strings of "symbols" that could mimic human cognition and language. While far from a complete failure—it was this kind of machine that IBM eventually used to beat Garry Kasparov at chess—"symbolic" AI largely foundered following a number of spectacular failures with expensive and clunky robots in the 1970s and 1980s. In 1973, an influential report from the United Kingdom that dismissed the idea as close to a joke almost buried symbolic AI completely.[14]

However, in his canonical papers Turing had also suggested that machine intelligence could be created by building a more basic machine that "learned" like a child. The idea was then given formal shape in a groundbreaking paper coauthored by the psychologist Warren S. McCulloch and the neuroscientist Walter Pitts in 1943, which suggested that the nervous system, and therefore the mind, might function in a similar binary way as formal mathematical logic.[15] The idea of building such a "neural network" had been included in the original Dartmouth proposal, and Frank Rosenblatt in 1958 had offered a tantalizing glimpse of a learning machine in his "Perceptron," but the idea of building brains lost significant interest and funding in the wake of a book coauthored by Minsky in 1969 that questioned neural networks.[16] With both approaches stymied in the 1970s and 1980s—the "AI winters"—neural networks, or "connectionist AI," returned triumphantly in the 2000s, most spectacularly at the 2012 ImageNet competition, where a "neural net" built by a team from the University of Toronto trounced its symbolic AI competitors in vision recognition.[17] It is these connectionist nets, now armed with extraordinary data harvested from the internet by technology companies, that promise to transform the world for good or evil.[18]

Though artificial intelligence has produced a robust literature—from technical guides to personal memoirs to deeply philosophical work to polemics against technology—most accounts largely rely on the narrative above and describe the history of AI as a somewhat fitful but ultimately progressive story of AI approaching the human level.[19] For example, in just one of the many "quest" narratives, the "grand goal" of "human-level AI" is presented as a project that should "develop artifacts that can do most of the things that humans can do . . . specifically those things that are thought to require intelligence."[20] Versions of this claim are repeated in many books on AI, and the idea of what has historically constituted those "things that require human intelligence" is either assumed or ignored. Yet, as historians of science have shown, the "human level" is not a static category. In the late 1800s, for example, "human intelligence" would have included the ability to construct a personal calendar, do basic rote mathematics, and measure the progress of celestial objects. For most of the twentieth century, "human intelligence" would have also meant the ability to take dictation or review legal documents, tasks now taken over in large part by machines. As shown in one classic study in the history of science, even tracing the movement of microscopic particles was once the province of human intelligence rather than machines.[21] Though some of these developments were driven by smart machines, others occurred through shifts in labor practices or cultural norms. Indeed, as one historian has noted, the ability to perform advanced calculations went from being the highly intelligent "distinctive activity of a scientist" to the work of "an anonymous drudge" at just around the moment when such labor shifted from men to women.[22] To judge by this history, then, the "grand goal" of the AI quest has already been accomplished several times over, often centuries before the term was coined.

While recent histories of artificial intelligence have provided a clear and triumphant narrative, the historian of science Stephanie Dick observed in a recent critique of histories of the field, "There isn't a straightforward narrative of artificial intelligence from the 1950s until today."[23] In her view, neural networks and generative AI—or what Dick calls "inhuman artificial intelligence"—might in fact represent a *retreat* from the original aims of Turing and the Dartmouth Conference, as the early dreams of AI and the modern realities seem related "in name only." Indeed, the proliferation of terms deployed by experts in the field—weak and strong AI, artificial general intelligence, and good old-fashioned AI—seem to indicate a splintering rather than a synthesis of ideas. While the original Dartmouth organizers had assumed AI would imitate human behavior, many modern researchers have dismissed the idea that their machines might be doing anything that resembles human thought. On the one hand, AI is often described as the same project developed at Dartmouth, for example when Eric Horvitz, director of Microsoft Research, testified to the US Congress in November 2016 that "AI is . . . aimed at a shared aspiration, the scientific understanding of thought and intelligent behavior, and in developing computing systems based on these understandings."[24] On the other hand, the limitations of neural nets as a model of human cognition have been pointed out in a series of recent articles and books asking whether the field is now a "one-trick pony" that offers flashy demonstrations but no legitimate explanation of how the human mind works.[25] As Margaret Boden claimed in 2016, representing some but certainly not all research, AI does not even try "to mimic human intelligence" anymore, opting instead for useful tools.[26] Confirming Boden's point, the enthusiasm in 2023 for generative AI further transformed the idea from a model of the human mind to simply one more profit-making exercise for Silicon Valley investors.[27] Regardless of which "trick" wins out in the contest for the dominant online chatbot, it can be argued that such confusion over AI occurs because its history has been separated from a longer history of thinking about human beings. As Dick concluded in her broadside against the quest narrative, "Artificial intelligence belongs in the history of *human intelligence*."[28]

Why a Long History of AI?

Expanding the scope of what counts in the history of AI has a number of important consequences. As seen above, the first consequence of a longer approach is to challenge the standard narrative at the heart of many philosophical and political debates about AI: that there exists a fixed standard of the "human level," which machines are approaching and will, eventually, surpass. Once restricted to dystopian novels, science fiction films, and the

Unabomber manifesto, this vision emerged in serious literature in the second decade of the twenty-first century, imagining the concept of human as indelibly fixed while machine intelligence is gradually "rising" to the point where it matches and exceeds the human.[29] Such a vision—often referred to as transhumanism—holds sway too in popular influencers in the world of computers and technology, where human life is either existentially imperiled or rapturously transformed.[30] Unsurprisingly, much of the concern and the enthusiasm has come from individuals heavily invested (in multiple senses) in machine technology and who harbor a view that historians and philosophers call "technological determinism," a vision of history where new technologies emerge out of a socioeconomic and cultural vacuum to transform unwitting and passive peoples.[31] Yet, as many of these same historians and philosophers have been at pains to point out, the determinist view misreads the historical complexity and agency of people themselves, as human goals, behavior, and intelligence have been in constant flux for centuries, just as often shaping the machines as they are shaped by them.

The second consequence for a history of AI told from the perspective of the sciences of man and social sciences is that it redirects attention to the amount of human labor needed to help build and run the machines. As the historian and sociologist of science Simon Schaffer noted, if "machines look intelligent," it is because "we do not concentrate on where their work is done."[32] Indeed, the artificial intelligence scholar Kate Crawford's recent *Atlas of AI* performs exactly this task, noting that an "abstracted analysis" of AI has blinded us to the "embodied and material" consequences of AI work.[33] As David Alan Grier noted in his survey of "human computers," during the nineteenth and early twentieth centuries, astronomers, mathematicians, surveyors, and many other "gentlemen of science" were forced to substitute "brawn for brain," using "immense labor" to complete scientific work.[34] Based on the principles of division of labor, social scientists of the nineteenth and twentieth centuries were also conscious about the need to create institutions and work practices that had proved successful in the natural sciences. As the resources and manpower devoted to astronomy, chemistry, and physics increased, so too did resources to study people using scientific tools.[35]

In spite of what has often been heralded as the creative spirit of modern technology companies, the labor practice in much AI work today too seems drawn from many aspects of "industrial science," the form of hierarchical labor first used in observatories, university laboratories, and other social scientific institutions of the past.[36] As one 2014 headline put the matter, most of the data processing that is central to AI today requires heroic "Janitor Work" to sift, mold, and shape the data, with 50 to 80 percent of the work dedicated to the "mundane task of selecting and preparing unruly digital data."[37] The 2018 German documentary *The Cleaners* also makes it clear that warehouses full of low-paid immigrant laborers in Europe are needed to sift through the enormous amount of racism, pornography, and violence on the internet that would otherwise render most AI machines as spiteful and hateful beings.[38] As a recent spate of books have shown, a rigidly disciplined hierarchical labor of "human-fueled automation" is at the heart of the supposedly "productive, nonhierarchical, and playful workplace" described in the earliest reports from tech campuses.[39] Recent studies have also connected AI success to the rampant poverty of contributors to sites like the Mechanical Turk and LeadGenius, to say nothing of the unpaid labor of the billions of people who provide the data necessary to train AIs.[40] In Heike Geissler's novel *Seasonal Associate*, for example, the roots of success for a massive digital marketplace is found not in the algorithms themselves but rather in the processing of human labor through constant monitoring and surveillance.[41] And in celebrated memoirs of recent years, even the supposedly glamorous world of tech work becomes a slog, recalling the dim awakening of a collective conscious seen in the factories of England in the 1830s.[42] Indeed, in their custodial metaphors, recent accounts of AI labor practices give new meaning to the historian of science Thomas Kuhn's famous invocation of the "mopping up operations" at the heart of modern science.[43]

The third consequence of reframing AI as the product of a long history of social scientific attempts to reduce human thought and behavior is that it can help explain one of the most fraught aspects of AI today: bias in the machines.[44] While it is plainly obvious that AI algorithms exhibit bias against women, minorities, and the poor, a philosophy of technical determinism has left technical "fixes" from the "coding elite" as the only solution for the "invisible men and women" who help produce AI.[45] As historians have shown, however, ideologies of European cultural superiority were co-produced with machines in the age of the scientific revolution and European exploration.[46] So too have historians shown that data collection often involves the "invisible labor" of the most marginalized groups, and a long history of the social scientific roots of AI demonstrates that bias has been embedded from the beginning.[47] The book's focus on European and American thinkers, mostly men, is therefore not because this history is more interesting or because the conclusions reached by these thinkers are any more true than other attempts to understand humanity through scientific means. Nor is it because much, though certainly not all, historical work in the social sciences and in AI has been done by men of European descent. Indeed, it might be argued that because modern AI has in recent years become so transnational and attracted so many practitioners from a variety of backgrounds, a narrow focus on Western thinkers—and even archaic terminology like "science of man"—obscures the modern diversity of the field.

Yet, as seen throughout the book, the centrality of European and American thinkers to AI and the social sciences is part of a corollary argument, developed in the final two chapters, that the concepts and codes embedded into the earliest vision of AI were grounded in Western social scientific thought.[48] For example, historians have shown that it was not a coincidence that AI and the modern social sciences emerged in America at almost the exact same time as "modernization theory," the belief that a particular Western form of politics and culture would necessarily predominate throughout the world.[49] Given that these same assumptions are under attack today, and that AI seems remarkably rife with bias for supposedly neutral machines, it is worth noting at the outset that AI may represent a last attempt to preserve a culturally specific and historically determined definition of human thought and behavior against challenges from non-Western perspectives on the meaning and purpose of human life.

Histories of People, Data, and Machines

The justification for a longer approach to AI history has been based in research and work in the history of science, a field well positioned to add context and understanding to the teleological and internal approaches that have mostly dominated and distorted AI history. Indeed, it was something of a surprise to many historians of science when a 2017 *New York Times* opinion piece criticized academia for failing to contribute to the story of modern technology, blaming the "ivory tower" for "being asleep at the wheel" and having "essentially no distinct field of academic study that takes seriously the responsibility of understanding and critiquing the role of technology."[50] For those in the ivory tower who have labored in the "distinct fields" of history of science, philosophy of science, and science and technology studies, nothing could be further from the truth. Rather, these fields have been attempting to understand technology for decades, producing nuanced histories of past machines, transformative works on quantification, and remarkably detailed histories and ethnographies of the kind of data science that make up today's AI machines.

No book has systematically attempted to connect the development of artificial intelligence to a long history of social scientific methodologies, but there is nevertheless a robust legacy of studies that makes such a book possible. Boden's *Mind as Machine* provided a vivid history of thinkers who had tried to define thinking as a mechanical process, and recent works by Matthew Jones and Jessica Riskin have also explored the considerable overlap between machines and ideas since the scientific revolution, providing richly sourced accounts of the search for agency in matter.[51] In *Sublime Dreams of Living Machines*, Minsoo Kang traveled back even further to investigate how automatons have been imagined since the days of Hesiod and Hephaestus, emphasizing how intellectual and cultural contexts can shape the creation of machines.[52] Adrienne Mayor's *Gods and Robots* too explored the importance of *ideas* in directing ancient machine-making, noting that the "black box" of ancient machines—the precise mechanisms of the bronze giant Talos, for example—might seem a handy metaphor for our own ignorance of modern proprietary algorithms and corporate technologies.[53] Even in the modern age of machines and the sciences of man, more specialized prehistories of artificial intelligence are a reminder that human cultures and ideas shape machines rather than vice versa. In Adelheid Voskuhl's *Androids in the Enlightenment* and Kevin LaGrandeur's *Androids and Intelligent Networks in Early Modern Literature*, for example, the authors examine how European machine makers and consumers valued feeling, affect, and sentimentality over intelligence.[54] In an irony for the Age of Reason, it was their very ability to *appear* artificial that made machines intelligent. The concern then was not that the machines would become intelligent but that people would come to look like automatons.

While these histories have provided key demonstrations of how culture and ideas can shape machines, they have not been primarily concerned with the role of the social sciences in shaping the kind of "human" to which the machines might aspire. In contrast, the story that follows draws on a larger context of thinking about data and social science that has emerged in recent decades, dating back at least to a conference and 1981 book on the "Probabilistic Revolution."[55] In this work, historians of sciences were particularly attentive to how governments, social scientists, and technology companies have used data collection to reduce the broad and messy idea of a person to a scientifically intelligible concept. In the 1980s and 1990s, a number of groundbreaking studies emerged where individual intelligence was redefined as "calculation" and where people and societies were redefined as quantifiable entities.[56] Outside of these more specialized books, other works on quantification demonstrated how numbers became associated with objectivity, which has subsequently shaped social scientific research.[57] Decades later, the literature on the quantitative revolution has moved into popular business and trade books, as the "boring but necessary" world of data exploded onto the bestseller lists in the twenty-first century, providing a mantra for the fusion of data and capital.[58] As an indication that much work remains to be done, however, a recent popular book on data has been praised for its "mind-altering insight . . . that the numbers we use to capture the human experience are themselves a form of creative story-telling."[59] For those familiar with the works described above, such "mind-altering insights" have been available for decades.

While the major revolutionary studies of quantitative history are close to thirty to forty years old, appearing before machine personal assistants, self-driving cars, and predicative algorithms became widespread sources of excitement and fear, a newer body of literature has begun to emerge that explores the recent impact on human intelligence and behavior resulting from the reductive process of quantification in the social sciences. In these works, scholars show how recent data collection has morphed from the days of crude averages and bell curves, as American corporations have tried to create individualized quantitative "selves" for the purpose of manipulation.[60] This new "turn" in data and quantitative histories can also been seen in two recent journals dedicated to the history of data collection processes. In one series of articles on the "Histories of Data and the Database," the authors explored the resurrection of interest in data histories, specifically tying this interest to the new ways in which data collection shapes modern life.[61] In a recent special issue of the history of science journal *Osiris*, the power of quantification was reassessed by many of the same authors in the current era of data mining, data hacking, and large-scale accumulation of data by private companies.[62] This collective effort to quantify, the authors argue, has even led to the creation of new statistical doppelgangers that move throughout the internet, "data doubles" or "algorithmic selves" that more and more stand in for our actual selves. In their malleability and transparency, these new "people" seem to resemble the "mere agents" necessary for AI success.[63] Rather than data collection "revealing" some essential "human level," these stories demonstrate that different forms of data collection have historically *produced* very different kinds of people, often reducing the concept of human thought and behavior to a form where it could be more easily mimicked by a machine.[64]

This recent work is important in updating the long history of quantification and data collection, and it is now beginning to find more relevance in histories and ethnographies of AI.[65] For example, Stephanie Dick and Hunter Heyck have examined different elements of the life of Herbert Simon, the social scientist and AI visionary who combined a novel approach to data with the idea of thinking machines.[66] Two recent works by Rebecca Lemov and Jill Lepore have explored fantastical dreams from the 1950s to assemble complete knowledge in the social sciences, and Lemov's work in particular identified the link between data collection and modern selves.[67] Yet few of these transformative studies of AI go back before the twentieth century. Perhaps most surprisingly, even the flood of books on the "quantified self" movement barely mention data collection prior to the first electronic computers, focusing on modern anthropology and ethnography over historical links to the first quantifiers.[68] It is the argument of this book, however, that the reductive aspects of modern data collection in AI have far deeper roots and that the collective work of historians of data collection and the social sciences can provide the basis for an alternate history of AI. Rather than the "ivory tower" having "ignored" technology, the breadth of studies surveyed above provides a powerful challenge to the determinist narratives that have emerged from AI researchers and the first drafts of science journalism.

The Descent

When artificial intelligence is viewed through the lens of the history of science, the simple story of a "quest" begins to fade. As those familiar with the long history of social scientific theories of humanity are well aware, the history of the sciences of man has not been a steady climb to enlightenment but rather a long trail of disappointment and error, one that might better illuminate the future path of AI research than the triumphant histories of the past few decades. It is for this reason that the book is told as a story of *descent* rather than a quest. Like Charles Darwin's usage in his own fledgling science of man—*The Descent of Man* (1871)—the term is intended to be ambiguous, both denoting the literal description of how ideas from the social sciences were inherited by AI and also connoting a sense of perceived moral or social decline. As will be seen, the reductive aspects of the sciences of man and social sciences were repeatedly criticized by contemporaries. From the earliest religious critics of Descartes to the "interpretive revolution" of the 1970s, critics noted the dehumanizing or injurious consequences of reductive social science for the idea of intelligence. For some, "intelligence" itself managed to fall from the sky, as an idea once linked to divine providence became the possession of humans as a rote process of mechanical thought subject to imitation by machines.

Just as modern AI of the past fifty years can be seen either as a natural growth of ideas or a decline, the four hundred years from Descartes to modern machines can be viewed as either a story of progressive accumulation *or* degeneration. Indeed, even today researchers in AI warn that the most revolutionary approaches of the past few decades are approaching a moment of crisis and that the AI triumphalism of the late 2010s has waned, with the idea of building a mechanical mind replaced by the economic potential of generative AI. As Melanie Mitchell pointed out in a recent survey of the field, "the quest for robust and general intelligence . . . may be hitting a wall: the all-important 'barrier of meaning.'"[69] Although "meaning" would certainly be an important step for computer scientists in creating a true AI, it is also an elusive idea that has bedeviled *social scientists* for centuries.[70] Similarly, in a 2018 *New Yorker* article on the perils of "superintelligence," the MIT physicist Max Tegmark remarked on the challenges that await future work in the field, challenges that might seem quite familiar to those trained in the social sciences and humanities. Tegmark claimed, "to program a friendly AI, we need to capture the meaning of life."[71] If AI researchers who pursue this path need a guide to their future prospects, they would be well served to look into the long and troubled history of the sciences of man and social sciences to answer such questions.

This book will not quite attempt to capture the meaning of life, but it does cover a long history of efforts to reduce human thought and behavior to a state where they could be understood through science and mimicked by machines. In framing the history of the idea of artificial intelligence as the story of a descent, the book therefore traces what might be called a "genealogical" path rather than the "clear" and "tidy" history that is most common in internal accounts of AI.[72] Rather than a simple "origin" story for AI, it provides the history of a set of ideas and practices that have come to shape the context in which AI has been developed. Instead of a series of progressive discoveries, where each new thinker self-consciously builds upon the ideas of the past, the story told here is one of ironical and paradoxical fits and starts, where ignorance of the past, new discoveries in the natural sciences, and the stubborn barrier of human complexity combined to vex those looking to study human thought and behavior through the tools of the human sciences. In some cases, knowledge accumulates, but in the majority of the stories seen below, it collapses. What does appear to change, however, and what so concerned the critics of reductive social science, is that throughout this history, the broad possibility of human thought and behavior—and even the idea of intelligence itself—became dramatically circumscribed. Ironically (again), while no one particular form of social science triumphed in the four hundred years covered by this book, many argued that actual human lives were transformed through the labor of scientific work and exclusive education in scientific methodologies. As in the case of artificial intelligence today, even when the sciences of man failed at the level of theory, critics maintained that in practice they nevertheless remained a powerful determinant of human lives.[73]

Reframing the story of modern AI—from a short history of a few technological triumphs to a long history of discarded social scientific theories—therefore requires a significant expansion in the sources traditionally used. This "deep history" of AI covers a wide range of reductive ideas that might be classified under the heading of "sciences of man." Among the creators of these sciences, the book includes discussion of René Descartes; the French Enlightenment thinkers Bernard de Fontenelle, Étienne Bonnot de Condillac, Baron d'Holbach, Claude-Adrien Helvétius, and the Marquis de Condorcet; the "father of economics," Adam Smith; nineteenth-century European social thinkers Adolphe Quetelet, William Stanley Jevons, Harriet Martineau, Charles Knight, Herbert Spencer, Alfred Russel Wallace, and Francis Galton; and twentieth-century American academic social scientists Frank Knight, Frank Hankins, William Ogburn, Milton Friedman, John Watson, Clark Hull, and Warren McCulloch. At the same time, the book includes a number of contemporaneous critics of these approaches, including French philosophes Jean le Rond d'Alembert, Denis Diderot, and Jean-Jacques Rousseau; the unclassifiable Julien Offray de La Mettrie; French reactionary monarchists Louis de Bonald and Joseph de Maistre; the British novelist Thomas Love Peacock; groundbreaking scientific thinkers Pierre-Simon Laplace and Charles Darwin; and twentieth-century academics Thorstein Veblen, Earle Eubank, Charles Ellwood, Wesley Mitchell, H.S. Jennings, Talcott Parsons, and John Tukey. While the dividing lines were not always so clear-cut between critics and proponents of a science of man—with many of the critics arguing for their own sciences of man—the overall discourse reveals that reductive social scientific theories held the potential in both theory and practice to drag human activity down to the simplistic degree of a machine.

Some of these names are well known to most readers (Descartes, Smith, Darwin), but some may be known only to historians of ideas and science (d'Alembert, Martineau, Jevons) or disciplinary specialists (Knight, Eubank, Tukey). Such a wide scope was both intentional and accidental. Many of the sources were encountered in research for a book on the nineteenth-century statistician and social scientist Adolphe Quetelet—himself a key figure in the story—where it became apparent that the many debates that have arisen over AI in the past decade were foreshadowed in the contested histories of the sciences of man, particularly those sciences based in quantification, probability, and mathematical reduction. Rather than produce a study limited to the well-trodden paths of intellectual history, the book looks at lesser-known works and stories to reveal moments when the most reductive forms of social science emerged. In cases where the author is well known—like Descartes and Smith—the analysis goes beyond their most famous works to bring out ideas usually only discussed in specialist literature. For less well-known figures, it relies on letters, notes, and published work that has rarely received substantial discussion in *any* context outside of specialized monographs and academic journal articles. For the selection of critics, I have mostly looked at those figures who have emerged from *within* the discourse of the sciences of man, rather than the legions of artistic and literary challenges to these ideas that have stretched from the Romantics through Dada to the Situationists. While the selection was idiosyncratic and personal, the intent has been to locate and critically examine the practices and critiques of reductive social science that appear most relevant to the discourse of artificial intelligence over the past several decades.[74]

While the approach described above can certainly allow for insight into how human thought and action came to a point where it could be modeled by a machine, it does not allow for a full and exhaustive survey of the social sciences, or even capture every reductive approach to studying people in the past four hundred years. In order to keep the notes at a reasonable length, only secondary material most pertinent to the argument has been referenced, though many more works that influenced this book could have been included. As noted above, the book is predominantly concerned with ideas in Europe and America, but it contains relatively little from Germany, as science in the nineteenth-century German-speaking regions largely *challenged* reductive accounts of human action that relied on the tools of the natural sciences, following in part the warning of the philosopher Johan Georg Hamman (1730–1788) that "the sciences, if they were applied to human society, would lead to a kind of fearful bureaucratization."[75] Indeed, even by 1862 the physiologist and influential German methodologist Hermann von Helmholtz (1821–1894) could still note that in his country "the name of science was often denied" to the physical sciences.[76] And in the first few decades of the twentieth century, the anti-reductive methodological practices of the "Germanic school" in economics and sociology was prioritized in American universities, whose first professors had often been trained in German seminars.[77] It was just these practices that needed to be overcome for professional social science to emerge in America. Similarly, anthropology—one of the primary social sciences of the twentieth century—is barely mentioned in the book, as the field spent much of the twentieth century *resisting* reductive accounts. While forerunners to anthropology like anthropometry, phrenology, and psychophysiology all contain startling reductions of human action and quantification, these stories have been well covered elsewhere and thus are only hinted at in what follows.[78] While the inability to include a broader range of stories is a liability for nearly any history with a long scope, the hope here is that those readers who know part of these stories are able to discover something new. Or, in the case of more recognizable names, they are able to see a new aspect to what seemed a familiar story.

It should also be noted that the approach described above is unlikely to reveal that modern AI researchers of the past seventy years self-consciously *based* their ideas on the sciences of man. In fact, the opposite may be true: modern AI successes and frustrations might be due to large swaths of ignorance in engineering and computer science departments about such histories. Historians and philosophers of science in particular have argued that the modern "scientific method" proscribes more historical and cultural approaches, instead offering what the historian of science Henry Cowles has recently called "an artificial, algorithmic scientific method."[79] For the philosopher of science Michael Strevens, science has in fact made great strides *because* it has been reduced to a simple set of narrow empirical theories, as he compares modern science to a process where a "simpleminded strategy" denies "students . . . the ability to think philosophically, theological, or aesthetically at all."[80] Although Cowles and Strevens do not mention AI and are largely focused on the natural sciences, their historical descriptions of the process of scientific methodology mirrors contemporary accounts of AI research. For example, in a recent critique, the entrepreneur and computer scientist Erik J. Larson noted that a "simple" approach to human intelligence is today a prerequisite for the field, claiming that "a dangerous simplification in the philosophical ideas about man and machine" is one of the chief obstacles to achieving a true AI.[81] Criticizing his fellow engineers for their limited approaches toward building thinking machines, Larson laments in *The Myth of Artificial Intelligence* that through "pulling down human intelligence, tying it to a definition more amenable to computation, current thinking about AI jettisons a richer understanding of mind."[82] Although Larson does not look beyond the past half century to tell this story, his recent insider view of the field confirms the central thesis of this book of a long descent of reductionist and simplistic ideas in the social sciences as the key to the idea of thinking machines.

The process of "pulling down human intelligence" occurred in a wide variety of social scientific fields across several centuries. In order to provide some narrative coherence, the book has been organized chronologically and geographically, exploring simultaneously how both social thinkers and their critics articulated the process by which the sciences of man simplified its subject. Part 1, which largely covers France in the seventeenth and eighteenth centuries, begins with the influence of mechanical philosophy and an increasing interest in applying techniques in the mathematical and natural sciences to the study of human beings, an idea dubbed the "geometric spirit" by Fontenelle, the head of the French Académie des sciences. Such an idea inspired a host of Enlightenment sciences of man that assumed that social thinkers could provide the kind of order to human life that Isaac Newton had brought to the heavens. Chapter 1, "Intelligence Lost," explores the depths of Descartes's revolutionary mathematical and geometric approach to understanding human action, tracing its influence on French thinkers like Fontenelle and Condillac, as well as highlighting criticism from French philosophes like Rousseau, Diderot, and d'Alembert. Chapter 2, "At the Bleeding Edge," explores how the geometric spirit was applied in mechanical sciences of man developed by the French *salonistes* Helvétius and d'Holbach, as well as efforts by Diderot and d'Alembert to blunt the sharper edges of a mathematically influenced science of man. As seen by the end of the chapter, the certainties of a Cartesian science of man would be complicated by the probabilistic conclusions of Laplace, who imagined a "great intelligence" that could predict all future events based upon certain knowledge of the past. The final chapter of this section, "Warnings of a New Barbarism," recounts the severe criticism of the geometric spirit leveled by counterrevolutionary figures Bonald and Maistre. Although these thinkers had their own idea of a science of man, they railed against what they believed to be the denial of divine intelligence among the French Enlightenment *sciences humaines*, as well as the mundane mental labor it portended.

In the second part, the story expands to Europe in the nineteenth century, exploring how prominent sciences of man like social physics, social mathematics, political economy, social Darwinism, and eugenics emerged in tandem with the growth of industry and machines in Europe to produce a new vision of the human subject, one which often ignored the Enlightenment focus on progress and equality that attended previous social thought. Chapter 4, "Progress to the Mean," explores how the idea of the "average man" was transformed in the statistical work of Condorcet, Quetelet, and Jevons. While Jevons combined the idea of a "social physics" with a mathematical approach to political economy, his work also revealed that the sciences of man did not merely *reveal* the workings of the human subject and mind but could *shape* the subject and mind as well. In chapter 5, "Tuning the Mind," Jevon's idea of an interventionist social science is explored in the work of earlier popular political economists Martineau and Knight, who sought to use the laws of the sciences of man as tools to instruct a population that did not always seem to follow the paths laid out by social physics or the "iron laws" of political economy. Through studying the early ideas of Smith and their deployment in Knight's and Martineau's popular pamphlets, the chapter explores how the interconnected ideas of political economy and real-life machines redirected attention to the *minds* of the workers at the machines. Chapter 6, "The Descent of Man (and Intelligence)," explores the fusion of social physics, political economy, and Darwin's *Origin of Species* in the emergence of social sciences developed by Spencer, Wallace, Galton, and Darwin himself, each of which would further challenge the Enlightenment focus on equality, progress, and human freedom that had motivated many of the earliest sciences of man. Here the geometric spirit was reborn as a form of biological reductionism, with even the idea of intelligence being reduced to a simple inheritable trait.

The final section of the book examines the development of formal academic social sciences in America in the twentieth century, many of which were strongly influenced by the evolutionary ideas of the previous century. While many eighteenth-century sciences had seen the abstract simplification of the human subject as the key to moral and political progress, and nineteenth-century social sciences suggested the need for political intervention and education in order to help people conform to these new laws, the move toward professional social science in fields like sociology, political economy, and psychology had at least three important consequences for the emergence of artificial intelligence in America. In the first place, it introduced a new kind of simplification and reduction: that of the social scientific researcher himself. Second, by projecting this newly professionalized ideal onto the human subject under examination, social scientists created forms of artificial intelligence decades before the term itself was born. Last, by creating an ideal-ized role for the researcher that stood outside of moral and political thinking, the model professional social scientist abandoned the early assumptions of the Enlightenment science of man that were explicitly guided by notions of equality, progress, and human freedom. At just the moment when the social sciences finally discarded their relationship to the broad methodologies and motivations of the Enlightenment, artificial intelligence was born.

In order to trace this transformation in the social sciences, chapter 7, "The Sacrifice and Rebirth of 'Man,'" examines the process of discipline formation in the world of sociology and economics in the first half of the twentieth century in America. It begins with the little-known story of Earle Eubank, a sociologist trained at the premier graduate school at the University of Chicago who, along with many others, struggled with accommodating his moral and political commitments to a new field of study. At the same time sociologists simplified the scope of their investigations, economics similarly became a "science" through the dismissal of methodologies that incorporated (explic-it) moral thinking, a process that its critics worried reduced social scientific work to mental and spiritual drudgery. Chapter 8, "Social Science by Other Means," examines the consequences of such visions of social scientific work for the development of the idea of "intelligence" at the heart of AI. While two new prominent American social sciences of the 1930s and 1940s—behavior-ism and functionalism—seemed to provide a new route to a science of man, their greater legacy would be in providing the central concepts of the human subject and environment for the pioneering AI work of thinkers like Turing and McCulloch. Not only did the earliest ideas of AI appear as the reflexive vision of the idealized professional (social) scientist, but such ideas also were able to thrive as the resources devoted to large projects of social science re-search were redirected toward computers in the competitive environment of the Cold War. Finally, chapter 9, "Second-Rate Mathematicians," examines a last attempt by the Princeton statistician John Tukey to guard against the increasing mathematical and quantitative reductionism in much social sci-ence research through the introduction of what he called *data analysis*. Tukey, one of the most neglected thinkers of the Cold War era, not only offered a vision of statistics and quantitative reasoning as autonomous subjects, warning against allowing data "to speak for itself," but he also offered an alternative to the simplified vision of the social scientist as quantifier. Years before the late philosopher of science Ian Hacking noted that ideas could "loop down" upon their creators, Tukey saw a similar process where mathematical and quantitative simplification could have a recursive effect on social scientists and engineers.[83] If statisticians did not broaden their outlook, Tukey warned, quantitative thinkers themselves risked being reduced to a kind of artificial intelligence.

After examining how artificial intelligence emerged out of the same milieu as a host of failed and discarded social scientific theories, the book concludes by returning to the current fractured and uncertain world of AI, where the field seems closer to the final days of Cold War social science than it does to completing its quest for a truly intelligent machine. In its naivete, public excitement, financial backing, and philosophical positivism, AI has also ironically left the scientific study of human thought and action as fractured as the days when Alan Turing first questioned the ability to understand people through scientific induction.[84] As A.O. Scott, film critic for the *New York Times*, explained in a survey of AI films, even the hype surrounding innovations in generative AI in early 2023 revealed that AI at its best revealed only "the banality of sentience," as the modern version has failed to live up to the fevered dreams of futurists and science fiction writers.[85] Yet, as this book aims to show, the very real labor practices and biases inherent in centuries of social science may provide more relevant villains in modern AI than dystopian machines that attain some "human level." In our hopes and fears about artificial intelligence, we would do well to remember that the modern world of twenty-first-century AI—with its endless questions of meaning and purpose, hierarchy of human labor, and endemic bias—owes it most distinctive traits to its descent from the social sciences.

PART I

Ideas *in* France

Intelligence Lost

The Enlightenment Response to a Mechanical Universe

IN 2019, ERIC SCHMIDT (1955–), THE FORMER CEO OF A MAJOR TECHNOLOGY company, claimed that his company could "more or less guess what you're thinking about."[1] Two hundred years before Schmidt, the great French mathematician Pierre-Simon Laplace (1749–1827) declared, in somewhat more vivid language, the possibility of "an intelligence which could in an instant know all the forces by which nature is animated . . . an intelligence so large that it could analyze all this data, [and] would know in this analysis all of the movements of the great bodies in the universe and those of the lightest atoms." For Laplace, "Nothing would be uncertain" for such an intelligence, so that "the future—like the past—would be seen by its eyes."[2] And almost two hundred years before Laplace, the seventeenth-century theologian Henry Oldenburg (1619–1677) pronounced that the Royal Society of London believed "all Nature's effects are produced by motion, figure, texture and various combinations, and that there is no need to have recourse to the inexplicable forms and occult qualities, the refuge of ignorance."[3]

It is a fun and sometimes dangerous game to play, mining the historical record in search of ideas and concepts that seem to anticipate the modern world. In the case of Oldenburg, Laplace, and Schmidt, it might even appear only a few short leaps from the scientific ideas of the 1600s to statistical determinism and modern data collection. Oldenburg, secretary for the Royal

Society in England and one of the most well-connected men of science in seventeenth-century Europe, was explaining to the philosopher Baruch Spinoza (1632–1677) the central concepts of the "mechanical philosophy," the belief that the natural world operated like a machine and therefore required none of the "occult" qualities that had marred scientific investigations for millennia. Laplace, who deployed probability theory to smooth out the disturbances in nature's mechanism, had imagined a demon ("an intelligence so large") that could evaluate and predict every future act in natural and human life, leaving the role of a divine intelligence in doubt.[4] As Laplace may or may not have said to Napoleon after being asked how God fit into his magisterial *Traité de mécanique celeste* (1798): "I do not need that hypothesis."[5] Finally, Schmidt's comment reflects a twenty-first-century belief common among technological utopians and proponents of artificial intelligence, that perfect knowledge of human behavior might be accomplished with sufficient data collection and mathematical manipulation. In the words of Oldenburg, Laplace, and Schmidt, the four hundred years between the introduction of mechanical philosophy and modern claims for artificial intelligence seem to disappear, revealing in an instant how nature, society, and the human mind can be reduced to quantifiable and predictable behaviors.

Similar as the sentiments of Oldenburg, Laplace, and Schmidt might seem, the first step in this story of reductionist science—from a world of willing beings inhabiting an intelligent universe to a predictable world of predetermined thoughts—was not quite so straightforward. While the majority of this book examines how the sciences of man and social science reduced human intelligence and action into a form that could be mimicked by machines, this chapter explores how early modern mechanical philosophy, particularly in France, prepared the foundations for these efforts. It covers a period roughly between 1620 and 1750, which overlaps with eras once conventionally known as the Scientific Revolution and the Enlightenment. While the correspondence of Oldenburg and Spinoza shows the wide reach of mechanical philosophy across Europe, and the Enlightenment clearly crossed many borders and oceans, French mechanical thinking and the resulting "geometric spirit" of the French Académie offer the best example of how a science of man could be built from the reduction of nature to simple principles.[6] As Natania Meeker claims in her analysis of French materialism, the world of the French Enlightenment in particular offers "an intense preoccupation with the objectification and rationalization of matter" and was "a crucial staging ground for the modern construction of human beings as objects of knowledge."[7] Free and unpredictable minds were certainly transformed into determined "objects of knowledge" elsewhere, but the clarity and explicitness of French ideas might provide the best introduction to the formation of the modern sciences of man out of mechanical philosophy.

To tell this foundational story, the chapter begins by investigating some of the neglected corners of the most influential mechanical philosopher of the age, French or otherwise: René Descartes. Descartes is best known for formalizing the deductive method of science in service of proving the existence of God and for marking a clear dividing line between matter and immaterial mind. Yet much of his earlier and less-read works indicate that his particular mechanism initially reduced *all* human action—body and soul—to mechanical processes and mathematical rules. While many theologians objected to such a vision, calling it the "doctrine of the automata," the following section examines the influence of Descartes's account on key institutions and thinkers of the French Enlightenment, including Bernard Le Bovier de Fontenelle, longtime director of the French Académie, and Étienne Bonnot de Condillac, the house philosopher for much of the French *Encyclopédie*. Here, the belief emerged that a mechanistic and mathematical approach to natural philosophy—or what was called the "geometric spirit"— not only could be used for a science of human actions but could also produce moral virtue. Finally, the chapter looks at critiques made by three of the most important French philosophes: Jean-Jacques Rousseau, Denis Diderot, and Jean le Rond d'Alembert. Though all three found fault with certain aspects of the geometric spirit, their biggest concern was not that mechanism infringed on a divine power. Rather, they worried that the new methodological approach was unsuited to most minds, producing a stifling form of intellectual work for many and an unintelligible nature for all.

Descartes and the Rise of Mechanics

Despite what might seem obvious in historical hindsight—that the deterministic consequences of a mechanical worldview could lead to the denial of individual free will and intelligence—few founders of mechanical philosophy would have embraced such a drastic conclusion in the seventeenth century.[8] While all rejected the medieval and ancient vision of a "purposeful" nature that acted toward a set goal, most believed the human mind and spirit retained a level of freedom from nature's machine. For example, Oldenburg's letter to Spinoza referred to banishing "occult qualities" only in the natural world, not in something as seemingly immaterial as human thought. Oldenburg has been called a kind of "philosophical merchant," and few at the Royal Society would have believed that *divine* intelligence or human action could be completely reduced to the "effects" of "motion, figure, texture, and various combinations."[9] Mechanical philosophers like Pierre Gassendi (1592–1695), for example, who had been influenced by the spontaneous "swerves" of the ancient Greek philosopher Epicurus, never could have imagined human thought as a lifeless and unintelligent machine; as the historian Lisa Sarasohn noted, Epicureans "spent thousands of words . . . proving that man is free."[10] The mathematician and philosopher Gottfried Leibniz (1646–1716) also insisted on a *spontaneous* free will that existed in human beings no matter what the makeup of the material universe.[11] Even Leibniz's great rival Isaac Newton (1643–1727), who did more to banish "occult qualities" from scientific investigations than anyone, did not think his laws applied to human action and divine intelligence.[12] Newton famously wrote that the world was guided by "the counsel and dominion of an intelligent and powerful being" rather than by "mere mechanical causes."[13] In the words of Newton's good friend and spokesperson, Samuel Clarke (1675–1729), "All things that are done in the world are done either immediately by God himself or by created intelligent beings."[14] With a few notable exceptions, such as Spinoza or Thomas Hobbes (1588–1679), most seventeenth-century mechanical philosophers could imagine the world as a machine while still insisting that human action and divine intelligence were outside of the mechanism.[15]

One ostensible believer in the distinction between the substance of a mechanical natural world and free human action was René Descartes (1596–1650), who set out his influential dichotomy of machine world and immaterial soul in *Meditations on First Philosophy* in 1641. As his biographer Adrien Baillet (1649–1706) noted, Descartes's belief that there was some form of "Genius" outside of "human intelligence" was first articulated much earlier, after a set of three fevered dreams in November 1619.[16] Descartes had detailed these dreams in a series of scattered writings known as the *Olympica* and later described this period in his life in the *Discourse on Method* (1637), but the incident is not well documented, with only Baillet's summary and some notes Leibniz took on the *Olympica* testifying to Descartes's inspiration. In this account, in the course of serving as a soldier in the Thirty Years' War (1618–1658), a youthful Descartes spent the winter of 1619–1620 quartered away from the fighting and found himself "alone in a small room, heated by a stove."[17] Uncertain of his future and forced to reflect on his existence without the help of books or much human contact, after three profound dreams

Descartes began to imagine that his destiny was to discover as much truth as possible about himself and the world through reason and introspection. As he later outlined in the *Meditations*, the first step was the "radical skepticism" of all prior beliefs, or what has been called a "mental purging" of all previous bases for knowledge.[18] Leibniz called the *Olympica* "a little chimerical," but one of his copied fragments noted that Descartes had placed both the "imagination" of poets and the "reason" of the philosopher as paths toward higher knowledge, indicating that Descartes had indeed reserved a role for intelligence outside of the mechanisms of nature. According to Leibniz's notes on the *Olympica*, in the higher realm Descartes had imagined that "God is pure intelligence."[19]

The accounts by Descartes, Leibniz, and Baillet have often reinforced the view that pure intelligence and immaterial soul were protected from the later rigid mechanisms of Cartesian nature, which notoriously viewed all living matter—including animals—as mere machines. In this view, the division of mind and body retained a role for religion, insulating Cartesians from the kind of reductive mechanisms that prompted Galileo's trial in 1616 and led to charges of heresy against Spinoza and Hobbes. Intelligence, as it had since the Middle Ages, was allowed to rest in the heavens while Cartesian science and mechanism described the working of God's creation.[20] While the story of Descartes's winter in 1619 provides a compelling origin story, recent historical work has elevated the importance of *actual machines* in Descartes mechanistic vision, revealing that a clear line between matter and soul was not always entirely maintained.[21] As J.A. Bennett noted in a seminal paper, "mechanical philosophy was not solely an intellectual construction" produced in "Descartes' stove-heated room."[22] Daniel Garber, too, has emphasized the role of machine making in the formulation of Descartes's scientific methodology, quoting Descartes as believing that "my entire physics is nothing but mechanics."[23] Historians have also highlighted the influence of the mathematician and natural philosopher Isaac Beeckman (1588–1637), who provided Descartes with dozens of books on technical machine making while also questioning much of his received philosophy and theology.[24] For example, after reviewing some works on the "art of navigation" in 1618, the young Descartes wrote to Beeckman to laud the simple rules of mechanics, marveling at his "discovery" that an important tool of navigation could measure distance simply by observing the stars. In an early indication of his search for clear and distinct laws, Descartes wondered why no one before had made a discovery "that required so little thought."[25] Decades prior to his most famous work,

Descartes seemed to imagine the tools of machine making, rather than pure reflection, as the key to a scientific method.

While all mechanical philosophy concerned the reduction of complex phenomena, what set Descartes apart from his contemporaries—and made his work so crucial to the sciences of man—was that his vision of a mechanical world was tied to a *universal method* for understanding all machines, people included. Though the *Discourse on Method* is Descartes's best-known attempt to formulate a universal scientific methodology, relying on pure deductive reasoning, his earlier *Rules for the Direction of the Mind* (1628) provides a rougher and more reductive grounding of scientific methodology in the mechanical arts.[26] Written in Latin and intended for a small audience, the *Rules* was less guarded than the much more famous and popular *Discourse*.[27] While the *Rules* has sometimes been dismissed because the work was published long after Descartes's death and was relatively unknown for most of the seventeenth century, it nevertheless provides the best summary of the kind of Cartesian thinking adopted by many French sciences of man during the Enlightenment, as well as those social sciences to follow.[28] As Martin Heidegger once claimed after reading Descartes's earliest draft for a scientific methodology, anyone who wants "an inkling of what is going on in modern science" needs to study the *Rules* "down to its remotest and coldest corner."[29]

While pure epistemology and reason guided Descartes's later works, the *Rules* was based on elevating the tools of machine making—geometry and arithmetic—to the level of science, or what he called "knowledge which appeals to the mind" (*OPD*, 80). Descartes believed there had been a "false comparison" between mechanical skills ("the arts") and the sciences, and he intended the *Rules* as a means to extend the simple and clear rules of machine making to abstract knowledge about the universe. Descartes's solution was literally revolutionary because, like Galileo, he upended the traditional hierarchy of sciences by suggesting that the lowly tools of machine building could be used for high-minded philosophy and even theology. In the *Rules*, Descartes believed that geometry and arithmetic were crucial because they alone were "exempted from all the vices of falsity or uncertainty" (*OPD*, 82). Where other sciences could be led astray through the introduction of what we might call observational error, compounded in the endless "dialectic" of medieval philosophy, arithmetic and geometry offered "much more certainty than other disciplines" because "they only treat objects simple and pure while allowing absolutely nothing from experience" (*OPD*, 84). The skills of machine making simply offered the best tools for philosophical and scientific investigation.

The extent to which the *Rules* extended arithmetical and geometrical reasoning into philosophy can best be seen in Rule IV, where Descartes proposed a "method by which one can gather [*quête*] the truth of things" (*OPD*, 90). Descartes began the section with an attack on the "disordered research" and "obscure meditations" that marred most science, but he claimed, "The human spirit possesses a certain divine something, where the first seeds of useful thought have been deposited . . . and [which] have produced spontaneous fruits" (*OPD*, 93). Not leaving any doubt as to the identity of that "divine something," in the next sentence Descartes praised "arithmetic and geometry" as the rare sciences that bore "fruit" (*OPD*, 93). In combination with Descartes's other claims in the *Rules*, Rule IV seemed then the clearest statement of reductionism possible, as *all* sciences—from philosophy to theology to medicine to what would become the sciences of man—could be investigated through arithmetic and geometry. In Rule III, Descartes had even used the same terms to describe mathematics and "intuition," an early version of the "clear and distinct perception" that grounded the *Meditations*. Intuition was, like geometry and arithmetic, a means for Descartes to avoid the "unstable testimony of the senses" and the "deceiving judgment of the imagination." Anticipating the *cogito* ("I think, therefore I am") from *Discourse on Method*, Descartes in the *Rules* claimed that "each can see by intuition that he exists, he thinks, that a triangle is defined by three lines only, a sphere by a single surface, and other similar things" (*OPD*, 87). So intertwined were math and method that Descartes used the terms *énumération* and *induction* synonymously throughout the later *Rules*, indicating that the act of quantifying was itself a means of thinking (*OPD*, 110). As Pamela Kraus put in her study of Rule IV, only the *Rules* "shows how method can found a new science of the world."[30]

While much historiographical controversy over Descartes's intent surrounds Rule IV, there was far less ambiguity to be found in *L'Homme* (1664), a work composed between 1629 and 1633 but not published until after Descartes's death.[31] As part of a two-volume plan designed to explain everything in the universe through mechanics, Descartes imagined *L'Homme* (and its counterpart, *Le Monde*) as a rigorous application of the *Rules* to all sciences. Though suppressed because of Descartes's concern about Galileo's heresy trial, it demonstrated how what Descartes called the "mechanisms of man" could be explored through a geometrical and arithmetical analysis. He claimed, "All the causes of motion in material things are the same as in artificial machines. . . . The body [is] nothing but a statue or machine made of earth" (*OPD*, 379). In page after page of detailed descriptions and images,

Descartes articulated precisely (if not accurately) how parts of the human body operate as a machine, overlaying geometrical figures onto sketched anatomies of human bodies.

Descartes began *L'Homme* with the qualification that he was not talking about *actual* people but rather hypothetical machine "men" who "are made, like us, of a soul and a body" (*OPD*, 379). To understand these "men," Descartes posited that mechanics was a better guide than a more traditional science like anatomy, which could only describe the existence of various "bones, nerves, muscles, veins, arteries, [etc.]." Unlike great anatomists, mechanical explanations could show how and why machines work as they do. In one of the most famous passages from *L'Homme*, Descartes explained how the metaphor of machines helped him articulate his own vision of the human body: "We have clocks, mechanical fountains, mills, and other similar machines that move by themselves in many ways, and these are made only by men. It seems that it would not be hard to imagine what kind could then be made by the hands of God, or to limit the artifice to those movements that we could think of" (*OPD*, 380).

From this startling comparison of "men like us" to machines, Descartes launched immediately into how blood flow, digestion, air circulation, and various other processes could occur in "this machine." For example, Descartes explained how blood flow to the "brain" could create either a "very subtle wind" or a "great flame" similar to "animal spirits" (*OPD*, 388). In describing the nervous system, Descartes made the machine analogy most clear, writing that "one must compare the nerves of this machine that I have described to the pipes of fountain machines," adding that "respiration and other such actions" are "like the movements of a clock, or a mill where the ordinary course of water continuously flows" (*OPD*, 390). Labeling different "points" of the human body with geometrical letters, Descartes explained in exacting detail how various forces are heated, decomposed, forced through the body, and processed from one geometrical point to another. In one diagram, Descartes imagined how "nerve A, whose exterior skin is like a great pipe which contains several other pipes (b, c, k, l), is composed of an interior skin more delicate" (*OPD*, 392). A typical passage, describing the circulation of "animal spirits," was as follows: "If the spirits which come to the brain tend to flow with more force by bf than by cg, they will raise [*ferment*] the small skin g and open f. . . the spirits continue in muscle E which will then enter muscle D by the canal ef" (*OPD*, 392).

Although Descartes's mechanical "man" may seem lacking in spirit, had this been the extent of the comparisons of man and machine, it is likely

L'Homme would not have needed to be suppressed. It was a work of anatomy explained geometrically, but hardly the challenge to the Catholic Church that Descartes feared. However, Descartes moved into trickier territory when he began to explain *mental* faculties in the same way, showing no evidence of his later division between material body and immaterial soul. When discussing how music operated, for example, Descartes began with a typical geometric depiction of "sounds" a person might hear and how they might be made either "sweet" or "rugged" (*raboteuses*) by the means though which they reach the ear. What followed, however, was a new claim: "And so it seems to be sufficient that I can show how the soul, which will be in the machine that I will describe, can receive pleasure from music following our same rules" (*OPD*, 414). As if to emphasize that there is no difference between the soul's appreciation of music and a bodily function, Descartes noted that the soul does not always want the sweetest sounds, "in the same way that salt and vinegar are often more agreeable to the tongue than sweet water" (*OPD*, 415). In his conclusion, Descartes summarized that "all of the functions that I attribute to this machine . . . common sense, imagination, the retention and importing of these ideas in our memory . . . are nothing less than the movements made by a clock or another automaton" (*OPD*, 479).

In its claim that there is no difference between a hypothetical man machine and a clock, Dennis Des Chene has argued that *L'Homme* went so far as to "eliminate the living as a natural kind."[32] While often overlooked in general discussions of Cartesian philosophy, it was the lessons Descartes drew from machines that would revolutionize philosophy. As famously put it in his *Principles of Philosophy* (1644), which depended as much on machine studies as it did on pure philosophy, "I do not recognize any difference between artifacts and natural bodies except that the artifacts are for the most part performed by mechanisms which are large enough to be perceivable by the senses."[33] While other mechanical philosophers had reached similar conclusions, many of them (like Hobbes and Spinoza) had been declared deranged and heretical, and none had Descartes's later direct influence on the sciences of man. Additionally, while neither the *Rules* nor *L'Homme* reached a large audience, much of this work found its way into his more famous books. For example, Descartes's "clear and distinct" formulation in the *Meditations* is drawn directly from the *Rules*, and nearly one-fifth of the *Mediations* is simply reprinted from *L'Homme*. By combining the idea of men as machines with a belief in mathematics as a methodology in itself, Descartes had created an enduring template for future sciences of man to follow.

The Doctrine of the Automata: Cartesian Critics, the Geometric Spirit, and the Path to a Science of Man

Even in its more cautious form, Descartes's contemporaries still understood the radicalism of his position. Most famously, Blaise Pascal (1623–1691) used the term *esprit géométrique* as a means to describe a particularly rigid and mathematical approach to the sciences, in comparison to the *esprit finesse* that Pascal likened to "intuition."[34] The English churchman Richard Baxter (1615–1691) also expressed concern with mechanical philosophy, joining a group of theologians and natural philosophers concerned that Cartesianism "involved the reduction of motion to local motion and the corresponding evacuation of intrinsic principles of motion from active natures."[35] Such a "reduction" had occurred because mechanical philosophers had elevated a science used in *machines* above all else, creating a "mechanization of the soul." As Baxter's biographer David Sytsma put the matter, "the transition to mechanical philosophy involved the transposition or re-imagination of a subdiscipline, mechanics, as a model for the entire natural order."[36] Or, in Baxter's more colorful language: "You may call these men New Philosophers, or Cartesians, but for my part I shall take them for proved fools, fitter for Bedlam than for a Schoole of Philosophie."[37]

Few, however, recognized the mechanism at the heart of Descartes's methodology as well as the Jesuit historian Gabriel Daniel (1649–1728). In Daniel's vision, "Cartesianism" became "the doctrine of the automata, which makes pure machines of all animals . . . taking from them all feeling and consciousness."[38] Perhaps even worse, Descartes's mechanism did not even accomplish the main goal of mechanical philosophy: to banish substantial forms and occult qualities. Daniel argued, "[Descartes] only explains physical effects by certain elements, certain assemblies of parts, certain movements and certain figures; this is hardly different from [previous] certain *entities*, certain *forms*, certain *virtues*, and certain *qualities*." Noting that both medieval scholasticism and Descartes had relied on abstraction, he claimed in jest that "by rights of seniority, the scholastic *I know not what* should win out over the Cartesian *I know not what*."[39] Supporting Daniel's claim that mathematics were no more real than metaphysics, Descartes had even claimed in the *Meditations* that "arithmetic, geometry and other studies of the simplest and most general things—*whether they really exist in nature or not*—contain something certain and indubitable."[40] More important than his mind-body dualism or deductive method, Daniel feared that Descartes had elevated a set of tools of dubious reality to the status of natural philosophy.

In spite of the philosophical and moral concerns of Pascal, Baxter, Daniel, and other critics of the mechanical philosophy, the methodological program of Descartes proved powerful, as nearly all instruction in natural philosophy in France moved from the Aristotelian to the Cartesian approach by the end of the seventeenth century.[41] While many teachers led the charge, the idea of mechanism as a guiding principle for scientific practice reached its apex at the French Académie Royale des Sciences, headed by Bernard Le Bovier de Fontenelle (1657–1757). The Académie has often been held up as the "last stronghold" of Cartesian physics against the rise of Newtonian ideas, and Fontenelle recognized that Descartes's ideas would find their best purchase through avoiding the philosophical questions raised by Daniel and instead focusing on the practical value of mathematics and mechanics.[42] For example, in his "Préface sur l'utilité des mathématiques et de la physique" (1702), Fontenelle even felt it necessary to assert that math and physics had been almost completely ignored.[43] Perhaps surprising to a modern reader, Fontenelle expressed concern that because "mathematics and physics are generally unknown, so they seem to be useless."[44]

For Fontenelle, the key to advancing the role of math and physics was to emphasize their practicality in constructing machines rather than their basis in reality. Despite requiring the kind of "painful observations and fatiguing calculations" that might alarm Pascal, Fontenelle praised the "immense observation" that had been made possible by the "great array of working instruments" in navigational tools and telescopes.[45] Ironically, Fontenelle's justification for relying on mathematics was almost exactly opposed to Descartes's expressed rationale in the *Rules*. While Descartes had seen mechanical sciences like mathematics and physics as disciplines perfectly suited for the clear and precise world of philosophy, Fontenelle praised them because they avoided philosophy completely.[46] Unlike the early Descartes, who had seen technical skills as limited and discrete, Fontenelle privileged the practical aspects of mathematics over Cartesian reason, writing that "if one wishes to limit [*renfermer*] math to that which is useful, one must only cultivate it in an immediate and sensible rapport with the arts, and let all that remains to vain theory."[47] As Fontenelle noted, tremendous improvements had been made in navigation and weaponry through math and physics, but "the utility of their progress was [still] invisible to the majority of the world."[48] For Descartes, "vain theory" had been exactly the point, but in the "Préface" Fontenelle made it clear that he was more interested in bringing "science" down to the level of machines than in elevating the lowly arts of mechanics to the realm of the mind.

The move from Descartes's justification for the study of math and physics to Fontenelle's had significant consequences for future sciences of man. As historians have long noted, Fontenelle did far more than report on the science of the day during his astonishing forty-two-year run as permanent secretary of the Académie Royale des Sciences. As part of his duties, Fontenelle had also composed hundreds of *éloges* to Académie members that amounted to a "scientific discourse" and coherent "theory of knowledge."[49] Rather than a "vain theory," Fontenelle created a philosophy of science *work* that reframed the pursuit of knowledge in moral and practical terms. Instead of a God-given intuition that allowed humanity to perceive the world "simply and clearly," mathematical training became a "painful and difficult" sacrifice, a theme that would endure for centuries in both the natural and social sciences. Fontenelle's Académie *éloges* are famous in the history of science for reframing scientific pursuits as "right conduct" and for refashioning science work as a selfless and tireless process rather than the result of isolated or divine genius.[50] In doing so, he was able not only to deflect the criticism of theologians who believed mechanical reductionism threatened the soul but also to provide a universal method to train new science workers. As Charles Paul claimed in his study of over one hundred *éloges*, the message to fellow *savants* was "incontrovertible. . . . Science is not so much a natural as a moral philosophy."[51] For the properly trained *savant*, the truth of science was reflected not just in the functioning of the machine but in the morality of the machinist.

Fontenelle's transformation of Descartes's "universal method" into the "geometric spirit" of the French Académie involved more than the reformation of science work. It also, crucially, extended the scope for which such a methodology could be useful, effectively creating the first articulation of a potential "science of man" based on mathematical reasoning. Fontenelle did so by updating Pascal's mildly pejorative idea of a geometric spirit: "The geometric spirit is not so attached to geometry that it cannot be displaced and transported [*tire et transporté*] to other forms of knowledge. A work of morals, of politics, of criticism, perhaps even of eloquence . . . can be made by the hands of geometry."[52] Though Descartes had imagined geometry and mechanics as a means to understand sensation, memory, and other intellectual faculties, in Fontenelle these approaches became a tool for understanding a broad range of human action and behavior that far exceeded anything imagined by the ancients. In the same way that Descartes had overlaid a geometrical grid onto the human body and brain in *L'Homme*, Fontenelle's "geometric spirit" was grafted onto the world of human affairs. Creating a clear overlap between geometry and the proper conduct of a *savant*, Fontenelle noted that the "order, cleanliness, precision, and exactitude which reign in our good books for some time have their first source in the geometric spirit."[53] Making the final case for how physics and mathematics could transform all knowledge, and all knowers, Fontenelle noted that "physics, as far as possible, understands and disentangles [*suit et démêle*] all the traces of intelligence and the infinite wisdom which has ever been produced."[54]

While Fontenelle and the French Académie institutionalized the geometric spirit and instructed hundreds of *savants* in Cartesian mechanics, Descartes's actual mechanical account of the universe did not fare as well in the eighteenth century. Though he had begun with clear and precise rules, his actual theory of "vortices" as the key to celestial motion was surpassed by the less mechanical—but more mathematical—Newtonian account.[55] As the French philosopher Étienne Bonnot de Condillac (1714–1780) pointed out, Descartes had erred in departing from the original reductive vision of the *Rules* and *L'Homme* through the addition of a complex "system."[56] In fact, although Descartes had been an inspiration to many French thinkers, by the middle of the eighteenth century, Cartesianism was included in Condillac's *Traité des systèmes* (1749), a work intended to expose "the artifice of abstract systems."[57] Inspired by Newtonian mechanics and the philosophy of John Locke (1632–1704), Condillac suggested that the grand ideas of Leibniz, Spinoza, Descartes, and other mechanical philosophers were a mere continuation of the "sophisms" of the scholastics, agreeing with Gabriel Daniel that they had merely provided a new terminology for another hopeless set of occult qualities.[58] He noted, "Descartes, Malebranche, Leibniz, and many others" had "worked only to increase the number of abstract principles" and to create a "fecundity of maxims which had not been seen before."[59] The fewer principles and "obscure ideas" like "being, substance, essences, [and] nature," the better.[60] Like many in the French Enlightenment, Condillac imagined Descartes as a model skeptic, admirable for destroying the superstitions of old but unreliable as a guide to future science. As Condillac and others had noted, Newton had surpassed Descartes by reducing "several phenomena" to a single "principle," as the layers of substances, metaphors, and dichotomies that Descartes had added in fear of the Inquisition only complicated his key insight that nature was written in the language of geometry and arithmetic.[61] Newton's great genius was to simply describe the *effects* of gravity rather than hazarding a guess about the nature of the cause, and as Condillac and others recognized, such an approach could be a template for all scientific knowledge.[62]

Like Fontenelle, Condillac stressed that a more simplified methodology could even be useful for a science of human action. As he wrote in *Traité*, "the military and political sciences have their general principles, just like all the other sciences." Was it not possible, Condillac asked, for grand strategies to be reduced to mathematical principles, and for anyone to "become, after some time of meditation . . . a Richelieu"?[63] Some might laugh at the idea of using the geometric spirit to compete with the "good statesmen or generals," but Condillac reminded his audience of the absurdities of the scholastics, who had once taught that one could become "a scholar" for whom "nature held no secrets" simply by the "charm of two or three propositions."[64] If basic axioms could lead to metaphysical insight, Condillac asked, why could basic mathematical principles not lead to good government? Condillac also shared with Fontenelle a moral objection to what he believed were the shortcuts of metaphysics. Decrying the laziness of "innate ideas," Condillac lamented that "at the birth of philosophy" people were "content with notions less exact" because "observation appeared too slow." Those who looked toward substantial forms or innate ideas were "impatient to acquire knowledge."[65] Though not as focused as Fontenelle on right conduct, Condillac's own philosophy was similarly laden with moral implications for science work, insisting that the lessons of "sensations" could be learned only through patient and detailed observation. While Fontenelle had imagined training people to observe, Condillac saw his "geometric metaphysics" as a process of observation that itself trained the mind.[66]

While Condillac has been described as "scarcely a major philosopher," perhaps his greatest contribution was to elevate the more simplistic Cartesian vision of the *Rules* and *L'Homme* over other potential interpretations. In fact, though he was critical of Descartes, Condillac reserved most of his scorn for one of Descartes's most influential interpreters, the philosopher and Augustinian priest Nicholas Malebranche (1638–1715). Malebranche was no scholastic, and Condillac had credited him with a "beautiful spirit," but his ideas were attacked in the *Traité* because he had nonetheless kept alive innate properties, active wills, and other complications to mechanical philosophy. In part, this was because Malebranche took Descartes's ideas on God seriously, perhaps more seriously than Descartes, asserting that "the spirit of man was not material."[67] Malebranche's first mistake for Condillac was to assume there was therefore an immaterial world of "ideas" that guided human actions, that "all our inclinations, all our passions, and all our loves" lead toward this notion of "the good."[68] For Condillac this language merely recalled the teleological approach of the medieval and ancient world, where acorns became oak trees because it was their purpose to do so, and it was foolish to believe that "idea[s] and inclination are to the soul what features and movement are to the body."[69] Like Descartes, Malebranche had seen divine power as the first and only cause of all action, but Malebranche had further argued that each successive action was filled by an active God eternally creating and infusing the human will, providing a constant—and, for Condillac, needlessly complicated—system of exchange between the material world of nature and the divine world of the mind. Instead, following Locke, Condillac imagined that sensations alone, impressed upon the blank slate of the human mind, taught one how to feel and act.

In retrospect, it may seem surprising that Condillac spent even more time on Malebranche than he did on "Spinozism," Leibniz, and Descartes; however, it may have been the theologian's popularity at the time that was most threatening. As David Cunning has shown, Malebranche left far more room for human *action* than the stereotypical Cartesian automaton, and he represented a different path for a Cartesian science of man.[70] Opposed to Malebranche's dense and complex immaterial mind acting in concert with a divine intelligence, Condillac offered his famous "statue man," who learned through experience alone, a concept that had an enduring influence on theories of the mind.[71] Yet Condillac's attack on Malebranche was necessary to return Cartesianism to its brute origins in the *Rules* and *L'Homme*, before Descartes had added God and an immaterial universe to hedge against potential critics. While Malebranche would become a hero for French anti-Enlightenment figures who sought to rescue God and soul from mechanism and scientism, his influence in French social thought was limited by Condillac's criticisms and standing among French philosophes. Perhaps more importantly, Condillac's suggestion to worry about practical utility alone meant that sciences of man need focus only on effects rather than causes, as crucial works like the *Encyclopédie* were explicit that "science" be used for the practical goal of human happiness.[72] Last, by framing the argument as a moral one, Condillac produced an ideological framework to complement Fontenelle's vision of science work. Rather than Malebranche's approach of active wills and the constant interplay of mind and matter, the "geometric spirit" in France would be built on the dual foundations of rejecting metaphysics and a new vision of the *savant*. In Condillac and Fontenelle, what Daniel had dismissed as the "doctrine of the automata" now had both its feet on the ground.

Analyst over Geometer: Rousseau, Diderot, d'Alembert, and the Autocritique of the Geometric Spirit

While Fontenelle and Condillac had used the geometric spirit to rid institutions and popular philosophy of lingering scholastic and Cartesian "mystery," not everyone connected to the French Enlightenment was convinced that the pursuit of the sciences led inexorably toward the moral good. Although criticism from religious figures was to be expected, the concerns of three prominent philosophes suggested that the worry over reducing human behavior to mechanical and mathematical rules—or even *teaching* such a methodological approach—was more widespread. Rather than a religious concern that education in math and physics had the potential to dispel the mysteries of divine intelligence, the critiques of Jean-Jacques Rousseau, Denis Diderot, and Jean le Rond d'Alembert were more grounded in the "intelligence" and "morality" of those trained in the sciences. While all three stressed the success of science and technology in improving the material lives of human beings, each worried in different ways that exclusive immersion in the geometric spirit reduced the scope of the human mind and soul. Even if people did not truly operate like machines, these authors warned of the danger of even teaching such a doctrine.

Perhaps the most famous critique of the spread of education in scientific methodology occurred in 1750, when the Académie de Dijon offered the following prompt for their prize essay: "Has the restoration of the sciences and arts contributed to purifying our morals [*épurer les mœurs*]?" One respondent, a then unknown Swiss author named Jean-Jacques Rousseau (1712–1788), ended up winning the prize with a resounding "no." Rousseau, the rhetorical champion of the French Revolution, nevertheless held a complicated position in French social thought. As one chronicler has noted, Rousseau managed to be the Enlightenment's best champion *and* its best critic.[73] Like most of Rousseau's writing, his essay response, *Discours sur les sciences et les arts* (1750), has been subject to a number of interpretations. Often called the *First Discourse*, it has sometimes been considered a misstep or false start before his more subtle and influential work on inequality. At times, the *First Discourse* certainly can appear as a factually dubious history of empires laced with fervent militarism and essentialist nationalism. Yet what makes the work so interesting—and so relevant to the story of the mechanization of intelligence and science work—is that it challenged the fundamental claim of Fontenelle and Condillac that science work and right conduct were mutually reinforcing. The center of

Rousseau's vision was that while the cultivation of the sciences undoubtedly led to numerous benefits, the very qualities necessary for Fontenelle's ideal *savant* were unsuited to the development of the soul. As Jeff J.S. Black put it in his exhaustive survey of the *First Discourse*, Rousseau's critique of science was not based on whether it accurately captured the reality of nature but was rather "primarily . . . a critique of the popular dissemination of scientific knowledge."[74] In Rousseau's essay, which won the Dijon Académie's "prix de morale," Fontenelle's right conduct and Condillac's mathematical approach become a stifling and dangerous form of intellectual machinery, one that was in danger of spreading far beyond the Académie.

In the opening of the *First Discourse*, Rousseau offered little beyond a conventional critique of "luxury," arguing that in their practicality the sciences and arts produce an enervating wealth and that the cultivation of idleness led to a concomitant loss of courage, morals, and equality.[75] Such a critique would fit into many a sermon of the age and was far from a challenge to the geometric spirit. However, in the second half of the *First Discourse*, Rousseau attacked the practice of scientific research on the very grounds of moral virtue which Fontenelle had praised. He wrote, "Today, where subtlest research of the finest taste has been reduced to fixed principles, there reigns in our customs a vile and deceiving uniformity, and all minds seem to have been struck in the same mold" (*OCR*, 2:54). Initially, Rousseau built his argument through selective (and often questionable) interpretation of historical evidence, believing that those nations trained in the sciences had achieved little, while the great nations of the past had mostly ignored education in the sciences. Applying the Académie's question to the past, Rousseau claimed, "If the sciences purify our morals, if they led men to shed their blood for their country, if they animated courage, then the people of China would have become sages, free and invincible" (*OCR*, 2:55). Conversely, in Rousseau's telling, Rome had been built on virtue rather than talents, and only fell because "the sciences . . . prevailed." Putting science squarely in opposition to morality, Rousseau claimed that "science and virtue are incompatible." As evidence for the moral degradation of Rome, Rousseau quoted Seneca, who claimed that "since the scientists [*savants*] have begun to appear among us, the good men have been eclipsed" (*OCR*, 2:57).

Though polemical in tone, Rousseau's *First Discourse* did not dispute the findings of the sciences or argue (like the theologians) that Cartesian ideas had infringed on God's creation. Instead, Rousseau's attack on science was far broader, condemning almost all intellectual work as enervating to the soul.

No matter where one looked, "where the commodities of life are multiplied, the arts are perfected, and luxury spreads, true courage fades and militarism vanishes." As an example, Rousseau claimed, inventively, that the Goths allowed the Greeks their libraries to keep them "amused with sedentary and idle occupations" (*OCR*, 2:62). Nor did he follow Condillac in critiquing the system builders, instead arguing that scientific research and knowledge *should* remain in the hands of an elite class. In one of the most novel arguments from the *First Discourse*, he lamented the very attempt to spread scientific education and work to a broader public. Describing those who might seek to join an institution like Fontenelle's Académie, Rousseau claimed, "One would hope that . . . they would be thrown [back] into arts useful to society. Such would be the life of a bad writer or a lowly geometer to perhaps become a great maker of fabrics" (*OCR*, 2:68). For Rousseau, it was not just that scientific accounts of human actions threatened the breadth of human potential; rather, simply learning about scientific methodology might lead people away from more vital pursuits.

Rousseau was not alone in his warning that limiting thought to "fixed principles" might create "a vile and deceiving uniformity" in France. In fact, he had seen the Dijon Académie notice while on his way to visit his friend Denis Diderot (1718–1784), who had been recently imprisoned for scandalizing polite society and the Catholic Church by expressing his atheist beliefs in *Letter on the Blind*. Diderot, a cofounder of the *Encyclopédie* who often promoted knowledge and technology for the good of human happiness, did not exactly share his friend's interest in moral virtue. Instead, Diderot offered a more specific attack on the problems of mathematical reductionism and the geometric spirit. Though certainly enthusiastic about the ability of the natural sciences to undercut the claims of the church and to increase human happiness, in *Pensées sur l'interprétation de la nature* (1754) Diderot managed to combine a thorough skepticism of science education with a distrust of mathematical reasoning.

Like most of the philosophes, Diderot had a complex and at times paradoxical approach to the natural sciences, with one historian claiming that "Diderot's mobile mind, like the weathervanes of his native Langres, shifted freely in the winds of thought."[76] As coeditor of the *Encyclopédie*, he was a champion of science who nevertheless rejected the two main streams of Cartesian rationality and Baconian experimentalism. As he claimed in the preface to the *Pensées*, a collection of almost random asides on eighteenth-century science, there were two extremes in the "phenomena which appear to occupy our philosophers." Concerned about a dual fetishization of machines and mathematical dogmatism, he warned, "On one hand, it appears, there are a lot of instruments and few ideas. On the other hand, there are a lot of ideas and no instruments at all."[77] For Diderot, this meant that scientific work often led to either mental drudgery or pointless abstraction, and he reminded his readers of one of Pascal's more astute paradoxes: "Throughout one's life, one searches for rest in battling certain obstacles, and if one surmounts these obstacles, the rest becomes unbearable in the boredom [ennui] it engenders."[78] In sum, he believed in "the grand revolution in science" but was not always happy with the results (*DD*, 287).

Although he disagreed with Rousseau about the value of the material benefits of science, Diderot echoed his friend's concern that great effort and human labor were needlessly dedicated to observing esoteric parts of an ultimately unfathomable whole. Granting that the world may be one giant machine, Diderot questioned the point of trying to understand "the variety of the same mechanism in an infinity of different manners," pointing out that "to collect and read the facts are two very painful occupations" (*DD*, 295). Later in the same work, he noted the "painful and difficult" process of "large observations," implicating the modern mechanistic approach and the laborious process of observation and experiment Fontenelle and Condillac had championed (*DD*, 324). Unlike Descartes, for example, who believed the *Rules* could lead to a complete understanding of the world and humanity alike, Diderot remained skeptical that the research would ever end, believing that scientific knowledge "could never be completed, and even if it were, it would be beyond the reach of human intelligence" (*DD*, 288–89). In a revealing analogy that echoed Rousseau's proposed hierarchy in the sciences, Diderot explained how this reorientation of scientific labor "will stand for centuries to come like the Egyptian pyramids, where we see overburdened masses in hieroglyphs, revealing to us a frightening idea of their power, and the resources of the men who had elevated them" (*DD*, 287). Here was the concern of Rousseau—of frustrated "geometers" dedicating their life to reducing the world's diversity to "uniformity" rather than cultivating their own virtue or happiness.

Like critics before and after him, Diderot was specifically concerned with the effects of reducing the study of individual diversity of nature and humanity to a single methodological approach. Bypassing Rousseau's concern with the wanton luxury and atheism provoked by Cartesianism, the greater crime for Diderot was to remove "intelligence" from the universe. As he warned, the geometric spirit risked repeating the mistakes of the system-builders. In his words, it even had the possibility to create "a kind of general metaphysics where bodies are stripped [*dépouillés*] of their individual qualities." In contrast, Diderot praised the ideas of the French mathematician and philosopher Pierre Louis Maupertuis (1698–1759), who had suggested that "desire, aversion, memory and intelligence" could be found in even the smallest particles of matter.[79] Diderot had written the *Encyclopédie* entry for "Epicureanism" and had read Lucretius's *On the Nature of Things*, indicating that he believed certain elements of nature were simply irreducible. Rather than see intelligence as a mechanistic process confined to the mind, Diderot saw it everywhere. In later writings, Diderot even suggested that qualities like intelligence could be found in "grains of sand" as easily as they could be found in "elephants."[80] Diderot shared few concerns with Richard Baxter or the theologians who had warned of the hazards of a Cartesian world, but he did note that an inherent and active intelligence in both the world and individuals was under threat. Intelligence did not reside in the world machine but could be found in every living and nonliving part of matter.

Idiosyncratic thinkers like Rousseau and Diderot might have been expected to challenge the guiding principles of the geometric spirit. Neither were mathematicians, and both often found themselves outside of the main philosophic discourse of France. Yet the third major critic of the geometric spirit discussed here, Diderot's *Encyclopédie* coeditor, Jean le Rond d'Alembert (1717–1783), might be more surprising. D'Alembert was not only one of the most accomplished mathematicians in the world; he was also one of the central figures of the French Enlightenment, corresponding with nearly every key thinker from Voltaire to Condorcet and eventually becoming permanent secretary of the Académie Française. As a sign of d'Alembert's standing, the intellectual historian and philosopher Ernest Cassirer claimed that the opening lines of d'Alembert's *Essai sur les éléments de philosophie* (1759) were "the direct expression of the nature and trend of contemporary intellectual life" in the Enlightenment.[81] In the *Essai*, d'Alembert had even expressed enthusiasm for potential new breakthroughs in the sciences through the application of "Geometry" and the "lively fermentation of minds" that would follow.[82] In other words, d'Alembert should have been the perfect thinker to carry forward the geometric spirit in France. Yet a closer look at the *Essai* reveals that d'Alembert was not quite the champion of the geometric spirit that he appeared. The opening lines were not promising a triumphant account of the geometric impulse that began with the Cartesian revolution but offering a warning about its overreach.[83]

D'Alembert's *Essai* began innocently enough with an acknowledgment of what everyone had agreed upon: there had been "revolution in the human spirit" begun by Descartes, which had led to enormous progress in the study of nature. As d'Alembert acknowledged, Descartes's mechanical revolution had led to the "true world system being known, developed, and perfected." Geometry in particular had "expanded its limits" to encompass physics and causation, leading to a revolution in understanding everything "from the Earth to Saturn, and from the History of the Heavens to insects."[84] Yet the revolution had shown both "advantages" and "abuses" in the spread of geometric thinking. D'Alembert noted that there had been "a great enthusiasm which accompanied the new discoveries," but he worried it may have gone too far. Though many had offered up geometry as a solution to the unification of knowledge, d'Alembert found it unsuited to the task. As he claimed, the geometric spirit had become too reliant on classical axioms and principles, which allowed it to be just as reductive as any seventeenth-century metaphysics. He specifically chided Christian Wolf (1679–1754), the expositor of Leibniz, for his simple abstractions and deductions, claiming that "sterility and childish truths [were] the least of the problems with axioms" (*DML*, 715). D'Alembert further questioned Cartesian followers who had put forth principles like "one must exist simply before existing in such and such a manner" as absurd and lacking empirical validity. As he noted, the "idea of simple existence, without neither quality nor attribution, is an abstract idea which is only in our mind," wryly adding that "one of the great inconveniences of imagined general principles is to realize they are abstractions" (*DML*, 715).

While d'Alembert shared Fontenelle's and Condillac's desire to reduce "abstractions," he was less committed to the geometric spirit, noting that it was his goal to "unmask the Sophist disguised as a Geometer" (*DML*, 722). In a general sense, d'Alembert warned that the geometrical approach to a science of man—or *any* science—was impossible, claiming that "the majority of the sciences, like physics, medicine, jurisprudence, and History," contained "an infinity of cases where we are forced to act and to reason." In such cases where things could not be expressed with simple axioms, the evidence could never be "clear and convincing" (*DML*, 722). Even more provocatively, d'Alembert's stressed that research into these sciences required a more broad and varied set of skills than mathematics. In a statement that challenged Descartes's rationale for using geometry and mathematics to discover truths about nature, d'Alembert claimed, "The spirit which only recognizes the truth when it strikes them directly is inferior to that which recognizes . . . that which is fleeting and hidden far away" (*DML*, 723). For d'Alembert, the former spirit— which he derided as "l'esprit purement *géomètre*"—was far too "inflexible" to make sense of the contingent and ever-changing experience of nature, useful only in a "narrow and limited sphere" (*DML*, 723). While math and geometry had unlocked a number of important secrets, d'Alembert called on *savants* to "conserve the spirit of flexibility" by not always "turning towards . . . lines and calculus." In a remarkable conclusion from one of the best mathematicians of the era, d'Alembert suggested that science best proceeded by "tempering the austerity of Mathematics by studies less severe" (*DML*, 723).

Like Rousseau and Diderot, d'Alembert warned that *l'esprit géomètre* not only reduced nature to a few simple principles but that its application would also influence the conduct of scientific *labor*. He had written the largely positive "Geometry" article for the *Encyclopédie* and had praised the geometric spirit and Fontenelle in the *Preliminary Discourse*, but d'Alembert warned that the process of gradual accumulation of axioms Descartes had inherited from the Aristotelean scholastics was not well suited to modern investigations and that "one can regard the method of the Ancients as a torturous, difficult and upsetting path in which Geometry trains and tires its readers" (*DML*, 785). In contrast to the geometer, d'Alembert imagined a modern "Analyst" who had the "spirit of play" and could instantly seize upon a wide range of facts to see the relations between them. Such an analyst would need more than the practical training offered at the Académie royale des sciences. Like many French mathematicians, d'Alembert invoked the analogy of the gambler, a figure who could grasp instantly what it would take the geometer days to deduce. As he concluded, for this reason "it was not surprising that a great Geometer is often a very mediocre gambler" (*DML*, 788).

For d'Alembert, the "analyst" might do better than the geometer but would never approach the status of genius. Though he did not share the concerns of Diderot and Pascal about the boredom that would necessarily follow complete knowledge, he did reiterate their worry about the inability of a slow and tedious process to produce great truths, as even the best analyst could not capture everything. Rather, d'Alembert reserved the most important abilities for what he called a "supreme Intelligence." Like Diderot, he believed that an "immense labyrinth" of truth could only be appreciated by "a supreme Intelligence which embraced all of the turns in the blink of an eye." Geometry, he claimed, should not be mistaken for this "Intelligence" (*DML*, 714). Although d'Alembert's *Preliminary Discourse* to the *Encyclopédie* is often taken as one of the great statements of modern science, the *Essai* ended on a more ambivalent note, claiming, "The sciences are a kind of great edifice in which many people work in concert. Some with the sweat of their bodies carry the stones from the quarry, some carry them with effort to the foot of the building, and some lift them with the strength of their arms and machines. But it is those who put them to work and in place who have the credit for the construction" (*DML*, 842). Although d'Alembert differed from his philosophe friends in temperament and training, this "great edifice" looked a lot like Rousseau's world of frustrated "lowly geometers" and Diderot's analogizing science to building the pyramids of Egypt. In this vision, the geometric spirit not only created an abstract and diminished vision of mind and matter, but it also threatened to shape the everyday lives of those caught up in the process. As seen in following chapters, the "autocritique" of Rousseau, Diderot, and d'Alembert would recur with each new attempt to model a theory of human thought and action on the methodologies of the natural sciences. Not only was the general and divine intelligence of the universe at risk of being reduced to a few simple principles, but the individual minds and souls of people were as well.

Conclusion

Beginning four hundred years ago with Descartes's first reflections on the precise rules of geometry and mathematics, French thinkers of the early modern era did significant work to transform nature and human action into subjects that could be analyzed through the methodologies of the natural sciences. From mechanical philosophy to the geometric spirit of Fontenelle and Condillac, the "occult qualities" of supreme intelligence, immaterial soul, and metaphysical reflection were removed as obstacles in the study of nature and humanity. Henry Oldenburg's dream had been realized, and while Laplace's all-knowing "demon" had not yet arrived, it was close at hand. Even more, as Fontenelle had championed and Rousseau lamented, the *savants* necessary to study the world were created during this era as well, as a new form of science work emerged in France to complement the new methodological imperatives.[85] Although the efforts of French thinkers were not full-fledged attempts to create a "science of man," the necessary theory and practical training for these sciences had been put in place by the middle of the eighteenth century.

The debates of the scientific revolution and Enlightenment over the consequences of the mechanical philosophy and the geometric spirit did not dissipate in the coming centuries. Rather, the move to reduce the study of nature and human behavior to a single methodological approach lived on in a host of "social sciences," seen over the next several chapters, including social mathematics, social physics, neoclassical economics, eugenics, behaviorism, and modern computer science. The idea of person as machine was, of course, also prerequisite for the idea that machines might gain intelligence or consciousness. While Cartesian vortices and other attempts at natural science became an embarrassment to even his most vocal defenders, Descartes's grounding of the analysis of living beings in mathematics and mechanics endured, becoming foundational for both the modern social sciences and our vision of intelligence today.[86] In many ways, as Margaret Osler asserts, Descartes did not just reduce living things to mathematics; he reduced even the idea of divine intelligence to the level of a machine.[87] Descartes in the *Olympica* had imagined "God as pure intelligence," but this intelligence was now dependent upon higher laws of geometry, arithmetic, and algebra (*OPD*, 63).

While critics like Diderot, Rousseau, and d'Alembert might have had less issue with dethroning the concept of divine intelligence, they likely would have agreed with the English clergyman and philosopher Ralph Cudworth (1617–1688), who declared that Descartes's world had become a "dead cadaverous thing."[88] Though they would have been surprised to find themselves in the camp of theologians like Baxter and Daniel, they too warned that mechanism had unleashed a more powerful occult quality than even the scholastics could have imagined, as the reduction of complex behavior to simple principles and methodological approaches became its own dogma. In doing so, these philosophes anticipated legions of critics, from counter-Enlightenment reactionaries to social reformers to American social scientists who believed that the methodologies of the natural sciences should not be used to study human thought and behavior. Even more, they recognized that the "spirit of play" had been lost and that the "dry and arid" work Fontenelle portended was unsuited for many *savants*. Not for the first time in this story, critics recognized that in the sciences of man, reductionist ideas did not just lead to a simplified portrait of humanity. They also entailed monotonous labor for those researchers who took up the task.

At *the* Bleeding Edge

From Mechanism to Probability in the Sciences of Man

> All has changed in the physical order, and so all must
> change in the moral and political order. Half of the world
> revolution is already achieved; the other half has yet to
> be accomplished.
> —Robespierre (Year II)

> I consider it possible to convert men into republican
> machines.
> —Benjamin Rush (1798)

WHEN MAXIMILIEN ROBESPIERRE (1758–1794) SPOKE TO THE FRENCH
National Convention in May 1794, some in the audience may have noted an
echo of Fontenelle's geometric spirit, where the great leap in mathematical
and mechanical methodologies provided the model and metaphor for a new
science of morality and politics.[1] In Robespierre's speech, which proposed
a new "Festival of the Supreme Being," he declared that the gulf between
monarchical Europe and Republican France was like "the distance between
the astronomical observations of the wise men of Asia and the discoveries of
Newton," and he implored his fellow representatives to complete the French
Revolution through similar progress in "morality" and "virtue."[2] As John
Carson traced in his history of intelligence, both French and American reformers attempted at this time to create what Benjamin Rush called "republican machines" through applying scientific methodologies to human action.[3]
Yet, as French crowds gathered around Notre Dame Cathedral one month

later to celebrate the "Supreme Being," a darker vision of a mechanical public began to emerge, and it is hard to believe that those assembled imagined their own lives as guided by stable laws derived from the sciences of man or that Newtonian physics could provide the model for a virtuous moral and political world order. The Reign of Terror had been in effect for over a year, and just a month earlier Parisians had seen the Revolution's most eloquent spokesperson, Georges Danton, executed by the guillotine, his severed head displayed before the crowd. In the years prior, French *citoyens* had seen forced conscription, the burning and drowning of counterrevolutionaries in Vendee and Lyon, and the murder of thousands of political prisoners in the streets and jails of Paris. If there was any intelligence guiding the world—divine or secular—it did not seem to share its graces with France in the spring of 1794.[4]

In France, as elsewhere, using science to simplify human action to the level of the machine had profound consequences. While mechanical philosophy, the geometric spirit, and the new sciences of man were only a part of the story of the French Revolution—and the Terror—the decades leading up to 1789 were marked by a number of *sciences humaines* that equated human actions with mechanical laws and were explicitly based on geometry and mathematics. Remarkably, even Robespierre believed the sciences of man had gone too far in reducing human thought and action to mechanical laws, and his call for a Cult of the Supreme Being was intended to reintroduce a sense of divine intelligence, a check on the excesses of the Cult of *Reason* started by his fellow revolutionaries on the Committee for Public Safety.[5] For those like Rousseau, Diderot, and d'Alembert, there was also good evidence that their worries about a hierarchical and mathematical science of man that privileged a select elite had been realized, as many of the most violent revolutionaries saw *savants* as the enemies. As one pamphlet entitled "Rethink the Terror through Science?" declared, France needed a *"sans-culotticized* science" because the *"savants"* were "employing a mystical language."[6] Not only did the new form of science alienate those at the bottom of Diderot's "pyramid," but increasingly mechanistic accounts of human behavior scandalized those at the top of French society as well.

Though the decades just before the French Revolution included a number of attempts to simplify the thoughts and actions of human beings to mechanical processes, this chapter begins with a comparison between two of the most scandalous sciences of man of the era: Julien Offray de La Mettrie's *L'Homme machine* and Claude-Adrien Helvétius's *De l'homme*. While La Mettrie's infamous "machine man" is regularly deployed in brief accounts of thinking machines and AI, cursory histories of the subject often miss that the work was intended as a parody of French social and intellectual life. Rather than act as the prophet of a new world of machine AI, La Mettrie used a series of reductio ad absurda to critique the artificial intelligence produced in real people by what he called the "mechanics of education." In contrast, in *De l'homme*, Helvétius imagined a *genuine* education machine that reduced human thought and action to invariable mechanical laws. Following La Mettrie and Helvétius, the Baron d'Holbach's *Système de la nature* was likely the most contested work among Enlightenment philosophes, as it seemed to deny even the idea of intelligence. While French thinkers from Descartes to Fontenelle to Condillac had reduced the role of a guiding intelligence in the complexity of human lives, Helvétius and d'Holbach threatened to simplify human action to the point where intelligence was eliminated altogether. As the historian of Enlightenment Jonathan Israel has noted of the major works of Helvétius and d'Holbach, "no aspect of the new eighteenth-century materialism seemed more menacing" to entrenched ideas.[7]

In many ways, the arc of the French Enlightenment sciences of man mirrored the Revolution itself. Just as Robespierre tried to push back against the extremes of the formal Cult of Reason, Diderot, d'Alembert, and others were again enlisted to temper the excesses of a mathematical and mechanical science of man produced by Helvétius and d'Holbach. And while the heated contests between various forms of republicanism eventually gave way to the stability and repression of the Consulate and Empire, by the turn of the century in France, the competing mechanistic sciences of man ceded importance to a science of man founded on probability.[8] While Napoleon of course provided the political solution, the mathematician Pierre-Simon Laplace finally dulled the blade of certainty that reigned in the mechanical sciences of man through his application of *probability* to the sciences.[9] Laplace was a *savant* rather than *un homme de lettre*, and he did not share his ideas in salons or in the impassioned settings of revolutionary politics like previous social thinkers. In the staider settings of Laplace's vision of professional science, intelligence in France could be rescued from materialist extinction by observation, data analysis, and probabilistic reasoning. As seen in the conclusion, Laplace's methodology for the sciences of man temporarily saved intelligence from the reductive oblivion of Helvétius and d'Holbach, but only through reconstituting the divine into the mundane.[10]

The Education Machines of La Mettrie and Helvétius

Before examining the sciences of man put forth by Helvétius and d'Holbach, it is worth reflecting on one other notorious project of mechanization in eighteenth-century France, Julien Offray de La Mettrie's *L'Homme machine* (1747). La Mettrie (1709–1751) was actively scorned by many philosophes and had almost no mathematical training, but he is a favored starting point for telescopic and anachronistic surveys of thinking machines nonetheless. One history even helpfully informs the reader that "the project of artificial intelligence has made great strides since the time of La Mettrie," while others see La Mettrie as the instauration of the very idea of a mechanical person.[11] To read *L'Homme machine* today, however, is to recognize that it was far from a precursor to the modern age of AI, thinking robots, or the many science fiction connotations that the title might inspire. La Mettrie's machine man looks nothing like an Enlightenment automaton, let alone an artificial intelligence, and actual mentions of machines in the text are rare.[12] Neither was it, as many critics contended, a simple continuation of mechanical philosophy, the geometric spirit, or a means to mechanize the soul in any modern sense of the word, as La Mettrie believed all those practices to be "rubbish."[13] As Kathleen Wellman put it in a long-overdue intellectual history of La Mettrie, "the analogy of the robot or even the computer . . . is quite foreign to the spirit of the text itself."[14] La Mettrie did not continue the mechanistic thinking of the geometric spirit, as many have claimed, but rather satirized its extreme conclusions.[15] As seen, instead of the first step in the advance of a great quest for AI, La Mettrie was one of its first critics, as his scabrous prose concealed a dire warning that human intelligence might descend to the level of the machine.

Though it caused a continent-wide scandal, *L'Homme machine* began conventionally enough with basic facts drawn from everyday experience that could hardly be refuted. For example, he noted that "the diverse states of the soul . . . are correlated to that of the body."[16] When the body slept, La Mettrie claimed, so did the soul, as the "brain fibers" become calm in the same way the "muscles of the body" relax during sleep. Both "rose" along with the sun, so it was no coincidence that upon waking both the mind and body struggled to function. Such was the "rule [*empire*] of climate," with great swings in emotion and physical state occurring because of temperature, hunger, and other external physical conditions (*HM*, 157). From these fairly banal examples of physical changes impacting the soul, La Mettrie moved into his true areas of expertise—anatomy and physiology—where it became clear that he was interested in demonstrating not so much that man was a machine but rather that at the level of matter man was nothing more (or less) than an animal. As he declared, "To know even the slightest bit about human nature," one must "open the entrails of man and animal" to see the "true parallel structures of one and the other" (*HM*, 158). Allowing that the human brain was more complex and larger than other animals', La Mettrie still claimed that "the Ape, the Beaver, the Elephant, the Dog, the Fox, and the Cat . . . are the Animals which most resemble man." Drawing on Thomas Willis's *Cerebri Anatome* (1664) and *De Anima Brutorum* (1672), La Mettrie then listed a series of correlations between brain size and "ferocity," noting that Willis's comparative anatomy had shown that larger brains led to docility. In humans, too, it was clear that the structure of the brain influenced personality, as La Mettrie quoted Fontenelle declaring that "imbecility" was not the fault of the person but rather the result of "a small fiber that not even the most subtle Anatomy could discover."[17] It was for this reason that apes could even be considered similar to childish or foolish humans, and La Mettrie noted that the orangutan—the animal that resembles humans so strongly—was often called the "Savage Man" (*HM*, 160).

La Mettrie's detailed knowledge of anatomy exposed a potential problem for the sciences of man based in mechanical philosophy. Descartes had attempted to wall off the mechanical world of animal life from the human soul, but if the founder of the geometric spirit, Fontenelle, could reduce human intelligence to a simple "fiber" in the brain—one that might be found in both humans and animals—how was human intelligence to escape the pure Cartesian machine? La Mettrie was not the first to note similarities between humans and primates, but he followed this discussion with a simple question that indicated the full scope of his satire: "Why, therefore, would it be impossible to educate an Ape?" Though it scandalized many readers, La Mettrie's presentation indicated that it was a natural conclusion drawn from simple anatomical studies. Perhaps even more astonishingly, he then speculated that simian education could be made possible through a medical procedure, equating the education of an ape to the restoration of hearing in deaf people. In his words: "Could not the same Mechanism which opens the Eustachian Canal for the Deaf be used to open it in the Apes?" (*HM*, 161). Taken at face value, this was an astonishingly stupid question, as it required La Mettrie to believe that the same physical defect that prevented deaf humans from hearing—a blocked passage in the ear—prevented apes from learning human speech. The logic indicated that La Mettrie believed the eustachian canal blocked both sound *and* understanding, as if the two were even remotely similar. (It is hard to imagine, for example, that La Mettrie believed that French-speaking people who do not understand Greek could literally not hear it, or that a medical procedure could allow a person to learn a new language.) Nor does

La Mettrie describe or suggest an analogous organ that might be "blocking" an ape from understanding language. The question then is patently absurd both as anatomy and as metaphor. Nevertheless, La Mettrie claimed that a mechanical vision of both human and animal meant that the "liberation of the organs of speech" in apes could give way to "intelligence" (*HM*, 162).

Taken seriously, it is easy to see why La Mettrie's ideas were a great source of laughter. Yet, taken as satire, they become quite serious. The perils of descending to mechanical analogies could be seen particularly in his discussion of learning apes, where he attempted to show how similar the French educational system was to animal training. La Mettrie first imagined an ape learning figures and colors like a child in school. After that, "Words, Language, Laws, Science, and the Fine Arts would come, and then at last the brute Diamond of the soul would be polished." Such a process was possible because, La Mettrie declared, in the French educational system, "One trains a man like an animal, and he becomes an Author like he becomes a Porter [*Porte-Faix*]" (*HM*, 162). Such a strange sentence becomes clearer when recognizing that the French word *portefaix* as a verb literally means to carry a heavy burden, but as a noun it can alternately mean a "rude and brutal person."[18] La Mettrie was imagining not that animals were capable of the wondrous intelligence of mankind, and could therefore take part in elevated dialogue, but that when viewed purely in mechanical terms, both animals and humans lived lives of stunning ordinariness and routine. Mocking the geometric spirit and "right conduct" of Fontenelle's *savants*, La Mettrie claimed that "a geometer learns to understand the most difficult Demonstrations and Calculations like an Ape learns how to take off and put on his little hat while he rides a trained dog" (*HM*, 162). Though it is not clear where La Mettrie got this analogy, his conclusion was no different from Rousseau's, who saw the "lowly geometers" as also practicing the most rote form of training. In a statement laden with sarcasm, and one that acknowledged the hierarchies of science noted by Diderot and d'Alembert, La Mettrie claimed, "One can see that nothing is so simple as the Mechanics of our Education!" (*HM*, 162). This was no praise of smart apes but mockery of dumb students, as the routine of mathematical education itself seemed to simplify the mind.

Though he created an astonishing title, La Mettrie was far from "setting a course towards . . . mechanism that proved to be irreversible."[19] Machines may well "be born with Intelligence," and a *savant* at the French Académie may be no more than a trained monkey, but this was not much of a foundation on which to build a science of man, and it was certainly not the first step in creating AI. Instead, La Mettrie's machine man revealed that the Cartesian *l'homme* was an empty vessel, one that could be filled to the brim with all sorts of nonsense by anyone with a few cursory observations and a good deal of imagination. Cartesian dualism had supposedly allowed room for an immaterial soul, but La Mettrie extrapolated to the most logical extremes what occurred when a science of man approached human thought and behavior as simply another form of matter. La Mettrie was not the natural continuation of the geometric spirit, and nothing in *L'Homme machine* portended the idea of creating an artificial intelligence or a thinking machine. Yet, in his crude and hilarious exposé of the foolishness of reducing all of life to matter alone, he demonstrated where the most reductive sciences of man might be heading.

The philosophes tried to ignore the absurd machine man of La Mettrie, worrying that that any engagement with him might discredit the whole idea of a science of man, but genuine versions of mechanical men entered mainstream French Enlightenment discourse in the following decades. To find perhaps the best example of the extension of the geometric spirit into a science of man, and certainly the most straight-faced attempt to refine the "mechanics of education," we must turn to the *salonnier* and philosophe Claude-Adrien Helvétius (1715–1771), who published his most complete ideas in the posthumously published *De l'homme, de ses facultés intellectuelles et de son éducation* (1772). Helvétius's earlier *De l'esprit* (1758) represented perhaps the rhetorical high point of French radicalism, and his later *De l'homme* presented the most systematic science of man of the age. In this work Helvétius had, incredibly, taken the punchline of one of La Mettrie's jokes—that education was just another machine—and turned it into a serious project for the betterment of humanity. For Helvétius, genius, talent, and moral virtue were exclusively the result of education rather than a person's innate ability or mental "organization," with genius in particular described as "the extended product of random events." For example, Helvétius argued that the great French inventor and machinist Jacques de Vaucanson (1709–1782) owed his genius not to inborn talent but to the good fortune of accompanying his mother every day to her work at a machinist's office.[20] Building from Condillac's "statue man," Helvétius proposed that a mind completely absent of internal qualities could be directed by determining which sensations resulted in which kind of intelligence.[21] His goal, put simply, was to "demonstrate rigorously . . . that all the operations of the mind can be reduced to sentiments" (*OCH*, 2:78).

Whereas La Mettrie used few mechanical analogies and spent a great deal of time on self-generated action and imagination, Helvétius argued in

De l'homme that there was no internal mechanism at all. For him, the human mind was a passive receptor of external experience, or what he called "random accidents." Helvétius has been called the "extreme radical thinker of his time," and his radicalism consisted mostly in the logical application of Condillac's "statue man" to a science of man, taking these theories of learning to their natural end. Unlike La Mettrie, Helvétius relied on almost no anatomical evidence and presented his theory of man as a set of demonstrable and abstract principles.[22] In another contrast to La Mettrie's machine man, which was based upon the organization of the physical body, Helvétius also made it clear that physical organization had nothing to do with education, cleverly dismissing "organization" and "temperament" as "occult qualities" (*OCH*, 2:70). There were no animated "brain fibers" or descriptions of the physical body at all in *De l'homme*, and certainly no suggestion that the structure of matter dictated human intelligence. For Helvétius, people differed not because they were "organized" differently, but because of the "accidents" of education to which they were exposed. Therefore, if all mental education were the result of accidents, Helvétius's science of man proposed to study the results of these accidents and to determine which outcomes were connected to which experiences. In the end, he argued, one could eventually "reduce the science of man" to a few principles of education (*OCH*, 2:608).

Like Descartes, Fontenelle, and Condillac, Helvétius also believed that the principles of reduction which succeeded in mathematics were the key to studying human action. Helvétius imagined *De l'homme* itself as a kind of geometrical proof for the possibility of a science of man, noting that "in geometry all problems not completely resolved can become the object of new demonstrations. It is the same in morals and politics" (*OCH*, 2:12). As he wrote in the introduction, "the science of man is part of the science of government," and the main problem of education was that it was "almost entirely reduced to the study of some false science, to which ignorance was preferable" (*OCH*, 2:4). While a geometric study of morals and politics did not necessarily need to challenge existing ideas, vast sections of both *De l'homme* and his earlier *De l'esprit* were given over to denunciations of revealed religion, church corruption, privileged hierarchies, and institutions that had claimed legitimacy from their "natural" standing in society. Yet in Helvétius's reading, they were simply accidents of unscientific education. Not only would a science of man reveal an alternate set of ethics, but it would show that established institutions actively hindered ethical progress (*OCH*, 2:121). Such conclusions shocked his contemporaries, and *De l'esprit* was called "unquestionably . . . the most radical form" of Condillac's sensationalism and burned by Parisian authorities.[23] While it has been claimed Helvétius's greatest sin was to have "rehabilitate[d] the dreaded La Mettrie" under the pretext of the "respectable" Condillac, it might be more accurate to say that Helvétius simply took seriously ideas that La Mettrie lampooned.[24]

For Helvétius's science of man to work, however, he needed to go beyond axiomatic deductions, and here he joined Condillac's philosophy with Fontenelle's vision of science work. Instead of pure Cartesian reflection and introspection, he proposed a more inductive and experimental approach, one that required an extensive level of observation and data collection. Because all men were "commonly organized," different levels of genius could only be recognized through patient and careful attention to the effects of different accidents, as "the principles of good morals . . . must . . . be established on a large number of facts and observations" (*OCH*, 2:150). Rather than something like the "law of large numbers" proposed by Jakob Bernoulli (1655–1705), where an increase in observation increased the likelihood of determining the truth, Helvétius had imagined such a process as simple trial and error. For Helvétius, such large-scale experimentation had proved successful in machine making, and he believed a "science of education" could proceed exactly in the same way that new machines were invented. As he noted, "A mechanic invents a new machine, and through calculating the effects and proving its usefulness, the science is perfected. The machine is not made . . . but it is discovered" (*OCH*, 2:598). So too could the education machine emerge through observing and "calculating the effect" of different forms of instruction. For Helvétius, this process of controlled trial and error represented a true "science of man" that he called "physical sensibility," and *De l'homme* presented a thorough excavation of how such an inductive science could be applied to education (*OCH*, 2:608). It was this expansive plan of experimentation that led Jeremy Bentham to later remark that "what Bacon was to the physical world, Helvétius was to the moral."[25]

Though Helvétius often used the Cartesian language of clean and distinct impressions and imagined his science of man as a series of geometrical proofs, he did not believe one could grasp complete and immediate truths intuitively. Furthermore, Helvétius followed Descartes in the idea of man as a machine, but he rejected the extensive *machinery*, leading to a muddled and inconsistent science that managed to borrow from both Condillac's empty "statue man" *and* Descartes's densely wired "man like us" from *L'homme*. Caught between Cartesian rationalism and Baconian empiricism, Helvétius chose both, creating a lasting tension that would provide methodological troubles for later sciences of man and the social sciences.[26] As the historian Ira Wade noted, these kinds of equivocations in Helvétius dated back to *De l'esprit*, which Voltaire found to be "systematic but unmethodological" and Diderot believed was "methodological but not very systematic."[27]

Part of the reason other French thinkers objected to Helvétius was that he often conflated one or another idea that each philosophe had separately championed and brought together so many different strands of Lockean, Newtonian, and Spinozan ideas that it was difficult to find any unifying principle in his work other than his distrust of existing institutions and the desire to study human beings scientifically. The result was a radicalism that was different from La Mettrie's, and one more threatening to the status quo.[28] Voltaire, for example, found it ludicrous that common people could have the same aptitude for intelligence as someone like Aristotle, and he denounced *De l'esprit* in letters to both Condorcet and d'Alembert. Rousseau objected for completely different reasons: because he believed in the inherent goodness of mankind in nature, he objected to Helvétius's belief that men are born neither good nor evil but could be created so through the means of their governments and education.[29] Rousseau's popular *Emile* (1762) was in fact a response to Helvétius, and it is notable that *De l'esprit* attracted criticism both from those who saw French hierarchical society as the hallmark of civilization (Voltaire) and from those who saw it as an extreme fall from an idyllic state of nature (Rousseau).[30]

As a fellow radical thinker, however, it was Diderot who felt the need to launch a line-by-line attack on Helvétius's theory of education, calling the work *Réfutation suivie de l'ouvrage d'Helvétius intitule "L'homme"* (1780). In Diderot's account, Helvétius's education machine had stripped people of all internal qualities. As he wrote of the oft intractable nature of intelligence, "stupidity and genius occupy the extremes of the scale of the human mind. It is impossible to remove stupidity. It is easy to remove genius" (*DD*, 457). Diderot agreed with Voltaire that natural aptitude, rather than the result of random accident, determined intelligence. For example, Diderot challenged Helvétius's account of the famous inventor Vaucanson, who had created a series of "automatons" that delighted European audiences. Helvétius had claimed that as a child the inventor had been directed toward his future career because his mother worked for a machinist, but Diderot joked that one could "Give me the mother of Vaucanson, and I could not create a better flute automaton" (*DD*, 463). Vaucanson, notably, had been the great creator of empty mechanisms, most famously the duck that supposedly ate and defecated, and Diderot might have imagined the pliable *l'homme* of Helvétius as a similarly unintelligent fraud.[31]

Although much of Diderot's "refutation" consisted of anecdotal evidence on the education of various historical figures, as well as a great deal of simple disagreement about the "nature of man," he also pointed out that Helvétius's "geometrical" education was trying to have it both ways. While Voltaire and Rousseau hinted at these problems, Diderot once again questioned whether a science of man could be reduced to either simple deductive principles or large experimentation. Refuting Helvétius's idea that all truths can be reduced to simple "terms," Diderot asked if "such a reduction is always possible" (*DD*, 528). For Diderot, geometrical certainty did not result from "the results of a single step, or even in their totality, but in their linkage [*enchaînement*]" (*DD*, 529). For example, Helvétius had claimed that all actions were the result of simple pleasure or pain, but Diderot objected to this as an absurd simplification, asking, "Is it true that pleasure and pain are the only principal actions of not just animals, but of men as well?" (*DD*, 482). As Diderot noted, Helvétius had to at least acknowledge the existence of different innate talents as well, as human beings clearly have "hidden" motivations.[32]

Although Diderot was sympathetic to the idea of mechanism as applied to human experience, he still believed that each human "machine" was unique. For Helvétius, however, the very existence of mathematical and mechanical laws *required* that the machines that operated according to those laws had to be identical as well. As he put the matter, "If . . . each individual was a different machine, how could the heavens, or even earth, require the same effects of different machines? . . . Therefore all commonly organized men must have equal abilities of the mind given to them by the Divine" (*OCH*, 2:604). Helvétius's stirring vision of radical equality was unique even among French thinkers, and he may have been the one philosophe most disposed to converting *all* men and women into "republican machines."[33] This may have accorded well with the republican vision, but for the more politically conservative Diderot, the machine analogy offered another possibility: that the range of inherent mental abilities pointed to the fact that diversity was *itself* part of the laws of nature. In Helvétius's *De l'homme*, Diderot had seen his earlier concerns about the extent of reductionist thinking fully realized, where nature and humanity were reduced alternately to mathematical laws and trial and error. While his 1754 attack on the geometric spirit had been mostly a critique of a *hypothetical* science of man, the popular success of Helvétius's works had shown that such sciences were now spreading among the masses.

The Sandstorm and the Revolution: The Denial of Order in d'Holbach's *Système de la Nature*

Though his refutation of Helvétius was thorough, Diderot was considerate enough to delay publication until after his own death in 1784, thus sparing any direct conflict with Helvétius's widow, the prominent *salonnière* Anne-Catherine de Ligniville, Madame Helvétius (1722–1800). Social niceties too may have kept Diderot from writing a similar refutation of the works of his great friend Paul-Henri Thiry, Baron d'Holbach (1723–1789), whose own salons provided Diderot with unequaled access to French politics and ideas.[34] In particular, d'Holbach's extraordinary project of mechanization, *Système de la nature* (1770), should have been subject to a similar scathing review, as the Baron violated nearly every principle of judgment Diderot had laid out in the *Pensées* and in his refutation of Helvétius. Instead, it has been suggested that the Baron's most notorious work was inspired, or even written, by Diderot himself.[35] Yet such a connection is unlikely and unnecessary, as d'Holbach published dozens of books and hundreds of articles, only publishing his most scandalous work anonymously to spare his name and fortune.[36] *Système de la nature* could not have been further from Diderot's belief that intelligence could be found in every particle of matter. For the Baron's science of man, even the concept of intelligence was an illusion.

Système de la nature opened with one of the more direct statements of eighteenth-century materialism, the claim that "the moral man is only this physical being considered from a certain point of view."[37] This was a step beyond the arguments of Fontenelle and Condillac, which had merely suggested that the methodologies that had proven successful in the physical sciences should be used in the "moral sciences" as well. For d'Holbach, morality, the soul, and other seemingly unique aspects of humanity that Descartes and others had walled off from mechanical descriptions were only illusions born of ignorance. Yet perhaps, the Baron offered, man viewed mechanically was not such a bad thing. As he explained, the goal of his work was "to restore Man to Nature" in order to bring about the "bliss" hidden behind the "shadows" of the "imagination" (*OPH*, 166–67). For d'Holbach, a scientific "system of nature" was necessary because all previous religious and metaphysical approaches to understanding humanity had failed. Although he acknowledged that these systems had ordered experience in creative and novel ways, they had been based on the fatal illusion that some form or substance existed beyond matter. As d'Holbach concluded in a passage suppressed in the 1781 edition of *Système de la nature*, religion "was only a science of illusions" (*OPH*, 169).

Many attempts to reduce human actions to mechanical principles had been based on the spontaneous "swerves" of Gassendi's Epicureanism, but d'Holbach's reduction of humanity to nature retained no such unpredictability.[38] Instead, what made the Baron's vision so stark was his belief that people operated under infallible and strict mechanical laws. For him, all human actions were "equally natural, the necessary consequences of its proper mechanism and of the impulse it receives from the beings which surround it." This meant that "all our ideas, our desires, and our actions are necessary effects" resulting from the "qualities that nature has placed in us and . . . the circumstances by which we are obliged to experience and to be modified" (*OPH*, 168). In other words, people were perfect machines subject to modification only through external causes they could not control. This was a mechanism that surpassed even Helvétius, who believed the causes of supposed "accidents" could at least be studied and therefore reproduced in education. For d'Holbach, however, *all* education—good and bad—was equally natural, and there were no such things as accidents. For humanity, the chain of wills and actions went back to the very origins of life and the universe itself. Or, in d'Holbach's own words: "All the steps that we take to modify our being can only be regarded as a larger consequence of causes and effects which are only developments from the first impulses that nature has given to us" (*OPH*, 169).

To demonstrate just how determined nature and mankind really were, d'Holbach even presented an early (and even more deterministic) version of Laplace's Demon, the all-knowing intelligence that could predict future outcomes given the proper data.[39] As a thought experiment, d'Holbach presented two analogies, "one physical and the other moral." The physical analogy was that of a "sandstorm," which appeared as a vast "confusion to our eyes," with all matter of particles—large, small, and invisible—flying about. Though it would be impossible for any one person to trace or predict the path of each particle, d'Holbach stated that "there is not a single molecule of sand . . . which has been placed at *random*, and which does not have sufficient cause to be placed where it is found" (*OPH*, 196). Similarly, in political revolutions (the moral example), among the revolutionaries "there is not a single action, a single word, a single thought, a single wish [*volonté*], a single passion . . . which is not necessary" (*OPH*, 196). Anticipating the claims made by many participants of the French Revolution and the Reign of Terror, d'Holbach said all the agents of his hypothetical revolution "act as if they must act . . .

following from the place each person occupies in the moral vortex."[40] In both the sandstorm and the revolution, d'Holbach argued, such a necessity could only be seen by "an intelligence which would be in a position to grasp and appreciate all the actions and reactions of those minds and bodies who have contributed to the revolution" (OPH, 196).

Though the idea of an "intelligence" that could "appreciate" all the events of a revolution might suggest that d'Holbach believed in some unifying principle of nature, the fifth chapter of Système de la nature confirmed that even the idea of intelligence was an illusion. An intelligence like Laplace's Demon *may* have been able to "grasp and appreciate" the location of every particle of sand or the motivations of every revolutionary, but that did not mean that such an intelligence actually existed. In a chapter titled "On Order and Disorder, Intelligence, and Randomness," d'Holbach denied the existence of all four of the titular concepts. Notions of order and disorder, which were necessary to the definition of intelligence, were instead found only "in our minds," as the residue of fruitless attempts to sift through the confusion of sense experience. It was not just that ignorance led to incorrect ideas about nature, as Descartes and Helvétius held, but that the *concept of ignorance itself* was an illusion. For example, d'Holbach believed that a true account of nature would not convince people that "monsters," "marvels," or "miracles" were false but would instead completely remove the *idea* of monsters, marvels, and miracles (OPH, 201). Even illness—imagined since antiquity as a breakdown of the natural order—was for d'Holbach merely a different kind of order. As he noted, "That which occurs in a state of illness . . . the movements which are provoked in the human machine . . . are necessary, and are subject to the same certain, natural, and invariable laws which produce a healthy course." Only death, it seemed, could bring freedom: at "the moment of the cessation of a human existence . . . [the] machine becomes an inanimate mass by the subtraction of the principles which made it act in a determined manner" (OPH, 202).

For d'Holbach, the idea of "intelligence" was no different from the ideas of miracles, monsters, or illness, a foolish concept that arose only because of an ignorance of nature's mechanism. If intelligence had previously been seen as the mental faculty that was able to understand order in chaos, removing the dichotomy of order and chaos meant that the faculty that perceived them became an illusion as well. Even worse, d'Holbach believed intelligence was a particularly pernicious concept because humans had used it to (falsely) separate themselves from nature. D'Holbach claimed that we see intelligence only in "those that have organs and a goal similar to us" and deny it to those things that we simply do not understand. Randomness too was an illusion, or what the Baron called "a word lacking in the sense that we always oppose it to intelligence without attaching to it any certain idea" (*OPH*, 204). Order and disorder, and therefore intelligence and randomness, were just more occult qualities to be banished from a proper science of man. As d'Holbach wrote, "As soon as we see, or think we see, order, we attribute that order to an intelligence, a quality similarly borrowed from ourselves and from our own way of acting" (*OPH*, 204).

For those who understood d'Holbach's shifting use of "intelligence" as a concept and "an intelligence" as a being, it was clear that the true target of his attack was, like Helvétius, the notion of a supernatural deity. The Baron was not a subtle thinker and, as Louis Dupré has recently claimed, *Système de la nature* was "the most elaborate *summa* of atheism until the Soviet encyclopedias of the twentieth century."[41] It may have been just this vision of a world lacking order and intelligence that caused Robespierre to create the Cult of the Supreme Being in response to the Cult of Reason. After all, if a rational science of man saw *all* actions as equally natural, and even health and illness were merely different descriptions of natural processes, then the actions of both the counterrevolutionaries *and* the agents of terror were equally necessary and therefore indistinguishable as moral acts. Robespierre, like many members of the Committee for Public Safety, imagined himself as an instrument for a cause greater than himself, a mere cog in machine of public order, but ideas like d'Holbach's threatened to undercut the whole purpose of the Committee. D'Holbach asserted that every single action of a revolution was "necessary," but necessary for what? If order itself was an illusion, and there really was no guiding intelligence that structured and ordered the course of the revolution, then it would be impossible for any act to be morally good or bad. Though d'Holbach died six months before the storming of the Bastille, and long before the *terroristes* declared order a virtue, he had exposed to many how far the ideas of mechanism could go before it eliminated intelligence completely.

D'Holbach's science of man was reviled in equal measure to Helvétius's, despite d'Holbach himself being a popular salon host. Voltaire, on cue, claimed that his was "a work of shadows, a sin against nature and a system of folly and ignorance. . . . I have never seen anything so degraded in our century as this work of idiocy." Voltaire wrote of d'Holbach's next work, *Le bon-sens* (1771), a largely abridged version of *Systéme de la Nature*, that "it was terrible" and "emerged from the same store [*boutique*]" as his previous work. Voltaire had sent his complaint to d'Alembert, who responded: "I agree with you that *Le bon-sens* is an even more terrible work than *Système de la Nature*."[42] Frederick II (1712–1786), the Prussian king who had employed Voltaire and provided the eulogy for La Mettrie's funeral, found the book so insulting to common sense that he wrote his own refutation of the work, joining a number of proper attacks that emerged in the German-speaking world.[43] Frederick was hardly a reactionary, and it is notable that the seriousness and popularity of d'Holbach's works had caused him far greater concern than La Mettrie's buffoonery. In 1770, Frederick even asked Voltaire and d'Alembert if they knew the author of d'Holbach's *Essai sur les préjugés*, which had been anonymously published earlier that year.[44] Both likely did but professed ignorance of the actions of one of the best patrons of philosophe thought.

Just what was it about d'Holbach that so concerned his fellow philosophes? To offer a final summation of why so many French Enlightenment figures rejected the extreme mechanization and mathematization of his science of man, it is helpful to return to an earlier work by d'Alembert, who was, as always, ready to caution his colleagues on the excesses of mathematical enthusiasm applied to nature and humanity. Like Frederick II's denunciations, d'Alembert's *Éclaircissements sur différences endroits des eléments de philosophie* (1767) arose out of a concern about using mathematical reasoning to express a form of materialism. Frederick had been impressed by d'Alembert's *Essai*, which had warned of using only "geometric" reason, and in 1767 he commissioned a more ambitious work, one that would explain once and for all why mathematical methodologies did not apply to human nature or society. Though published three years prior to *Système de la nature*, d'Alembert's clarifications represented the final and most substantial attempt of the philosophes to try to contain the reduction of human nature to mathematical rules.

D'Alembert began *Éclaircissements* with a simple point about scientific theories: in spite of their best hopes, *savants* always needed to express their ideas through language. As a mathematician, he lamented "how Philosophers are always obliged to submit themselves to the tyranny of figurative expression," admitting that "the majority of the expressions are figurative." Therefore, as he noted, the "imperfection of language . . . is the source of intimately false judgments" (*DML*, 952, 953). The figurative nature of language had led philosophers like Descartes, Leibniz, Spinoza, Hobbes, and the mechanical philosophers to try to get around the problem through reducing ideas and truths to their smallest parts, a process d'Alembert called *decomposition*. In doing so, these authors believed they were following the clarity of geometrical reasoning, where even the most complex shapes and ideas can be reduced to points and axioms. In the process of what d'Alembert called *générDalisations*, these simple points and axioms could then be combined to form precisely the same shape that had been "decomposed" (*DML*, 946). It was just this simple process of "decomposition" and "generalization" that had so attracted d'Alembert to mathematics in the first place.

The problem with using decomposition and generalization outside of mathematics was that language was far more complex and malleable than geometric proofs. According to d'Alembert, a word like "matter" could be decomposed, without any loss of meaning, to three simpler words like "extension, boundedness [*bornes*], and impenetrability." One could reverse the process, too, and generalize extension, boundedness, and impenetrability back into "matter." However, d'Alembert noted that not all words and concepts could be decomposed and generalized so easily. A verb like *voir* could be subject to simple decomposition but *not* simple generalization. "To see" could be reduced to "sensation and existence," but these two latter words could also be generalized to encompass far more than simply "to see." In this case, many words, especially those that represented complex concepts, took on new meanings as one generalized and decomposed. While many mechanistic authors had sought to reduce ideas to their smallest parts through decomposition, d'Alembert warned that something very different may emerge when they were brought back together. Similar to Diderot, who held that geometric proofs rested on the "linkage" of one idea to the next, d'Alembert insisted that most words and concepts could only be understood through *enchaînement* (*DML*, 952).

Beyond the linguistic problems with applying geometrical certainty to a science of man, d'Alembert also pointed to the problem that most sciences relied on sense experience, which was random and accidental. Helvétius and d'Holbach had imagined that randomness was only a word for ignorance, but d'Alembert believed it had real ontological standing.[45] As he did in the *Éléments de philosophie*, d'Alembert claimed that mathematical truths are only certain because they are abstract; as soon as they are applied to nature, they can only be taken only as probable or contingent. For d'Alembert this meant that intelligence—in the sense d'Holbach had dismissed—was real, because there was often no way to get beyond the randomness and accidental nature of much experience. Ironically, randomness entailed the concept of order, which kept alive the possibility of intelligence that could recognize such order. If randomness was real, almost all knowledge therefore had to come through what d'Alembert called the "art of conjecture," which meant that truths in nature could only be generalized and not decomposed into smaller and more universal truths (*DML*, 987). If true, this was a substantial blow for any science of man that sought geometrical precision, as d'Alembert reiterated that there are only two "objects of our knowledge" that did not need to be "submitted to the art of conjecture: the mathematical sciences and the truths of religion." As he noted of geometry especially, "one smiles at the Geometer who wishes to employ probable arguments to prove the propositions of Euclid" (*DML*, 987). Left unsaid, but implied, was the inverse of this claim: that d'Alembert would laugh at anyone who used geometrical propositions to understand the necessarily probable truths of mankind and nature.

The combination of linguistic imprecision and the contingent nature of experience led d'Alembert to invoke a metaphor cautioning those who would attempt to create a science of man out of mathematics. After noting that the "human mind" had been searching for first principles for a long time and had taken a "thousand paths . . . and found nothing," he claimed that humanity was like a "criminal locked up in a dark recess [*réduit*], looking around uselessly in all directions to find an escape. Seeing only a glimpse of a weak light through a narrow and crooked crack, they try in vain to expand [*agrandir*] it" (*DML*, 955–56). For d'Alembert, the "narrow and crooked crack" was mathematics in general and geometry in particular, and it was easy to see this as a rebuke to anyone—Helvétius and d'Holbach included—who would use this "glimpse" to light the future course of a science of man.

"An Intelligence So Large": Laplace at the End of the Machine Man

D'Alembert's thoughts on materialism and the mind/body distinction take up relatively little space in his overall work, and it is for good reason that he is mentioned only as a footnote in most histories of materialism, as the happy interlocutor in Diderot's *D'Alembert's Dream* (1782) who ends up spouting inanities and absurdities after falling asleep.[46] In part, this was because d'Alembert believed strongly in the existence of an abstract world through which mathematics operated but was far less interested than eighteenth-century mechanists in the ability of that abstract world to have much say in the truths of the real world. What he was left with, and what he devoted many of his last works to, were the "probable" truths that could come about through repetitive observation. As will be seen in chapter 4, one of his best students, the Marquis de Condorcet, would use such probabilistic theories to try to construct a "social mathematics" to reveal the collective will of man, a science of man considerably more nuanced than d'Holbach's. At the time d'Alembert wrote, however, probability theory was still in its infancy, and he worried that it might supplant the geometric spirit he was concerned with in the *Essai*. As he noted, "One commonly complains that the formulas of mathematics, applied to nature, too often are only in error. People nevertheless do not seem to see, or don't believe they see, this inconvenience in the calculus of Probabilities. I have dared to propose some doubts on some principles which serve as the basis for this calculus" (*DLM*, 1063).

Pierre-Simon Laplace (1749–1827), the great French astronomer, mathematician, quintessential *savant* of the French imperial era, and *another* prominent student of d'Alembert's, did not appear to listen to his mentor.[47] Rather, Laplace's use of the "calculus of probabilities" to correct the "errors" of the Newtonian cosmos famously transformed the reigning mathematical approach in the sciences of man, redirecting the efforts of "social scientists" away from the strict mechanism of d'Holbach's sandstorm to the more statistical and observational sciences of man of the nineteenth century. Though Laplace was famous for his deterministic "demon" that seemed to imagine the world to be as predictable as the sandstorm, a review of his most famous work on probability reveals that his contributions to the development of the sciences of man were far more subtle than the demon might allow. In the realm of ideas, Laplace's focus on probability rather than pure Cartesian mechanism allowed for some level of indeterminism—and even freedom—into the lives of human actors. As will be seen, in a direct challenge to the idea of mechanical philosophy, for Laplace even nature itself seemed capable of free and independent action. Perhaps more importantly for this story, however, Laplace's scientific methodology—patient and painstaking observation by well-trained observers—seemed to extend and confirm the practical side of Fontenelle's geometric spirit. While La Mettrie had mocked the mechanical education of French "geometers," and D'Alembert and Diderot called for the return of *esprit finesse*, Laplace's success in mechanics and astronomy confirmed that science seemed to work best through the large-scale accumulation of data by science workers trained in the techniques of proper observation. Though he was neither a proud proponent for a science of man nor a fierce critic, Laplace exerted a powerful influence on both scientific ideas and methodology for the future sciences of man. Even if humanity could escape the theoretical determinism of Laplace's Demon, critics would contend that actual workers could not escape the confines of scientific practice and education.

Laplace may seem an odd figure in which to locate the end of mechanism and the instauration of a new kind of science of man—and an even stranger figure to call for a numerate science in plainspoken language. His *Théorie analytique de probabilité* (1795) was a notoriously dense work understood only by specialists, and he was first author and director of the massive collaborative work *Mécanique céleste* (1799–1825), which famously *diminished* the role of chance (and God) in astronomy.[48] Laplace's scientific work also represented the kind of hierarchy that Diderot and d'Alembert had envisioned, where great discoveries were brought about by a few individual geniuses while a great many labored in obscurity.[49] Though Laplace was only six years younger than the "last *philosophe*" Condorcet, and was helped in his career by d'Alembert, he had little connection to the *gens de lettres* who had sought to order the world of politics and morals along Newtonian lines, refusing to enter the salon life of d'Holbach and Helvétius and, according to one biographer, "shunning . . . the Enlightenment fraternity all around him."[50] He seems in fact to have worked in a parallel universe, so it must have been surprising when he proposed one of the greatest thought experiments in the history of philosophy: his all-knowing "intelligence," or what has come to be known as Laplace's Demon.

Though Laplace is often seen as a strict determinist, his demon was not the unerring perfection of the machine that might be imagined. As he noted, "the most important questions in life . . . are questions of probability" (*Essai*, 31). Moving further away from the certainties of mechanical philosophy and the geometric spirit, and embracing the lessons of his mentor d'Alembert, Laplace also claimed that "all our knowledge is only probable" and that "the few things that we can know with certainty . . . are founded on probability." Unlike geometric proofs, which were "pure science" based on timeless and self-evident truths, probable truths required what Laplace called "induction and analogy," terms he repeated often in the *Essai*.[51] Yet Laplace was also quick to mention that "probable" did not mean the same thing as "random," and he explicitly denied that there were events that were the result of the "blind chance of the Epicureans."[52] Instead, although Laplace did not attempt a science of man on his own, he offered a new methodology for later sciences, one that deserves to be quoted in full. After delivering his famous thought experiment about "an intelligence so large," which could "analyze all . . . data" and for which "nothing would be uncertain," Laplace followed with these less-referenced lines:

> In the perfection that it has given to astronomy, the human mind offers a weak sketch of this intelligence. Discoveries in Mechanics and

Geometry, joined to universal gravity, have put within reach the ability to understand with the same analytical expressions the past and future states of the world. In applying the same method to some other objects of knowledge, it will be able to achieve and reduce the observed phenomena to general laws, and to predict that which those circumstances should produce. All these efforts in search of the truth tend to approach without end the intelligence as we have come to see it, but which remains always infinitely far away. The tendency to progress, proper to the human species, is what renders it superior to the animals. It is also what will give distinction to nations and ages and will constitute their true glory.[53]

What was so remarkable about Laplace's claim was not its novelty in calling for a use of mathematical tools to study "other objects of knowledge," as this had been central to the geometric spirit since Fontenelle. Many philosophes had talked about applying the tools of mathematical sciences to other sciences, and many too had linked this process to national "glory." Instead, the striking aspect of Laplace's formulation was the turn made in the middle of the passage, where he noted that the progress of trying to "approach intelligence" was "without end." Not only did Laplace undercut the possibility of actually realizing his "demon" one paragraph after proposing it, but "endless progress" was a strange point to make for a mathematician who believed he had found universal and timeless harmony in the skies. None of the promoters of the geometric spirit, for example, had claimed that a final understanding of the laws of nature and human action would remain "infinitely far away," and a hypothetical Laplacian science of man could never reach the certainty of those offered by d'Holbach and Helvétius.

While Laplace's "intelligence" might seem like d'Holbach's sandstorm, allowing properly trained moral and political "scientists" to imagine the past and future states of human activity, such potential sciences of man never approached such certainty. In fact, in a short note titled "Application of the Calculus of Probabilities to the Moral Sciences," Laplace echoed Fontenelle's geometric spirit but only in the most mundane of fashions. He claimed, without much further comment, "Let us apply to the political and moral sciences the method founded on observation and on the calculus [of probabilities], a method which has served us so well in the natural sciences."[54] Laplace's statement influenced the later development of "social physics" and helped set the course of quantitative social science for centuries, but Laplace devoted all of two paragraphs of the essay to a subsection on the "moral sciences," and he hardly mentioned the application of the "calculus" to human actions again. For the sake of comparison, Laplace's two paragraphs on "moral science" were dwarfed by the section on applications of probability for the *natural sciences*, which stretched twenty times longer. While probability theory might mean that there was some transcendent intelligence that can know all actions, and that probability might be helpful to studying the actions of mankind, Laplace's vaunted demon offered little explicit threat to individual intelligence or free will.

As will be seen in chapter 4, "social scientists" like Condorcet, Quetelet, and William Stanley Jevons would use probability theory as a means to level the sciences of nature and human society, but Laplace likely would have warned against such measures. In part because of the skepticism of his mentor d'Alembert, and in part because he had witnessed the very *unpredictable* events of the French Revolution, Laplace made it clear that a rigid geometric spirit would not be applicable to the world of human affairs. Though not in the "moral sciences" section, Laplace had earlier explained why a deterministic science of man was simply impossible. In his "third principle" of probability, far less referenced than his great demon, Laplace insisted that calculating probability differed significantly depending on the number of "mutual combinations of events" and that it became more and more difficult to determine probability when events were "independent of one another." This mattered because it meant that probability in mathematics differed substantially from how it operated in the "moral sciences." In the "sciences purely mathematical," Laplace claimed, even the most faraway events "participate in the certitude of the principle from which they derived," and in physics, the "consequences have the same certainty as facts or experiences."[55] This was d'Holbach's sandstorm. Yet in the moral sciences, Laplace noted that probability could only be based on an inference of the previous state and, therefore, the further away from the original event, the higher chance of an "error" that "would surpass the truth."[56] Unpredictability was simply part of the moral sciences. Contra d'Holbach, the revolution was not the sandstorm.

Indeed, once the reader gets past the demon at the gates of the *Essai*, Laplace allows for far more room for "error" and indeterminacy than might be expected. While Laplace's idea of an "intelligence so large" is often considered "wrong" because of indeterminism in quantum physics, such an anachronism is not needed to rescue free will and human complexity, and Laplace seems to have learned more from d'Alembert than is usually supposed. As an example of a probable truth that could never approach mathematical certainty,

Laplace asked what a person would think upon coming across a series of tiles that spelled out the word C-O-N-S-T-A-N-T-I-N-O-P-L-E. The probability that someone had ordered the letters was indeed high, high almost to the point of certainty, because it is "a word in usage among us," and Laplace noted that it was "incomparably more" probable as the result of intention than that of "randomness."[57] Yet it could not be completely certain. In a section on the "illusion of probability," Laplace saw that in "the actions of the oceans, the atmosphere and meteors, the trembling of the earth and the eruptions of volcanoes," there appeared to be order, but it was "unlikely" that "the causes . . . would remain in exactly the same qualities." In a claim that seemed to undercut the certainty of the most mathematical of natural sciences—astronomy—Laplace wrote, "The sky even . . . is not unalterable. The resistance of light and other ethereal fluid and attraction of stars must, after a great number of centuries, considerably alter the planetary movements."[58] The inventor of the determinist demon, and the man who had removed God from Newton's universe, allowed a little freedom even in the heavens.

Conclusion

After reading Baron d'Holbach's *Système de la nature*, the Russian socialist writer Alexander Herzen (1812–1870) declared, "This book is the conclusion of French materialism, this is Laplacian 'J'ai dit tout'! After this book . . . it was impossible to go further."[59] When Herzen compared d'Holbach's *Système de la nature* to Laplace, it was likely in reference to Laplace's *Mécanique céleste*, the epic five-volume work which had "said everything" that needed to be said about the laws of celestial motion. Laplace had corrected even the great Newton, and it seemed to many that the sciences of the heavens had finished their task. The fate of the sciences of man, however, was less certain, and few today would hold up d'Holbach's social science as an authoritative equal to Laplace's account of the heavens. While the Baron may have indeed taken materialism to its logical limit, the decades following his work proved that the sciences of man had no Newton or Laplace. To many later commentators, in fact, the endless eighteenth-century French debates about materialism and the role of mathematics in understanding human beings were some of the most absurd ideas ever proposed, sounding more like scholastic ruminations than a science of human thought and action that could achieve the certainty and clarity that Newton and Laplace had brought to astronomy. As the French Revolution and Reign of Terror made clear, the other "half" of the revolution that Robespierre had called for—the moral and political revolution to accompany that made in the physical world—was not soon forthcoming.

The sciences of man of this era therefore did little to bring harmony to political and moral affairs, as the "weak light" of the geometric spirit grew no brighter. A leading expert on human anatomy and physiology, La Mettrie used his knowledge to mock the education of French men of science, and those philosophes who knew their mathematics—like d'Alembert—warned against mechanism. Added to this, the century ended with a non-philosophe mathematician like Laplace questioning the certainty on which Descartes, Fontenelle, and Condillac had placed so much hope. And the two most popular writers who had taken up the geometric spirit—Helvétius and d'Holbach—had been forced to publish anonymously and were criticized by their friends. In perhaps the most concrete challenge to extending the geometric spirit into the sciences of man in France, after taking power, Napoleon moved for the "suppression" of the moral and political sciences at the recently formed Institut de France, branding them "ideological" and replacing them with military training.[60]

This does not mean that the sciences of man based in the enlightened geometric spirit did not exert an influence over later theories of human thought and behavior. Laplace had made thinking and judgment less mechanical and axiomatic than Descartes, and the process of thinking through "induction and analogy" had in some sense saved the idea of intelligence from being swallowed up in d'Holbach's sandstorm. Similarly, d'Alembert's "Analyst" and "generalization," Diderot's "linkage," and the "art of conjecture" all seemed hopeful efforts to bring back Pascal's *esprit finesse*, offering to the moral and political sciences a balance to a rigid geometric spirit and an escape from La Mettrie's parody of the education of "geometers." Yet, as seen in the rest of the *Essai*, Laplace's vision also offered a potential catch to what otherwise might be imagined as the free play of scientific research: a science of man based on probability rather than axioms required large amounts of data to consistently observe and compare. Laplace may have saved intelligence in theory from the simplification of the geometric spirit, but probability had opened the door to a form of statistical reasoning that required a project of massive labor much larger than that which produced the *Mécanique céleste*. Rather than showing hypothetical beings trapped in a deterministic universe, critics echoed La Mettrie in the complaint that science work and education left actual people—and minds—trapped in predictable routines. As historians have noted, in the age before mechanical and electronic computation, data collection along Laplacian lines meant heroic levels of tedious work by *savants*, science workers, and people known as "computers."[61]

Laplace had managed to temper the strict mechanism of Descartes and his heirs, but he did little to combat, or even discuss, a complementary aspect of the geometric spirit that had been proposed by Fontenelle and mocked by La Mettrie. By 1800, many assumed that all sciences—of man and nature—would proceed through an analysis of mass observations. Such an idea united as disparate figures as Napoleon, Laplace, Helvétius, d'Alembert, and Diderot. Like almost every philosophe on all sides of the debate—except, of course, Rousseau—Laplace believed in the idea of a hierarchy of well-trained science workers who would contribute to science through patient and attentive observation. The simplification of science work, it seemed, might have been a more pressing everyday concern for the people of France than the bogeymen of machine men and demons that take up so much space in histories of artificial intelligence. As seen in the next chapter, however, there was one particularly vitriolic group of thinkers who *were* paying close attention to the consequences of the new form of science work and education. United as they were by a seeming opposition to all things modern, historians have branded them the "enemies of Enlightenment."

3

Warnings *of a* New Barbarism

The Defense of Intelligence in Bonald, Maistre, and the Enemies of Enlightenment

> We have descended, step by step, a staircase of systems,
> one that is certainly not a staircase of giants.
> —Jules Barbey D'Aurevilly, *Les prophètes du passé* (1851)

FOR MANY FRENCH ENLIGHTENMENT THINKERS OF THE EIGHTEENTH century, the mechanization of natural philosophy and the extension of the geometric spirit suggested a new approach to studying human thought and action, one that could simplify the human subject to the point where its actions could be as easily predictable as planetary orbits. Early modern mechanical philosophers like Gassendi, Leibniz, and (ostensibly) Descartes had allowed for an area of human will and action well beyond the reach of the natural sciences—seeking explanations for this behavior in theology and philosophy—but philosophes like Fontenelle and Condillac imagined using the tools of the natural sciences to study directly the moral and political sciences of human actions. And for d'Holbach and Helvétius, the hope for a science of man was based particularly on the progressive narrative of accumulated knowledge borrowed from the natural sciences. As Newton famously wrote in his letter to Robert Hooke (1635–1703) on the progress of science, "If I have seen further, it is by standing on the shoulders of Giants."[1]

Yet, as seen in the last chapter, not everyone agreed that a revolution in the moral and political sciences should follow the Newtonian revolution in physics, or that a science of man would be able to reach such heights. Indeed, the nineteenth-century provocateur Jules Barbey D'Aurevilly (1808–1889) went so far as to invert Newton's accumulative and progressive metaphor for science.[2] Whereas many French thinkers had seen an ascent to perfectibility, happiness, and perpetual peace based on the applicability of scientific knowledge to humanity, Barbey saw only a decline. Citing thinkers that were heroes for the philosophes, his descending "staircase of systems" led "from Spinozism and Cartesianism to Bacon's experimentalism to the materialism of the eighteenth century."[3] Rather than lead society toward a higher calling, Barbey believed the philosophes and their systems had plunged humanity back into a "new barbarism," one that replaced the "living blood" of the old barbarians with a "cold blooded," "exhausted," and "corrupt" vision. Even worse, Barbey claimed, "the barbarians of the future will have in their hands means of destruction created by the scientific materialism of the modern world."[4] In short, new technologies combined with impoverished theories of human thought and action represented a fall from the civilized world, creating a simple "man" living in a world void of meaning.

Barbey launched his attack on modernity in *Les prophètes du passé* (1851), a book notable at the time for both its searing intensity and its seeming impracticality. It was published, for example, just a decade before Proposition 80 of the Catholic Church's Quanta Cura had declared that even the pope himself was not opposed to "progress, liberalism, and modern civilization."[5] Even more, the "prophets of the past" Barbey celebrated—Louis de Bonald, Joseph de Maistre, and René Chateaubriand, among others—had themselves been dead for decades, the last considered a great hero of French literature but the others relatively minor figures who left little in the way of influential ideas or enduring philosophical systems.[6] Barbey was no intellectual historian, and his recent conversion to Catholicism heavily influenced how he wrote *Les prophètes du passé*, which deployed the prophets as hidebound reactionaries against all things modern.[7] In his treatment of Bonald and Maistre, and in particular their writing on science and technology, polemics and outrage took the place of textual analysis. To see the prophets from Barbey's perspective was to see the sciences of man only through the lens of reaction and to imagine that the "descent of systems" was the inevitable result of embracing science and technology. It was, quite rightly, an easy book to dismiss.

Yet the thinkers Barbey chronicled, who worried about a world of "artificial" intelligence created through the scientific study of man, are important to this story beyond what a summary of Barbey's tiresome book might suggest. While debates over AI today often direct historical accounts to discussions of past scientific discoveries or novel machines, the focus of the "prophets" on the consequences of deriving rules of human action from mechanical theories of science points to a different kind of artificial intelligence. Just as La Mettrie and Rousseau had seen science education as a rote process by which man becomes a machine, and Diderot and d'Alembert warned of restrictive hierarchies in scientific life, these very different thinkers worried that reducing human actions to scientific laws—and the widespread diffusion of such an idea—was far more of a threat to human intelligence than the rise of smart machines. Instead of dogmatically opposing every aspect of science, the "prophets" attacked what they felt was the reigning *scientism* of the era, or what we might call the exclusive adoption of scientific methods for understanding nature and humanity.[8] They saw scientism most clearly in public education, which was awash in enthusiasm for science and mathematics but offered little support for literature, history, the fine arts, or traditional instruction in the humanities and "moral sciences."[9] For them, the greatest threat to human intelligence was not "machine learning" but rather the simplification of actual education to the level where students were taught only a mechanical picture of the world.

As historians of the French educational system have noted, by the late 1700s, there was no shortage of educational reforms for a conservative-minded and traditionalist thinker to critique. Even prior to the Revolution, the decision of Louis XV to banish Jesuits in 1764, and the eventual closure of all Jesuit colleges by 1773, ended what has been called a two-hundred-year "war" between the University of Paris and the Society of Jesus over the foundations of Catholic education.[10] After the Revolution, the University of Paris, along with all other Catholic universities, was closed, as the revolutionaries sought to replace a "patchwork school system" with a unified and formal system of schooling, even suggesting at first that all male students be instructed in boarding schools.[11] Most galling to some, in 1798 revolutionaries had even led an investigation into private primary (elementary) schools, with agents showing up unannounced at local schools to ensure that revolutionary texts and the "geometric spirit" were being taught.[12]

At the level of higher education, the revolutionaries had more success with the creation of two new institutions in 1795: the École normale supérieure and the École polytechnique, the latter of which became the most important site of mathematical and scientific education in the country.[13] The École Polytechnique has been the subject of significant scholarship, but most historians agree that in the decades from the Revolution until the French Empire, the school went from being "rather free-wheeling to meticulously bureaucratic," as students were expected after 1805 to learn in a "military-style" environment.[14]

Unsurprisingly, such a turn occurred after Napoleon's rise to emperor and the subsequent influence of Laplace, who had attended the Corsican's coronation ceremony and began to reorganize the school as early as 1799.[15] By 1819, engineers who graduated from the school were even expected to practice "arithmétique sociale," a form of objective quantification that ensured their status as public experts.[16] While there has been disagreement over whether the École polytechnique became a breeding ground for narrowly educated "technocrats," it certainly appeared so to many contemporary observers who called for a return to the broad-based education of the Catholic colleges.[17] Remarkably, in just one sign of how rigidly mathematical French education had become, when Bavaria finally removed the yoke of Imperial France, it dissolved the Real-Institute for modern mathematics and science instruction that had been set up in 1808, eliminated nearly all math and science instruction, and fired the entire mathematics faculty.[18]

Such was the state of educational reform in France during the age of the "prophets." While Bonald, Maistre, and the counter-Enlightenment may seem a departure in telling the story of the creation of the sciences of man, their radical, reactionary, and often bizarre condemnation of the Enlightenment embrace of science provides one of the clearest windows onto that world. As Gérard Gengembre claimed in defending studies of such forgotten political thinkers, "the Counter-Revolution permits the best understanding of the Revolution; it is a pure negative."[19] So too does their profoundly negative vision of the new sciences of man illuminate the transformation that took place during the nineteenth century, or what Justin Smith has called the "tragic arc of science."[20] As historians have noted, their critiques were based in much more than simple religious dogmatism. Both Bonald and Maistre, for example, were well educated in math and science, and exemplified what Darrin McMahon has called the "enemies of Enlightenment," a form of "anti-philosophe discourse" that stands apart from the more formal counter-Enlightenment and Romantic movement in Germany and England that takes up so much space in intellectual history.[21] Though McMahon does not spend much time discussing their opposition to scientism, focusing instead on the role of the church and monarchy in politics, he does note the forward-thinking critique of Bonald and Maistre, claiming that their "defense of tradition was not traditional." The historian Zeev Sternhell too highlighted the forward-looking theories of these thinkers, arguing that "the enemies of Enlightenment . . . were not turned towards the past."[22] While it may be too much to say that "progress was as much a part of the agenda of the counter-Enlightenment as of the Enlightenment," historical revisionism has helped rescue such thinkers from Barbey's anachronisms.[23] In defending traditional education and a broad concept of intelligence from the reductive sciences of man, they were searching not for a return to the Middle Ages but for what Sternhell called "another modernity."[24] As will be seen, the legacy of the prophets of the past for this survey is that in their alternate "modernity" it would have been impossible to imagine the artificial intelligence of today, where neural nets are presented as discrete beings with no gender, race, religion, or lived historical past. For these writers, "intelligence" was instead inextricably bound to a nexus of cultural, social, and historical institutions and simply could not exist on its own without a connection to traditional society.

To explain how these authors defined and defended a different kind of intelligence against what they felt was an encroaching scientism, the chapter begins with Louis de Bonald, who was emblematic of the complex approach many counter-Enlightenment thinkers took toward a science of man. Bonald did not condemn the philosophes for looking to science to understand humanity; rather, he thought they were not scientific enough. Though his vision of science looks quite strange to the modern reader, many at the time shared his distrust of quantification, applied technology, and mathematics. In particular Bonald worried that science education had overridden French society, an idea stoked in the pages of the French counterrevolutionary journals *Mercure de France* and *Journal des Débats*. In these periodicals, Bonald was inspired by strong critiques of science education, where the authors revived Rousseau's concerns from the *First Discourse* and complained of "entire generations" receiving only a technical education with no appreciation of literature, virtue, or morality. Writing from his exile in Saint Petersburg, Bonald's fellow prophet Joseph de Maistre agreed, believing that Russia offered a form of education to ward off the enervating study of science he had witnessed in France. Maistre's anti-scientific writings have been largely overshadowed by his apocalyptic warnings and anticipation of reactionary conservatism, yet, seen in conjunction with Bonald and the journal writers, his screeds against science—in particular his critique of Francis Bacon's definition of "intelligence"—seem just as relevant today as his politics. In ways the philosophe critics only hinted at, the prophets of the past claimed that exclusive education in science and science work had reduced intelligence to what Maistre would call the "rank of a simple faculty."[25]

A Sad and Savage Solitude: Bonald's Attack on Scientism

In the world of politics, Louis-Gabrielle-Amboise de Bonald (1754–1840) likely deserves every epithet he has received. A legendarily poor writer, he was also a political opportunist. In David Klinck's excellent survey of his life and ideas, it is clear that family pressures and the status of noble privilege shaped Bonald's opinion of the French Revolution more than profound philosophical reflection. When, for example, Bonald was elected president of his *département* in 1789 as a consequence of the Revolution, divided sovereignty between the crown and people seemed a workable solution. When, however, he lost his next election, popular will became a beast.[26] Marginal even to histories of conservative thought, Bonald has been called a "quasi-unknown," whose works are "rarely studied" and represent "a phantom passing in the desert" of intellectual history.[27] As early as 1831, a poet had recorded in his diary: "What to say about the political books of Bonald? What will remain? Nothing, absolutely nothing."[28] Part of this may be due to his style of writing, what one anthologist called a "turgid prose style that betrays the influence of his education."[29] Perhaps damning Bonald forever to the margins of even the counter-Enlightenment, Isaiah Berlin echoed the condemnation of Bonald's style, calling it "ponderous, and remorselessly monotonous," the product of a "pedantic" and "unoriginal" thinker. Summing up his legacy in just a few paragraphs, Berlin concluded that Bonald's "works, and to some extent his personality, seem today . . . to be deservedly forgotten or ignored."[30] Perhaps this was the legacy Bonald wanted, as he indicated in a letter sent to his good friend Joseph de Maistre that "I am only truly happy in my sad and savage solitude."[31]

While it is impossible to argue that Bonald's prose sparkles, or that his antisemitism and patriarchal views were not odious even for his time, part of the reason for the many dismissals of his writing has been the political focus of much counter-Enlightenment history. For good reasons, political and intellectual historians like Berlin have tried to tease out proto-Fascist or right-wing ideologies in authors who criticized the Enlightenment, and for such a project Bonald is of little help as a serious thinker.[32] Yet Bonald's writings on science, and in particularly on society, might seem to have more resonance today than his defense of Louis XVI or his successful effort to bar divorce in France for seventy years.[33] For example, Bonald was one of the first writers to articulate that individuals were unable to form societies *as individuals*, and that society itself might be an autonomous entity that structured individual action.[34] As Bonald put it in his first major political work, "not only does man not constitute society, but it is society that constitutes man."[35] Bonald also knew his science. Calculus, geometry, trigonometry, and algebra formed the "centerpiece" of his last year of study as a student, and the dryness of his works owes to the "mathematical" approach he took to philosophy and history.[36] Indeed, in summing up Bonald's thought, Klinck claimed that "science provided a second type of universal truth to which French counter-revolutionaries could turn," noting that in 1800, there was no separation in Bonald's mind between "philosophical reflection" and "the empirically-based search for regular patterns of human behaviour."[37] Even his great champion Barbey had to admit that Bonald was no Romantic and that he approached his studies with "the rigor of a mathematician who resolves the terms of an algebraic equation."[38] Bonald was a spiteful monarchist, but he was also a genuine *savant* who took the science of society seriously.

Bonald even had some sense of the geometric spirit, referring to the "grand machine of the universe."[39] In the preface to *Théorie de pouvoir publique* (1796), an attempt to defend the monarchy as a historically rooted and necessary institution, however, he explained that "the physical sciences had made great progress" but had been stunted when applied to man. Although people could "observe the heavenly bodies and their movement," such research into a science of man faltered when reason extended itself too far. Bonald's goal in *Théorie du pouvoir publique* was to rescue humanity from what he believed to be the abyss of mechanical amorality and to situate a science of man firmly within the social character of human beings. As Gengembre has noted, for Bonald, "only social man [was] worthy of scientific discourse."[40] Unlike the philosophe sciences of man, with their astronomical and physical analogies, Bonald claimed that he was "daring to search for the fundamental laws of society" outside of nature.[41] While *Théorie du pouvoir publique* was rooted in tradition, Bonald's traditionalism was not biological or even natural. In a belief that anticipated the sociological methods of Durkheim and Weber by a century, Bonald argued that because the social world was distinct from nature, and constituted an irreducible reality in itself, the science that studied it needed a distinct methodology as well.

In addition to extolling the virtues of an autonomous social world, Bonald's proposed science of man was also noticeable because it embraced *littérature* as the primary means through which to study the past and present of human societies. Though Bonald often criticized the physical and mathematical sciences, it was almost always only in comparison with what he believed was the general neglect of *littérature* and *belles lettres*, fields that would overlap with most of the humanities today. He had even written an essay titled "On the War of the Arts and Sciences," which posited that the physical sciences had succeeded in overtaking all other forms of knowledge.[42] In 1810, he equated the very idea of "progress" with the "decadence of literature," writing that "one had plainly seen that literature had been in descent for a long time among us" (*OCB*, 3:1106). For Bonald, Napoleon's triumph in reorganizing higher education around the technical skills taught at the École polytechnique only made obvious what had been occurring for a century under the rise of the philosophes.[43] While in retrospect, Bonald's prejudiced political ideas seem to be almost a caricature of reactionary right-wing thinking, his emphasis on literature as an antidote to reductive science anticipates a number of later critiques of mathematical approaches in the social sciences. In these writings, Bonald often sounds more like a modern professor arguing for the value of the liberal arts—or even a believer in the "cultural turn" of the 1970s—than a fire-breathing defender of feudalism, monarchy, and religious education.

For Bonald, the problem of literature's fall first arose in a "divorce" between "the sciences and *lettres*" (*OCB*, 3:1139). Past scientific thinking, which for Bonald included everything from theology to physics, had been based on a balance between the "foundation" of science and the "form" of literature. In other words, *savants* had presented scientific information in a way that unified what today might be called the technical content and the form of presentation. *Both* content and form had counted as scientific thinking, and Bonald saw that, in the previous centuries, "the distinction between the sciences and letters was less noticeable" (*OCB*, 3:1137). Yet, in a critique of increasing technical language, Bonald complained that modern writers had dropped the "form" completely and were communicating in an incomprehensible jargon that protected their ideas from scrutiny. The philosophes had "attached themselves exclusively to physical knowledge." As a result, "the physical sciences were therefore the only sciences," and those who pursued them were "*les savants par excellence*" (*OCB*, 3:1140). Mathematics in particular had "taken the reins of science under the name of the *high sciences* or the *exact sciences*" and had subsequently trampled over "men of letters," who were "satisfied with their part" and did not challenge the rise of *savants* (*OCB*, 3:1140). In "War of the Arts and Sciences," Bonald noted the irony that the philosophes, whose actual strengths were in *belles lettres* rather than science, had helped bring about their own irrelevance. For Bonald, the consequence of this usurpation was clear: "The name of science has therefore become close to exclusively to mean natural sciences, or rather mathematics."[44]

Rather than a mere aesthetic preference, the loss of *littérature* meant two things for Bonald. The first was that the debasement of literary form made communicating scientific ideas difficult. Referring to mathematics, he complained that "all the sciences . . . are speaking a *technical* language foreign to literature and employing a style previously unknown" (*OCB*, 3:240). Bonald wanted to create his own science of society, but the conflation of science and mathematics meant that any forthcoming science would mean people would become subject to endless classification and quantification. Sounding much like the Rousseau of the *First Discourse*, Bonald warned that a "dry and solitary" field of study caused the "spirit to act against itself, to wither, and to be consumed by mute abstraction of reason over the soul." Compounding the problem was a second consequence of the divorce: schools now stressed mathematics over literature. While "the sciences of measuring and calculating" were "useful for a small number of people," it was becoming the "foundation for all instruction." Rather than elevate knowledge, the subsuming of science into mathematics had made it "impossible" for students "to conceive of higher truths and the great sentiments of morality" (*OCB*, 3:1157). Predicting disaster for the future of France, Bonald concluded that it was "certain that the centuries where the sciences are exclusively cultivated will not be centuries of eloquence, poetry, religion, and morality."

To illustrate the process of degradation, Bonald focused on a particular faculty he saw as most in danger of abuse: intelligence. In *Recherches philosophique sur les premiers objets des connaissances morales* (1818), Bonald claimed, "Intelligence . . . was the most noble part of [man's] being," yet it was often treated as an afterthought in the new vocabulary of science. In one of his longest philosophical works, Bonald explained that his "definition of man" was simply "an intelligence served by its organs" (*OCB*, 3:149). Bonald stressed the difference between the "soul and spirit" of intelligence and the "body and material" that made up the physical substance of the organs, arguing that the latter "served" the former and was "clearly inferior." In a telling metaphor, Bonald explained just how *social* intelligence really was and how this social intelligence was nearly synonymous with the soul. As he argued, "In man, the *power* of intelligence is in the soul, and the organs are the administrators. In society, the soul and intelligence are in . . . the social body; the administrators . . . are the organs" (*OCB*, 3:154).

Bonald's clever use of a Cartesian metaphor meant human beings were "evidently made for society," because it was only in society that "intelligence" emerged. Many philosophes had emphasized *individual* rationality, but Bonald countered that *social* intelligence should rule over the bureaucratic organs of the state in the same way that *human* intelligence (the soul) should rule over the material body. For him, the superiority of the "social body" over the "administrators" in state governance was the essence and great strength of the ancien regime, and it is not hard to see why such a science of man might lead to monarchism. Bonald believed traditional society was the soul of politics and should therefore always guide the "agents" of government. In this way, he combined a critique of biological reductionism (reducing the soul to the body) and technocracy (reducing social intelligence to administrators). In both conditions, he feared a material body had overthrown an immaterial intelligence. To restore the order of the ancien regime, and to rescue humanity from the stupidity and drudgery of nature, Bonald required an immaterial, irreducible, collective, and social intelligence.

While Bonald saw the degradation of this social world all throughout France, in no place was the fall of intelligence more pronounced than in education. The contested status of intelligence was the reason education mattered, and education and the defense of literature were in fact the only two areas of Bonald's vast oeuvre where he focused exclusively on science. Bonald had argued in "Des Lettres" that if the tools of the physical sciences were taught exclusively, and applied to something like the soul or intelligence, they "would diminish intelligence among a mass of miniscule details" (*OCB*, 3:1158). "Moral knowledge" needed to be placed above "physical knowledge" in the same way that "society is superior to the individual man." The "sterile nomenclature" of the exact sciences could only succeed then if the "taste" of humanity for literature was limited. Intelligence, Bonald warned, had to be completely reframed as a concept in order to be subject to investigation in the physical and mathematical sciences (*OCB*, 3:1158). Gengembre has argued that Bonald believed that, through quantification, the "descent" of humanity "degraded from the authentic to the inauthentic" and that consequently, "society fell into artificiality" as the soul and intelligence were removed.[45] As Bonald's fellow counterrevolutionaries would assert, the greatest threat to producing an inauthentic and artificial society did not come from machines but the spread of a particular kind of education in French schools.

"Substituting Calculus for Discussion": Popular Critiques of Science Education in the French Empire

Bonald's anti-scientism was a singular blend of Catholicism, traditionalism, and personal resentment at the course of the Revolution, but he was not alone in sounding the alarm over the primacy of science education in France. In fact, the revolutionaries themselves had even proposed a law that would limit elementary schooling to reading, writing, and morality.[46] In defense of the law, one member of the National Convention—Gabriel Bouquier (1739–1810)—claimed in full Rousseauian fashion that "free nations have no need for caste of speculative *savants*" and that "pure science" was a "poison which infects, weakens, and destroys a republic."[47] Indeed, Bonald's lament was part of a much broader movement of writers concerned that traditional education in the moral sciences was withering away because of education in the physical, exact, and quantitative sciences. In particular, many Catholics at the time echoed his specific concern over the increasing power of mathematics and "technical language" in French schools, wondering "how a fully integrated society could nonetheless leave room for individual expression and autonomy."[48] While not every enemy of Enlightenment went so far as Bonald to imagine his own science of society, or believed that an "ultra-Monarchy" was the only way to maintain traditional society, numerous authors, pamphleteers, and critics linked the perceived failures of the Revolution to the problem of science education. Just as Bonald worried about the consequences of reducing human action to mechanical and mathematical laws in the abstract, these writers worried that an education focused solely on the mathematical and physical sciences limited the potential of human intelligence.

An early article in *Journal des Débats* from 1800 stands out in anticipating the "divorce" Bonald described between science and literature. The anonymous article included in the "Varieties" section began, as almost all critiques of scientism did, with a conditional statement of support for science education. The author pointed out how the "current generation" had "elevated the study of the exact sciences" and noted that schools were turning out extraordinary students.[49] Yet the writer was equally "amazed" that teenagers "as pale as the most consummate scholars" were being produced with little forethought as to how they could help the nation and "the spirit of society." The problem had been that the "exact sciences"—chemistry, astronomy, and mathematics—which were meant to be "only a branch of the general sciences," had become the "entire tree." Education in *belles lettres*, history, and the fine arts had been completely "neglected."[50] France, a nation born in the "urbanity" of culture and arts, had prospered due to a focus on exact sciences, but the change in education meant a profound change in the future of the country. In the words of the anonymous author, "if the study of the exact sciences continues to be carried out at the expense of other parts of knowledge . . . we must expect that the national character will experience a proportional change."[51]

What made the 1800 article so prescient was that it occurred when Napoleon's great renovation in the French educational system was barely underway.

Yet like Bonald, the author of the "Varieties" section feared that the changes Napoleon had proposed for higher and secondary education—encouraging knowledge in the exact sciences instead of *belles lettres*—were already filtering down to local primary schools. For example, the National Convention in 1793 had closed twenty-two French universities, and the new curriculum skewed heavily toward the natural sciences. Critiquing the new form of Napoleonic écoles, the "Varieties" author complained that governments were merely interested in creating "a child who is a learned fool" capable of doing "only what their profession makes of them." Echoing Pascal's complaints against the geometric spirit, the author argued that a child became a fool because the "the exact sciences, principally mathematics, exercise only one faculty; they dampen or kill all the others." There would be legions of "astronomers who could predict cold in winter," but no good government officials or jurists.[52] Anticipating Barbey's complaint in *Les prophètes du passé*, the *Journal* piece ended with a warning: if the study of languages had led humanity out of "barbarism," then it would be the "lack of study" that would "return us to those days."[53]

The anonymous article received a wide audience. A constant critic of the *nouveau regime* and a staple of the exile press, *Journal des Débats* was the most widely read paper in Paris, accounting for nearly 30 percent of the newspapers sold in 1803.[54] So strong was its defense of religion and *belles lettres* in the face of science that the piece may have even been responsible for influencing the direction of *Mercure de France*—a literary organ of Francoise-René Chateaubriand (1768–1848) that functioned as the popular mouthpiece of the counterrevolution.[55] Bonald's depiction of the great harms of science education in "Des sciences, des lettres et des arts" in fact cited two articles published in *Mercure de France* in 1807 (*OCB*, 3:1135, 1155). While Bonald eventually found common cause with the counterrevolutionary press, things were not always so pleasant. Both papers were critical of Bonald for his continued interest in a science of man and held a "hostility" toward him for being too like the "social" *savants*.[56] Yet by 1807, the *Mercure* "Prospectus" thanked Bonald for "compensating" for Chateaubriand's absence and for providing extra articles while its founder navigated imperial politics.[57] The writers in *Mercure de France*, lacking Bonald's systematic and plodding attacks, made a much more direct point, one that highlighted Chateaubriand's view that "modern science was the antithesis of the mysterious spirit of religion."[58] Rather than posit the existence of an independent and autonomous society or debate the nature of intelligence, they reiterated a single complaint: the French educational system prioritized education in mathematics and the sciences over traditional education in *belles lettres* and moral education. In doing so, the country risked a return to "barbarism" in their neglect of literature, history, and the traditional liberal arts.

The first article that Bonald referenced from *Mercure de France* was a commentary on a speech by François-Joseph Génisset (1769–1839) titled, appropriately enough, "Quelques réflexions sur les sciences et les lettres."[59] Though Génisset had spoken on the potentially hopeful ideal of "An Accord Between Sciences and Letters," the review, simply signed "C.," promptly departed from this conciliatory attempt.[60] A professor of humanities who had been active in the Revolution, Génisset had looked for common ground, but the review warned that the exact sciences, backed with mathematics, were crowding out all the other fields of study in higher education, especially as expensive equipment for observation and experimentation stretched limited budgets.[61] The reviewer argued that "science" should be defined more broadly to include political and moral sciences and therefore the study of great writers like Plato, Aristotle, Machiavelli, and Montesquieu. Like Bonald, they believed exact sciences like chemistry and astronomy could only be "disseminated" in proper language and were therefore more beholden to *belles lettres* than they liked to imagine.[62] While many institutions of the Revolution had separated these ideas, the review insisted that the perceived progress in the sciences was only a result of a narrowed scope and definition. If new knowledge could not be explained in clear language, it did not deserve the name of science.

Another argument that influenced Bonald was the idea that knowledge in poetry and literature endured far longer than scientific theories. Questioning the philosophe progressive narrative, the review argued that most science was ephemeral, as ancient ideas were swept away by the slightest new discovery. Homer's truths could be reconciled with those of Virgil and even survive a challenge from modern literature, but Archimedes's ideas could not compete with modern science. Even a genius on the level of Newton, who had "erased" the discoveries of Descartes and offered a "simple and clear theory, in perfect harmony with every observation," could not last. The author of "Quelque réflexions" foretold the possibility of a new theory that might one day replace Newtonian physics, presciently asking if "one can affirm that" Newton's ideas "will remain perfect" in the face of new observations. In contrast, the great pieces of history and literature were unified as complete works, unalterable throughout history.

Historical oblivion was a problem for the exact sciences, but there was a more prosaic objection: science education was boring. Echoing the concerns of d'Alembert and anticipating worries about science education discussed in later chapters, there was now a concern that it was being forced upon students to their own detriment. As the review claimed, everyone "reproached *savants* for the dryness of their language," and technical language made it accessible to only a few people.[63] Indeed, if the current trends continued, "in an era without great writers," the "physical and mathematical sciences" would become a self-contained unit made of a "few *savants*."[64] This was the very elite "caste" the French revolutionary Gabriel Bouquier had worried about, and this was the same concern of those revolutionaries from the previous chapter who had declared the need for a *sans-culotticized* science. While education should "prepare the student in how to form judgments, how to exercise reason . . . and how to nourish noble and elevated sentiments in the soul," learning about mathematics and science produced only "positive knowledge" and simple tools.[65] Though some argued that mathematics indeed helped students learn "a taste for order and method," it was contrary to the needs of everyday life. Echoing positions taken by Rousseau and Diderot, the reviewer argued that because it focused on narrow and simplistic answers, mathematics was no match for the "subtleties" of life. Mathematics only appeared progressive because "a false spirit could make rapid progress."[66] As Bonald had argued, the progress of science could only advance through a degradation in language and the moral sciences.

Like the earlier critiques of education by the French philosophes, the author of "Quelque réflexions" was able to recognize the benefits of science education while still wondering why so many had to study the subject. The "mechanical arts" led directly to material gains, but the practical benefits of technology could be dismissed by noting the sheer luck and contingency of most inventions. The telescope, for example, so crucial to the business of navigation and commerce, was "found by accident in the time of semi-barbarism."[67] Like the anonymous 1800 *Journal* article and Rousseau, the *Mercure de France* reviewer believed humanity had made great progress in civilization since the Middle Ages but that it had been built on the great truths of history and literature rather than technical training in the sciences. Citing the historian Edward Gibbon (1737–1794) on the "destructive habits" that were formed out of "judgments made through rigorous demonstration," the review warned that the exclusive teaching of experimental and mathematical science would have profound moral consequences. If the sciences were taught "with a prejudice against *belles lettres*," there was a danger that they would "discourage and extinguish the imagination" by "substituting calculus for discussion."[68] Without a single reference to God, the ancien regime, or the monarchy, "Quelque réflexions" argued that the progress of humanity was dependent upon a broad education in the humanities rather than the exact sciences.

Given such strong language, it was remarkable that it was inspired by Génisset's speech on an "accord" between the arts and sciences. The next article Bonald referenced, written by a Monsieur Guairard, had no such ambiguity. In an even more scathing piece, Guairard declared that an exclusive focus on mathematics was nothing less than the capitulation of the soul.[69] Reviewing a math textbook designed for students who wanted to attend Napoleon's new École polytechnique, Guairard felt the need to remind his readers that "mathematics has not always dominated over the sovereignty of all education." As he explained, the middle of the eighteenth century had brought about a "revolution in teaching" that created the "painful feeling of today."[70] Similar to the *Journal* article, Guairard noted that a "revolution" had taken place among the youth of France, who could no longer relate to teachers raised on Virgil and Homer. While this revolution took place in regional schools, it had spread to universities, where "old professors were humiliated" by having to discuss mathematics instead of classical Greek and Latin. When students were asked to discuss the principles of grammar or the ideas of Virgil, they could only respond in "theorems" and "equations." The "barbarous jargon" of technical science terms and equations meant that the young students and their older professors were "no longer speaking the same language."[71]

Guairard rued the "half-century of this terrible plague," but he was not against mathematics itself. Calling the achievements of the field "truly remarkable" and "among the most beautiful and useful" creations of the "human spirit," he extolled the many benefits it had brought to human society. Geometry in particular had the ability to bring about "true heart" and a "just spirit" in its practitioners, but the philosophes had exaggerated the ability for mathematical training to lead to a similar moral and social order through a science of man. Like the reviewer of Génisset, Guairard worried that the problem was not in the exact sciences and mathematics themselves but in how they tended to dominate education. "Our error," he wrote, was in "esteeming mathematics too much," and he was appalled that students had "sacrificed their entire youth and all of these years" in learning just one field among many.[72] There was no doubt that "mathematics is useful," but Guairard stressed, "Languages are too!" The only regret about teaching the sciences was that language, *belles lettres*, and history had been "sacrificed for the study of less useful things." In choosing to review a standardized textbook,

Guairard explained that students had been shaped by the exclusive focus on applied technology and mathematics, worrying that an entire generation would pass without knowledge of the "sublime sciences" which find their "principles in our hearts."[73]

Together, the articles from *Mercure de France* and *Journal des Débats* attempted to undermine, in polemical fashion, the many claims made for science in the name of technology and mathematical certainty. Chateaubriand was preparing to leave *Mercure de France* in the summer of 1807, but both articles bore the imprint of the great French Romantic's anti-scientism.[74] Most importantly, however, in criticizing the geometric spirit, these articles connected two separate but related ideas crucial to Bonald's defense of intelligence: not only did the exact sciences provide poor methodologies for studying the thought and behavior of humanity, but the exclusive study of such sciences restricted a fuller appreciation of human potential. In both cases, "intelligence" was lost. Guairard, for example, explained that mathematics was a "precious tool" for "the astronomer, the mechanic, and the engineer" but became "an instrument of damage" for "the lawyer, the doctor," and "all those" who searched for "principles in our hearts" rather than "in geometry."[75] At the same time, the anonymous *Mercure de France* review warned that while science had been useful in the Revolution, the establishment of national schools focused on science education only led to "a people [made] entirely of mathematicians and chemists."[76] As in most of *Mercure de France*, the articles were pleas for the importance of literature and poetry in society as much as they were criticisms of science itself.[77] Vitriolic as they seemed, *Mercure* and *Journal* authors maintained a respectful role for the sciences. The same could not be said about the final prophet covered in this chapter.

Joseph de Maistre and the Fall into Spiritual Anatomy

Bonald and the writers of *Mercure de France* and *Journal des Débats* may have been overlooked in the literature on anti-scientism, but this is hardly the case for Joseph-Marie, comte de Maistre (1753–1821). An impassioned critic of the French Revolution, Maistre attacked nearly every foundation of Enlightenment and revolutionary rule during his thirty years of exile in Sardinia and Saint Petersburg. As paraphrased by Graeme Garrard, Maistre believed that "the entire earth, continually steeped in blood, is only an immense altar on which every living thing must be immolated without end."[78] Maistre's evocations of innocents as sacrificial victims for a greater power led Isaiah Berlin to place Maistre at the "origins of fascism," and historians have noted that Maistre put science at the heart of his political critique.[79] As Stephen Holmes noted in his survey of anti-liberalism, Maistre believed nothing less than that "modern science was an immoral power that drags mankind into unhappiness" and "leaves human beings as morally unhinged."[80] If one were looking for a prophet of doom, or a rhetorical leader for Barbery's plan to drag humanity back to the superstitions of the Middle Ages, Maistre might have been the best fit.[81]

While Maistre's image as the arch-prophet of the past is well supported by the violent language and apocalyptic visions in works like *Soirées de Saint-Pétersbourg* (1819), such readings of Maistre have largely been made by political scientists who have tried to find intellectual coherence in hopelessly reactionary politics. When it came to science as a whole, however, Maistre was not always so dismissive, agreeing with Chateaubriand and Rousseau that science was a powerful force that could be useful for a few, but which needed to be tempered by religion. Garrard noted that Maistre had "a natural curiosity about modern science," and more specialized works have demonstrated that Maistre's critique of scientism was not a critique of science per se, with Richard Lebrun calling Maistre a "Cassandra" for his warnings of scientific overreach.[82] Jean-Louis Darcel has also claimed that while Maistre was in exile in Saint Petersburg, he was in the "city of refuge of men of science who did not adhere to the revolution."[83] Though it has been argued that "Maistre's status as a symbol of reactionary opposition to the spirit of modern civilization is probably beyond revision," this has not stopped a host of revisionary accounts in the fifty years since.[84] In particular, in recent years Maistre has been recast as a structuralist, postmodernist, or even a counterintuitive intellectual bridge from the Enlightenment to positivist social science. In the most sustained and provocative revision, Carolina Armenteros has tried to reorient Maistre as a profound historical thinker who provides the missing link between Enlightenment and socialism.[85] Directly challenging the idea of Maistre as an anti-intellectual with disdain for mathematics and science, Armenteros fashions a Maistre who builds his history out of "statistical reasoning" and who outlined "the transition from human nature to normality that accompanied the rise of statistics."[86] Rather than the most well-known French-speaking critic of Enlightenment, the revised Maistre becomes its savior, smuggling a host of assumptions into the social sciences of the future.[87]

Such debate likely reflects that Maistre was not the most systematic thinker. Much like Bonald, he offered few thoughts on science prior to being forced to flee from the agents of the French Revolution at age forty. During his exile, Maistre was sent to Saint Petersburg as ambassador to the king of Piedmont-Sardinia, a seeming relief because Russia had been spared from the "fanaticism" overtaking much of Europe.[88] Yet shortly after his arrival, Maistre became alarmed at a proposed overhaul of the Russian educational system, which promised to teach the exact sciences in the earliest years and reduced training in language, rhetoric, and writing. Reflecting on the nature of Russian life in a series of 1811 essays, titled "Quatre chapitres sur la Russie," Maistre claimed there were "two equally fatal ideas" to education: "The first is to put literature and the sciences at the beginning; the second is to teach all of the sciences together."[89] The Russian Minister of Public Instruction, Count Rasoumowski, to whom Maistre addressed the letters, was proposing to add a significant number of science courses, and Maistre worried "the same sophism" of France might spread to Russia, where the educational system "regarded man as an abstract being" (*OCM*, 8:161). In another series of five letters, written more in the spirit of frustration than fanaticism, Maistre explained why an exclusive education in science would doom Russia to the fate of the French.

In his typically bracing style, Maistre began his first letter with a set of extreme propositions, noting that "science renders man sluggish and incapable of dealing with . . . great enterprises" (*OCM*, 8:165). Referencing his experience in France, Maistre warned that such a "species of moral vegetation" had led other nations to "barbarism from civilization" (*OCM*, 8:168). Comparing Russia to the great civilizations of the past like Rome, Maistre insisted, like Rousseau, that scientific training had contributed little to the great nations and could even be threatening if left unchecked. Russia was particularly unsuited to scientific training, Maistre believed, and the country was fortunate to be "cut off" from other nations because of a language that was "undoubtedly beautiful" but "sterile." To support his argument, Maistre asked what role science had played in Russia's glorious past, noting that the country had achieved great things without training in the sciences. Rather, it was "the humanities which marched in front of the exact sciences" throughout history, and Russia was privileged to be one of the last European nations to be untouched by the new science curriculum. In their "wisdom," Maistre noted that "the ancients" had taught the exact sciences only at the end, believing that it was only specialized knowledge and useless to the general student. Summing up the country's great imperial and military past, Maistre informed Rasoumowski that "there is therefore nothing in Russia that requires science" (*OCM*, 8:173).

Maistre's worries were based on what he had witnessed in France, in particular the closing of all Jesuit schools in 1773. In a lengthy defense of the Society of Jesus, he noted that the dual independence of the Jesuits and the natural science faculty at the University of Paris had created "two excellent institutions." Mixing the two, however, had created "one bad one" (*OCM*, 8:229). The Jesuits had ignored "chemistry, natural history, botany, etc." while the university had ignored "religion . . . moral philosophy, and literature" (*OCM*, 8:229). Though seemingly ecumenical in his last letter to Rasoumowski, Maistre's concern about education in the sciences became more explicit in other, unpublished, writings. In an 1811 letter titled "Freedom in Public Education," signed as "Philalexandre," Maistre made his Jesuit sympathies clear, noting that the Society of Jesus had "demanded" nothing more than "a perfect freedom for their *régime intérieur*" in order to "reclaim the privileges of all human association" (*OCM*, 8:272). In his essays on Russia, which picked up on the theme of how science might threaten the "interior regime," Maistre restated his attack on the enervating qualities of science education, claiming that its study "made men useless for an active life, which is the true vocation of man" (*OCM*, 8:297). Unlike the Jesuits, who allowed for free association, science "tends necessarily to kill the public spirit and to harm society" (*OCM*, 8:298).

The loss of Jesuit schools in France meant for Maistre that scores of students would be trained in subjects for which they had no aptitude and no interest. On the other hand, religious education, while not tied to any particular profession, was necessarily beneficial for all, since all people entered into society and presumably had souls. Worried that all students, no matter their intended vocation, would devote their early years to study in the exact sciences, Maistre lamented, "Never had anyone dreamed that one must know chemistry to be a bishop, or mathematics to be a lawyer" (*OCM*, 8:179). Maistre warned too that "it was in vain for a government to make . . . one kind of knowledge necessary to obtain . . . distinction," as previous generations of Russian civil servants and religious bureaucrats had come from broad backgrounds. In its attempt to unify all education under the auspices of training in the exact sciences, Russia was therefore in danger of losing the services of those who had aptitude for religion, literature, or the fine arts. It was not that scientific training was inherently bad (although it could be in some circumstances) but that it ignored other forms of knowledge. As Maistre claimed, the Jesuits and the University of Paris differed "only in the coordination of the different sets of knowledge, on their respective importance, and on the time when it is most proper to study them" (*OCM*, 8:174).

Maistre's writings on education suggest why he believed the study of the exact sciences was so dangerous. Even if the reductive sciences of man based on mathematics and physics were failed attempts to reveal the true scope of human life, the exclusive study of these sciences could in fact reduce the scope of actual lives. Physics itself may not be able to capture the objective truth of human experience, but *learning* physics alone could drastically alter individual lives. He had earlier written to Rasoumowski that "science makes . . . [us] incapable of great enterprises," yet many had come to "believe that science education was [all of] education" (*OCM*, 8:165). For Maistre, this was the explanation for why the exact sciences and mathematics had defeated the ancien regime. Science, he complained, "continually exposes the state [to] men of nothing" and "was accessible for all" (*OCM*, 8:303). Like Bonald, who believed intelligence should rule over the "organs" of the state, Maistre believed that the proper place for most people was in society, not government. Rather than a victory of science, however, whereby *savants* rose to the level of the nobility, Maistre saw in the proposed educational reforms of the Revolution only a descent. Linking the nobility of privilege to the nobility of the human spirit, Maistre warned that scientific education reduced all to a common level. As he noted of France, "during the seventeenth century, the nobles gathered around them illustrious *savants* for the benefit of the state. In the eighteenth century, they descended, and had to serve them. What had happened? The nobility had fallen, and their sovereignty with them" (*OCM*, 8:303).

Taken by themselves, Maistre's vehement denunciations of Russian educational reform may seem overheated, but it was clear that he believed the exclusive study of science narrowed the possibility for a fully human experience. In fact, in his most extended attack on scientism, the posthumously published *Examen de la philosophie de Bacon*, the consequences of privileging science over all other forms of human understanding became clearer. While Francis Bacon (1561–1626) had been hailed by the French *Encyclopédie* for his new form of investigation based on experiment and induction, Maistre argued that prior to Bacon, "the fine arts and literature had been carried in the sixteenth century to an even higher point of perfection."[90] Notably, Bacon and his philosophe popularizers had not done any real work to understand human beings as they actually existed in the world, instead taking the easier task of defining humanity to fit a predetermined scientific theory. Sciences "remade human understanding" to fit its own ideas in the same way that "one remakes the human body for gymnastics."[91] Such "spiritual anatomy," as Maistre called it, could only work by inventing a new patient to examine.

As he later complained in *Examen*, echoing Bonald's concern about the restrictive language and impoverished "form" of language of modern science, Baconian science had "imposed" a "didactic order and regulated technique" on writing, which limited aspects of human existence and experience. "In place of adapting his system to man," Maistre countered, "[Bacon] invented a man which bends to his system."[92]

For Maistre, the heart of the problem was Bacon's reduction of science and knowledge to physics, which in turn reduced it to the quantification and calculation of mechanical causation. According to Maistre, Bacon believed that "all of *philosophy*, all of *science*, and all of *physics* . . . were synonymous," a problem that was compounded if the tools of Baconian science were used to study human action.[93] As he claimed in a note that would attract the attention of Auguste Comte, the founder of sociology, a "science of man whose goal is the knowledge of causes will be irreparably useless."[94] Sounding quite like d'Alembert's idea of the "Analyst," Maistre pointed out that great thinkers like Copernicus had used reason and a "freedom of spirit" in their physical investigations, borrowing from literature to make their arguments rather than reducing them to the "domination of the table." Rather than rely on "sterile calculations," Copernicus and others had studied "substance, movement, and the influence of celestial bodies according to their true essence."[95] The natural allegory was that people too must be studied according to their "true essence," not just as passive producers of trace data that can be tabulated and manipulated to prove causation. After noting that Newton and Descartes had both been attentive to purpose and meaning in their physics, he said Bacon "was the opposite of all."

Earlier critics had warned of the consequences of Cartesian mechanism, but for Maistre, the consequences of applying Baconian experimentation and induction to the sciences of man had even more devastating consequences for human intelligence. Maistre, reading an early nineteenth-century French translation of Bacon's *Great Instauration* (1620) by the former Jacobin Antoine de La Salle, imbibed a particularly strong dose of Baconian ambition. Writing enthusiastically about the project, in language sure to rile a traditionalist, La Salle introduced Bacon by saying that he "aspired to nothing less than the production of a new species of being and to transform the existing species."[96] Maistre agreed but noted that Bacon's accomplishment was not the elevation of humankind but rather the degradation of this "new species of being" into the realm of nature. After Bacon, the "great misfortune of man" was to have the "true sciences" of morality, politics, and civilization "hijacked [*le détournaient*] by physics." Baconian progress, he argued, was only possible through an extreme simplification of life to mechanistic causation, and Maistre saw Bacon's success as being due to the fall and even negation of the possibility of intelligence. As he wrote at the beginning of Book II of *Examen*, because of Bacon, "it is impossible to see any *intention*, and consequently any intelligence in the universe."[97] Therefore, sciences of man like those developed by Helvétius and d'Holbach had committed a "crime" against "depraved man" by substituting "*effects* for *intentions*."

For Maistre, intelligence was synonymous with free will (or intention) because it was a gift from God. Nothing was "comparable to intelligence" because "intelligence is the sole title of the definition of God which has been given to man."[98] As he claimed, "man is . . . purely and simply *an image of God*, that it to say *intelligence*."[99] Yet in *The Great Instauration*, Bacon had taken "intelligence" from the heavens and placed it in the same category as "appetite." For Bacon, intelligence was a simple "faculty of the soul" that could be understood merely with a "physical explanation." After listing the many "faculties" Bacon claimed made up the soul, Maistre criticized Bacon's "deception," where "he mixes [intelligence and appetite], confusing one for the other. He has placed *intelligence*, that faculty which all schools recognize as the distinctive character of the sensitive soul, or even the soul itself, in the rank of a simple faculty, uniting it with appetite."[100] Though Maistre did not note the connection, Bacon's reduction of intelligence to appetite recalled Descartes's *L'homme*, discussed in chapter 1, where the process of memory was likened to how the "tongue" processed the various sensations of vinegar and sweet water.

While Maistre's specific anger was directed toward Bacon's attack on the divine, his argument against the exclusive teaching of the mathematical and physical sciences could easily be adopted by an ancient Greek, a Renaissance humanist, a Romantic poet, or a defender of liberal arts requirements at an American university in the twenty-first century. For the purposes of the story of the descent of artificial intelligence, however, Maistre had located a key feature of the modern sciences of man and what would become the social sciences: even if mechanical, mathematical, or quantitative theories of human action failed to uncover a true explanation of human actions, such theories often had the corollary effect of "bending" humanity to fit the system. Were such ideas to receive widespread attention in education and in institutions of state and science, they could, Maistre worried, create that which they could not discover.

Conclusion

In many ways, it is not surprising to see Bonald, Maistre, and the counterrevolutionary press all arguing for one or another form of doom and descent. As Christophe Ippolito put it in an introduction to an anthology of "resistances to modernity," all anti-moderns are driven by a belief that "the modern world is a theater of de-civilization, marked by the collapse of culture and education, and of a catastrophic 'massification' of spirits."[101] This was what Barbey had meant by his metaphor of a descending staircase to barbarism. Bonald stated it more simply in a letter to Maistre: the ideas of the Enlightenment had substituted a "mortal indifference for all that is great, noble, and elevated."[102] Such is always the reactionary conservative position: to invent a supposed golden age, to propose a radical critique of the means through which a society has arrived at its modern perversion, and then to provide a remedy—usually violent—to this invented fall. Yet what makes these figures so compelling is that they departed from ancien regime thinking just as radically as the philosophes did in their vision of science. It is not necessarily the case that, as Gengembre wrote, "the counter-Revolution built upon the ruins of the Enlightenment" by reusing "their materials."[103] Rather, the novelty in their writing was to imagine a science of man based on irreducible collective intelligence in *society* and to recognize how education in literature, *belles lettres*, and the arts might be fundamental to such a science. The ancien regime, the nobles, the Bourbons, and even the church were not necessities in themselves, but rather served as protectors of a particular kind of social intelligence. As Bonald claimed at the time of Maistre's death, literature was "the proof of our origins and the stamp [*sceau*] of our intelligent nature" (*OCB*, 3:1106).

As the review of these thinkers makes clear, their fear of a "descent" from an intelligent nature was based on the idea that the reduction of humanity to generalized laws and mathematical formulas had severe consequences. Most importantly, for Bonald such an approach in the sciences of man "atomized" individuals, tearing them away from institutions like the monarchy and church, which were the historical guarantors of both society and intelligence. While the modern conservative Margaret Thatcher (1925–2013) famously claimed "there is no such thing" as "society," for Bonald's conservative science of man it was the *individual* that was the invention. In addition, as seen in the counterrevolutionary press, even if the sciences of man were wrong in their approach to studying humanity, the teaching of these methods exclusively at the expense of literature, poetry, and history could create the very kind of tedious and repetitive human behavior that reductive sciences of man sought to find. Maistre's critique of Bacon would not be the last time a social scientist would be accused of having "invented a man which bends to his system."

Without pushing the point too far, it should also be noted that these thinkers produced a few heirs and that their ideas returned at the same time when the quantitative study of humanity and the focus on science education reemerged in the middle of the twentieth century. In *Dialectic of Enlightenment* (1947), for example, Max Horkheimer and Theodor Adorno quoted Maistre at length, stating, "The reduction of thought to a mathematical apparatus . . . condemns the world to be its own measure."[104] Horkheimer and Adorno also appeared to recast Bonald's argument that progress in science can come only through a degradation of education, writing that "in the operations of modern science, the major discoveries are paid for with an increasing decline of theoretical education."[105] Rather than the originator of fascism, as he was for Berlin, in *Dialectic of Enlightenment* Maistre becomes a prophet warning of Arendtian banality, foretelling the dangers of an eventual "administered world."[106] While debates about the connection of an administered world to the horrors of the twentieth century are outside the scope of this study, it might be enough to say that Bonald, Maistre, and the journal writers all ultimately anticipated the critiques of scientism, bureaucratization, and atomization in the modern world, even as they failed to fully articulate a competing vision that led anywhere else but back to the chains of the ancien regime.

As this chapter concludes the first section of the book, it is worth reflecting on the connection between the "prophets of the past" and the overall concern with the decline in the concept of intelligence. For the enemies of Enlightenment, it was the erosion of interest in even the idea of intelligence—expressed through religion, literature, and traditional customs—that was most distressing, as sciences of man based on mechanical philosophy and the geometric spirit appeared to eliminate one of the great features of the natural world. Descartes's earliest critics recognized that intelligence had lost its connection to the divine in mechanical philosophy, but in the Enlightenment sciences of man, it seemed to disappear completely. Perhaps even more worrisome than the vision of man was the trend in education, as philosophes and anti-philosophes expressed concern that language, literature, morality, politics, and history were being crowded out by technical training or mathematics. As early as 1815, pamphleteers were arguing that French youth had lost their "true taste and love for literature."[107] This would be Barbey's nightmare, where a modern barbarism emerged as generations of students were denied education in the moral sciences, *belles lettres*, and fine arts.

Barbey's fears would not have been quelled by the introduction of more formal sciences of man in the nineteenth century, including social physics, political economy, and eugenics. Neither, of course, would the increase in *actual* machines have inspired much hope. As Bonald worried in 1810, "there are many machines to replace man, and there are a lot of men who are only machines" (*OCB*, 3:1152). The reviewer of Génisset's speech feared as early as 1807 that there would be an "ingenious machine which will multiply or replace the forces of man."[108] Decades before the fictions of Samuel Butler and the theories of Karl Marx, Bonald identified the moral threat machines posed to human intelligence: "The desire to invent machines that do as much work as possible, with the least amount of intelligence involved in operating them, has been extended to the moral world" (*OCB*, 3:1254). As will be seen in the practices of the nineteenth-century sciences of man, a transformation in mechanical technology was abetted by a new process of administration and observation, one that would begin with the first proposals to count and observe all the actions of humankind, including the quantification of intelligence itself. In seeing a world where machine-like intelligence was every bit as threatening as machines themselves, the prophets of the past had seen all too clearly what the future sciences of man might entail.

PART II

Numbers *and* Machines *in* Europe

Progress *to the* Mean

Social Mathematics and the New "Men" of Social Physics

In 1832, ANY REMAINING CRITICS OF THE GEOMETRIC SPIRIT MIGHT HAVE been disturbed by the conclusions of the influential Belgian statistician Adolphe Quetelet (1796–1874), who infamously declared that "the more the number of individuals observed grows, the more individual will disappear."[1] When Quetelet claimed that free will "disappears" as more individuals are observed, he was likely only referring to the "law of large numbers," the idea borrowed from probability and astronomy that particular "perturbances" in a given data set will tend to even out over time. Quetelet, who spent over five decades quantifying human actions and behavior (and much else), did not mean to say that observation *itself* lessened individual will.[2] Yet, as many critics had recognized, the very process of observing and calculating human actions—the heart of the probabilistic sciences of man since Laplace—seemed to require a further reduction of individual will. As seen in the past two chapters, French critics of the mechanical sciences of man—from La Mettrie and Rousseau through Bonald and Maistre—worried that particular forms of education in science and math had even constrained and narrowed the possibility of human intelligence. For the critic attuned to the real-life effects of practicing the sciences of man, Quetelet's note about the removal of free will in the process of scientific observation may well have taken on a more sinister cast.

The prophets of the past did not live to see the development of Quetelet's science of man—which he called "social physics"—but the development of the quantitative social sciences in the nineteenth century might have confirmed their worst fears. In the nineteenth century the most reductive sciences of man would become transformed from a handful of simple metaphors and suggestive ideas into a "relatively distinct intellectual field," relying on more sustained and literal applications of tools from mathematics, physics, and astronomy.[3] As a number of historians of statistics have demonstrated, the subtle "subjectivism" and freedom of Laplace gave way to a more "objective" vision of numbers, one that seemed to impose a form of statistical determinism upon the events chronicled, leading to profound consequences for both the practitioners and the subjects of social physics.[4] The shift occurred because of several interrelated developments. Not only was more data collected in the nineteenth century, but it began to "speak" in new ways, as probability theory replaced the certainties of geometrical and deductive reasoning, becoming the guiding light for researchers in all sciences. So too was there an increase in the number of researchers charged with counting and observing human actions. Last, the field of "physics" itself changed in the middle of the century, providing a radical new shift in the reigning metaphor for those trying to model a science of man on the theories of the natural sciences. By the end of the century, the sciences of man were not simply surveying or studying human activity but actively seeking to influence this action as well.

To examine how the transformation of the nineteenth-century quantitative sciences of man imposed further limitations on both practitioners and subjects of social science, the chapter begins with a brief survey of the work of Nicolas de Caritat, Marquis de Condorcet (1743–1794), the French mathematician who was the "apostle" for applying the tools of probability to the actions and behavior of human beings.[5] Condorcet called his science "social mathematics," and it was just one of several plans he had for a new kind of *science sociale*, a term he chose over what he believed to be the less precise fields of moral and political science.[6] Although Condorcet did not witness the flourishing of his new science, as both a government minister and perpetual secretary of the Académie des sciences, his influence on subsequent ideas was profound, as the French philosophe united the belief that a mathematical and statistical science of man required both sophisticated mathematics *and* a new class of *savants* to implement them. While the application of probability to the study of human behavior was most pronounced in France, this chapter also marks a geographical expansion of the story, as ideas that began in the French Enlightenment were exported (often by force) throughout Europe in the first decades of the nineteenth century. One young man schooled under occupation of the French Empire was Quetelet, the "father of statistics," who was just one of thousands of Belgians who had been educated in Napoleonic *lycées* that mixed Enlightenment political philosophy with practical science.[7] Quetelet was not the best mathematician of the nineteenth century—far from it—but he was the best and most enthusiastic *counter* of the age, dedicating his professional life to making a comprehensive statistical account of nearly every feature of the natural and social world, from subtle changes in barometric pressure to the causes of suicide. Although his handwritten journals and books on probability include shockingly reductive accounts of human action, suggesting that aggregated data and a mechanical science of man might inhibit free will, Quetelet's greatest legacy for the sciences of man was far more practical. Rather than check the free will of those under observation, Quetelet asked for the greatest sacrifices from those doing the observing.

Quetelet called his project social physics, but when he began his research in the 1830s and 1840s, the idea of "physics" hardly denoted a fixed set of theories and methodologies, let alone a standardized set of practices that could be directly applied to studying humanity. Such was not the case for William Stanley Jevons (1835–1882), a British political economist who credited Quetelet as the "true founder of social science" but who sought far more rigorous and lawlike explanations of human thought.[8] Not only did Jevons attempt to find a common methodology for the natural and social sciences, but he also built an actual machine that he believed could reproduce mechanically the process by which the mind reasoned. By the time Jevons began his work, the field of physics itself was on far sturdier ground, leading to even more certainty than the Newtonian and Cartesian metaphors that guided the first sciences of man of Helvétius and d'Holbach. The result: a new direction for the future social science of economics, which by many accounts would become the most influential "social physics" of the twentieth century.[9] Grounded in the political ideas of the Enlightenment, Condorcet had believed that education would drive progress in human affairs, and Quetelet had vacillated over whether social laws could be directed by *savants* and governments, but by the end of his life Jevons had concluded that interventions were necessary to make the seemingly free and stubborn actions of actual human beings more amenable to laws revealed by scientific methodology.

The Weakness of the Human Mind:
Condorcet's Plan for Progress

Although his life was tragically cut short during the Reign of Terror, the Marquis de Condorcet likely represents a high point in uniting a quantitative science of man with the idea of political progress in France. Similar to Fontenelle, Condorcet held the position of permanent secretary of the Académie des sciences and was as strong a believer in the geometric spirit as any Frenchman of the age, even after his death. In 1795, in a posthumously published work that famously attempted to sum up the entire "progress" of the "human spirit," he noted that "general laws . . . which regulate the phenomena of the universe, are regular and constant" and should be "applied to the development of the intellectual and moral faculties of man."[10] A decade earlier, in a more technical work devoted to probability theory, he went so far as to claim that a "great man" had taught him that "the truths of the moral and political Sciences were susceptible to the same certitude as those developed in the physical Sciences."[11] Condorcet has been a student of d'Alembert and taught Laplace, but as seen in chapter 2, neither of these two mathematicians had thought a probabilistic science of man could provide "certitude."[12] Instead, the "great man" who embraced the geometric spirit was not a mathematician but a politician: Condorcet's patron and mentor, Anne Robert Jacques Turgot (1727–1781), Louis XVI's chief economic advisor from 1774 until 1776. Neither Turgot nor Condorcet were the first to suggest that a science of man might achieve the "same certitude" as the physical sciences—Condillac, Fontenelle, Helvétius, d'Holbach, and Robespierre had said much the same thing—but Condorcet's approach was driven as much by an interest in political reform as it was in creating an entirely new science of human action and thought.[13] Indeed, for Condorcet, science and enlightened political progress were inseparable.[14]

Unlike Helvétius and d'Holbach, who proposed sciences with broad theoretical assumptions based on Cartesian and Newtonian mechanical analogies—and left the practical details to be derived later from first principles—Condorcet built his science of man from the ground up, investigating and quantifying how people actually lived. As a government minister and representative in the French National Assembly and National Convention, he offered hundreds of practical proposals he believed would benefit social life. Condorcet represented the "dream of public service under an enlightened minister," and his work allowed him "to demonstrate the indissoluble link between scientific and political progress."[15] His detailed plan for the 1791 Constitution—a "democratic utopia" which called for decentralized voting, the declaration of education as a natural right, and full suffrage for women—likely marked the apex of Enlightenment theory in practical politics in Europe.[16] Although Condorcet's vision of the human spirit as progressing from the "savage" to "Enlightened" anticipated the paternalism and racism inherent in the French "civilizing mission," his aversion to violence, severe condemnation of slavery, and promotion of women's rights have led to claims that Condorcet was "250 years ahead of his time."[17] Less well known today than major philosophes like Rousseau and Voltaire, Condorcet's last years were both dramatic and tragic, as he tried to "sketch" a complete methodological guide to the sciences of man *and* a schematic history of the past, present, and future of the "human mind." After resisting calls for violence during the peak years of the Terror, Condorcet was forced to compose two of his most famous essays while in hiding. His final work—a breathless argument for the possibility of perpetual improvement in human happiness—was written just prior to his capture and poisoning in 1794.

Any attempt to trace the implications of Condorcet's social scientific thought, even ignoring his political and personal life, runs into the limitations of space. As Keith Michael Baker demonstrated in his highly influential study of Condorcet's ideas, there were really three kinds of sciences of man wrapped up in his idea of a "social science": the concept that a study of the "ideas" and "sensations" of people could lead to a better form of education and practical reform; the idea of a "social mathematics" that relied on data collection and the application of probability theory to social questions; and the vision of a "social art" that could transform and "revolutionize" humanity.[18] Although the focus here is mostly on the second idea—social mathematics—as Baker noted, it is nearly impossible to understand any aspect of Condorcet's ideas without keeping the other two projects in mind. Indeed, Adolphe Quetelet would follow Condorcet in arguing that new social scientific methodologies required new kinds of scientific researchers, while William Stanley Jevons agreed that the subject of such studies would require the "social art" of altering society as well. While all three projects required a different kind of "social scientist," Condorcet was clear that new researchers would be needed to take up such an ambitious project.

Condorcet introduced the *sociale mathématique* in one of the two epic works written in the last years of his life, the *Tableau general de la science qui a pour objet l'application du calcul aux sciences politiques et morales* (ca. 1794). Though it has become one of his most enduring works, the *Tableau* was likely written in haste and partially in hiding, and it has been suggested that Condorcet may have written it in one draft, as most versions that have appeared contain significant errors.[19] Yet the overall message was clear: the sciences had progressed to a point of maturity where there could be "lines of communication" between them and "the application of one science to another" was possible (*OC*, 1:540). For Condorcet, fields like political economy could only progress so far without the "application of rigorous methods of calculating," as previous ideas had been based on ancient prejudices and "false certitudes." In particular, the ongoing revolution and reaction in France had proved that this method would have new "utility" in fixing the "disorders" that were "inseparable from all great movements" (*OC*, 1:542). Good ideas had clearly circulated before, but if the political and moral sciences were given mathematical rigor, it would "chain man to reason by the precision of these ideas" (*OC*, 1:543).

For a hastily sketched outline that sought to bring the lessons of all the sciences together, Condorcet's *Tableau* was remarkably clear in terms of why mathematics could be applied to human affairs. First of all, in the many decades since Descartes had introduced the "physico-mathematical sciences," Condorcet noted that mathematics had moved toward the probabilistic approach, noting that in all fields it was "useful to have average values of observations." Because people could be treated as individuals, or in relation to "things," social mathematics too required the "reductions of values to a common measure" (*OC*, 1:545). Here the recent applications of probability and the "average values" could be analyzed, and Condorcet outlined at least five kinds of mathematics that could be brought to bear on political and moral questions. Even at the level of the mind, Condorcet believed "one must count also . . . technical or even mechanical means to execute intellectual operations" (*OC*, 1:556). By "mechanical means," Condorcet did not mean the simplistic metaphors of the past, and he was certain to avoid any hint of the disgraced "machine men" from the condemned materialists. Explaining the problems of past sciences of man, for example, he compared previous efforts to understand the laws of wealth as a "simple" application of "mechanical principles" to describe a "great hydraulic machine" (*OC*, 1:567). Using probabilistic social mathematics, rather than inflexible and lawlike mechanics, would eliminate the problem of "too often adopting truths as absolute and precise," when there were "exceptions" and "even changes" to the most important laws (*OC*, 1:567). Just as Laplace would warn that even celestial bodies might deviate from prescribed paths, Condorcet noted that only through probabilities and averages could such "exceptions" be incorporated.

Social mathematics in the *Tableau* was clearly built on complex and technical ideas drawn from probability theory, though it was crucial to Condorcet's plan for a science of man that it be accessible to all. Returning to the language of the earliest mechanical philosophers, Condorcet promised that the social sciences would not be an "occult science" kept as a "secret among a few followers." While the dense works of Laplace would be incomprehensible to all but the well initiated, social mathematics relied only on "elementary knowledge . . . to understand the solution to the majority of the questions, to understand the theories, and to deduce the application the most immediate in practice." Social mathematics still required that it be led by "geometers" who are able to "deepen the social sciences," but in practice there would be a general knowledge deployed by all citizens in a rationalized France (*OC*, 1:550). For Condorcet, who may have been sensitive to the concerns of Diderot and d'Alembert, social science *required* broad adoption, because no single genius could capture every new truth. Ironically, progress would be based on what Condorcet called "the weakness of the human mind" (*OC*, 1:572).

Here it can be seen why Condorcet's three projects of social science were so interrelated and why he was not content to offer a simple mathematical or mechanical theory of man like the French materialists. Condorcet did not merely propose a set of mathematical tools through which one might learn about people and society; he also made his social mathematics contingent on his other two projects of human transformation, the idea of practical political reform and a "social" art that could "revolutionize" humanity. Although the "weakness of the human mind" may not be able to keep up with all the new discoveries in science, Condorcet believed that scientific advances themselves would lead to better instruction in the methods of social mathematics. As he claimed in his famous *Esquisse d'un tableau historique des progrès de l'esprit humain* (1795), "the progress of the sciences secures the progress of the art of instruction, which again accelerates in its turn that of the sciences; and this reciprocal influence, the action of which is incessantly increased, must be ranked in the number of the most prolific and powerful causes of the improvement of the human race."[20] While no single mind could know everything, everyone would be capable of knowing some part of the whole.

Condorcet's belief in the "reciprocal influence" of scientific discovery and education also meant that progress in science and instruction could itself transcend the limitations of human thought and morality. As he noted earlier in a speech to the Académie Française upon his induction in 1782, "The project to make men virtuous is a dream, but one day why should we not see enlightenment and genius create for happier generations an education system and a legal system which would make courageous virtue almost unnecessary?"[21] Notably, for Condorcet it was not in individual virtue and intelligence but rather in a collective "Republic of Minds" that progress would be guaranteed.[22] This was why his work to completely reform the French educational system was inseparable from social mathematics. Just two years prior to his fall from revolutionary leadership, in fact, he had been in the rare position to offer "nothing less than a fully integrated national education system from infant school to university."[23]

Condorcet's science of man had called for a *revolution* of human minds, but what distinguished his social science from later and more reductive forms of social physics was that he never imagined the individual at the mercy of larger forces. His plan for a system of "normal" schools offered precisely the kind of broad education that Bonald, Maistre, and the counterrevolutionaries had *wanted* (absent, of course, theological instruction). Further distinguishing social mathematics from the determinist path of French materialism, in his final two works there are few analogies with specific theories of physics or astronomy, and nothing like the deterministic education machine of Helvétius or the sandstorm of d'Holbach. Condorcet's vision of human thought also differed from the later abstracted "average man" of Quetelet and the "hedonistic calculus" of Jevons, each of which relied on far stricter applications of the methodologies and metaphors of the natural sciences. Remarkably, the main "scientific" theory in the *Esquisse* was a simple appreciation of history. In Condorcet's words, if "past experience allows us to anticipate future events with a great deal of probability, why should it seem an impossible undertaking to project the future destiny of the human species with some plausibility from the results of this history?"[24]

The *Tableau* had imagined the possibility for a fully mathematical social science, but Condorcet's account of the progress of man in the *Esquisse* depended on little, if any, actual mathematics. While his earlier work on voting procedures required sophisticated tools from probability theory in order to tease out the collective will of individual minds, there was no need to apply any particular mechanical metaphor from the natural sciences, just probabilistic reasoning. Condorcet's mentor Turgot began his career as a physiocrat, whose mechanistic political economy placed all wealth in agriculture, and Condorcet certainly seemed to leave much of social mathematics open to future development, never implementing the crude metaphors of a mechanical force. For Condorcet, social science was not just a simple application of physical tools to humanity but, as the *Esquisse* indicated, also a rich analytical and historical process. In the next two attempts to reduce human thought and behavior to a mathematical form discussed below, however, much of Condorcet's subtlety and vision for social science was lost.

Quetelet: The Average Administrator

As Baker noted in his study of Condorcet's social science, the most committed follower of social mathematics was Adolphe Quetelet, the Belgian mathematician and astronomer who encouraged generations of *savants* to collect and share data whenever possible. Quetelet, whose massive compilation of data on physical, moral, and intellectual traits—*Sur l'homme* (1835)—cited Laplace's admonition to "apply . . . the method of observation and calculus" to "the political and moral sciences," was perhaps the most dedicated follower of Enlightenment plans for a science of man.[25] Yet as important as Quetelet was for his intellectual contributions to applying probability to social data, it was his vision of science work that constituted almost a revolutionary new philosophy for the sciences of man. Whereas d'Holbach and Helvétius had hardly discussed ideas of observation and data collection, and Condorcet had hoped to train new generations at the Académie des sciences to practice the arts of social mathematics, Quetelet's social physics required a far greater sacrifice for science workers. As he noted in a history of Belgian science, the sciences could no longer progress through individual "genius." Instead, Quetelet created social physics with the belief that in "the sciences of the new era . . . *savants* have ceased to act as individuals."[26] While known to most historians of science, Quetelet's importance in the creation of the quantitative sciences of man has increased substantially in recent decades, as his core belief that numbers might be able to speak for themselves became more relevant in the era of "Big Data" and "the end of theory." Alain Desrosières, the late political thinker who helped develop the sociology of quantification, claimed that Quetelet's "mode of reasoning" might be more relevant to the modern age than the theories of Marx, Comte, or Tocqueville.[27] In advocating for data collectors over analysts—and mathematical tools over interpretive analysis—Quetelet's science of man helped shape the twin poles of future reductive social science.

In both intellectual histories and histories of science, Quetelet is today best known as a statistician, a dogged quantifier who claimed to have found invariable statistical laws in human behavior.[28] In his early career, however, he did not see statistics, especially statistics about people, as particularly important. Having taught astronomy in Ghent and Brussels and written an important dissertation on conic sections in his early twenties, he was only a few years removed from being a published poet and composing essays on Romanticism when the opportunity to coedit a journal in physics and mathematics changed the course of his studies. When his former mathematics professor at the University of Ghent J.G. Garnier suggested they begin a new international journal, the *Correspondance mathématique et physique* (1825–1830, 1832–1839), Quetelet had few other long-term options. His biggest dream at the time had been to build an observatory in Brussels, but this was more to elevate the scientific standing of the Low Countries than to attain any specific scientific aim.[29] Science in Belgium had been neglected under the rule of the United Kingdom of the Netherlands (1815–1830), which viewed the Catholic Low Countries as backward provinces, and Quetelet had made only fitful attempts to convince administrators to support an observatory up to that point.[30] Outside a vague desire to restore the sciences of *les Belge* to their previous glory, Garnier's offer to coedit a journal seemed like the next best option.[31]

In Brussels, Quetelet had hoped to create an institution like the Paris Observatory, where he had met Laplace and visited the "able generals" of François Arago (1786–1853) and Alexis Bouvard (1767–1843), but Garnier's offer of an international journal offered another path toward data collection.[32] The plan was simple—to contact *savants* from around the world to send in their most recent mathematical work, effectively replicating the "army" of researchers that Quetelet had witnessed during a trip to Paris.[33] In return, Garnier and Quetelet would share the most important research in Belgium, as well as create problem sets and equations for contributors to solve. Introducing a new "Statistics" section to the *Correspondance* in 1826, Quetelet called on his correspondents from "different countries" to "seriously occupy themselves" in the "statistical research that is so interesting and so useful." Again, he referenced the "administration of Paris" as a "good example to follow," one that would "guarantee the stability" of statistical research. To reach this aim, Belgium and other countries needed to "call for the creation of *savants* of the first order."[34] Quetelet even quoted the great French mathematician Siméon Denis Poisson (1781–1840), who lamented that "statistical science" up until then had been a "sterile science," which had, in his colorful language, been "reduced to simply understanding the Babylonian or the Carthaginian consumption of meat and lamb." Without applying mathematical techniques, statistics would remain a mere descriptive science that would create redundant work, equivalent to the "100 Gates of Thebes" (*CMP* 2, no. 3: 177).

Surprisingly, rather than the kind of "mathematization" of statistics that Poisson suggested, in the 1820s Quetelet embarked on an administration of numbers, prioritizing data collection above all else. Although Quetelet had been a math professor and had written a textbook on astronomy, in the next few years of the *Correspondance*, the theoretical and mathematical studies begun by Garnier were replaced with raw data. While simply compiling numbers might have seemed a poor way to begin a science of man, Quetelet was convinced that data collection was not just necessary for an inductive search for lawlike patterns, it might even be sufficient. In his new science of man, which he called "social physics," data collection *was* science. He claimed, "Statistics should first gather [*réuni*] numbers; these numbers gathered successively and with more care and caution, on a bigger scale, will reveal curious facts and make known the laws that govern the moral and intellectual world as they have governed the material world. It is in these laws which appear to me to constitute *social physics*, a science still in the cradle, but one which will irrefutably take precedence every day. It will in the end become among the sciences of which men will make the most use" (*CMP* 9: 485–86; italics in original).

While Quetelet might have seemed naive in his belief that gathering numbers alone would "reveal curious facts" and "laws" of the "moral and intellectual world," on multiple occasions in the *Correspondance*, they certainly seemed to do so. In one study, he combined dozens of reports to conclude (correctly) that a person's height is tied to environment, or what he called the "inverse reason of troubles, fatigue, and deprivations felt in childhood" (*CMP* 3, no. 3: 161). Similar comparisons between heights and weights would later lead Quetelet to the development of the formula that would become the body mass index (BMI), a simple equation based on a handful of studies that is (with great controversy) still the primary measure of obesity used today.[35] In another notable example from 1826, after observing only a few tables of data, Quetelet observed that year after year births in Brussels were at their highest in February and lowest in July, and that the graphical representation looked similar to the sine wave that described heat transfer in the work of the French mathematician Joseph Fourier (1768–1830).[36] In the most famous example of raw data "revealing" itself, Quetelet found that a group of chest measurements for Scottish soldiers contained the same level of "error" as astronomical observations. In the same way that the mathematician Carl Friedrich Gauss (1777–1855) had demonstrated that astronomical measurements clustered around an average—the true location of the star—so too did the soldiers' measurements point to a true "average" soldier. Such regularity led Quetelet to believe that there was, therefore, an "average man" of physical traits, around which all other sizes were "errors."[37] Although it was difficult to find as many statistics on moral and intellectual traits as there were on physical ones, Quetelet assumed that the only thing that stood between the determination of the average physical man and the average moral and intellectual man were practical means of data collection.

The idea that deviations from the average might constitute an "error" led many of the contributors to *Correspondance* to complain that "miserable numerical tables" led to "materialism," and it has been suggested that "fatalism underlay Quetelet's conclusion about the stability" of certain numbers.[38] However, Quetelet's interest in the practical, rather than philosophical, aspects of determinism can be confirmed by the recent publication of a handwritten journal Quetelet composed around the same time he created social physics and edited the *Correspondance*. Published and translated by the historian of science David Aubin in 2014, the work, "Principles of Mechanics that Are Susceptible of Application to Society," seems at first glance to be an astonishing document of geometrical reduction on the level of d'Holbach's *Système de la nature*, presenting a mechanistic social world that operated in a Newtonian void of laws and forces.

Quetelet had said that individual will "disappears" when aggregated in large numbers, but the notebook entries seemed to suggest that social physics left no room for free will at all. For example, he bluntly claimed, "Society [as] a moral being obeys moral forces, like raw bodies obey physical forces."[39] Even at the level of the individual, something like inertia might inhibit free will: "The man that always tends towards the same goal ultimately acquires an immense force like the stone that falls at the center of the earth." In what might seem like a vast denial of individual autonomy, Quetelet even questioned whether those who commit crimes are truly responsible for their acts, noting that "society will deliver the same number of victims to prison" because "when a system is in motion, its pace will stay invariably the same, unless the causes of the motion come to be modified."[40] Quetelet's recasting of criminals as "victims" of societal inertia would appear a few years later, when he wrote in *Sur l'homme* (1835) that "society itself contains the germs of all the crimes committed. It is the social state, in some measure, that prepares these crimes, and the criminal is merely the instrument that executes them."[41] Later historians noted that Quetelet's work in *Sur l'homme* was at the vanguard of modern criminology, but given such strong language, the publication of "Principles of Mechanics" would have only increased the charges of determinism, plunging Quetelet back into the determinist debates that haunted the French materialists.

While such language might appear to reflect an all-encompassing mechanical vision of the universe, like Descartes in *L'homme* or d'Holbach's sandstorm, a further reading of the notebook indicates that Quetelet was working out far more practical concerns. The "mechanisms" of society were not unrelenting metaphysical burdens but rather the practical gears of career advancement in the administrative state. As Aubin noted in the introduction to his translation, Quetelet was "grappling with ways in which principles of mechanics could be applied to society" at the same time he was "wrestl[ing] . . . with the issues of someone who had not yet achieved the position he aspires to."[42] It was possible too that Quetelet found these personal and individual metaphors too crude for publication or that they were frustrated notes of an astronomer who labored at an inferior observatory. He had even warned any prospective reader at the top of the notebook that the ideas had "been thrown on paper in a disjointed manner and without order."[43]

Aubin suggests that the undated work was written prior to 1831, a time when Quetelet had been named director of the observatory and had achieved limited success with the *Correspondance* but had yet to integrate Belgium into the growing world of international science. And indeed, the notebook does appear to contain as many examples of petty frustrations as it does serious philosophical reflection, the outlines of a career plan as much as a sketch of a social theory. For example, Quetelet explained in "Principles of Mechanics" a series of forces that "act on man as an individual," "act on man in society," and "act on nature like the development of the species." Each, in its turn, helped fashion the development of a person's life and career. In a strange analogy that foretold his "average man," he wrote that a man pulled in two directions by his own "self-interest" and "feeling for what is good" would be "in equilibrium between forces."[44] However, if for whatever reason, these ideas "act on him in different directions with unequal forces," he would be stretched at "a diagonal of the parallelogram."[45] For Quetelet, such a deformity in shape pulled a person away from their natural "center of gravity," which he located as almost a geometrical point in an individual's mind or soul. In another geometric analogy, he claimed therefore that the "wise man" should place himself at the center of the rotating circle of "society" in order to "eliminate all the forces that could divert him from his path or cause him to pivot around himself." Quetelet believed by doing so this "man" would "place himself at the center of motions; and when ambitions will carry some to the heights of guile, he will quietly position himself on the axis and will not be exposed, like them, to being squashed on the ground or dragged in the mud at the slightest turn of the wheel."[46] Rather than a grand theory that applied mechanics to society, Quetelet's notebook entry looked more like a guide for administrative advancement in a bureaucracy, as those who wanted to avoid getting "squashed" would learn to avoid any extreme positions. As strong (and bizarre) as this language was, given later controversies over determinism in social physics, it was wise that Quetelet never published such speculations.

Even without reference to Quetelet's "Principles of Mechanics," the project of social physics was still often hounded by complaints that it denied free will.[47] Although it would be helpful to know his position on determinism, metaphysics was not among the dozens of subjects Quetelet specialized in over his fifty years of professional service in the sciences. Objections that social physics denied free will did not bother him as a matter of philosophy, but he did worry that social physics would be undone by the fatalist charge that had attached itself to d'Holbach's sandstorm and Laplace's Demon.[48] When, in 1848, he attempted to defend himself against determinism in a series of talks before the Académie Royale de Belgique, he again noted that "free will only fades" when large numbers are brought together, as social physics was simply irrelevant to individual lives.[49] His friend Pierre-François Van Meenen (1772–1858) had offered an even more straightforward defense: God existed and granted people free will, and the numbers will say what they will.[50] Quetelet's and Van Meenen's arguments were more dismissive of the deterministic charge than engaging, but the third member of the debate, Pierre de Decker (1812–1891), offered a more nuanced approach that may have better captured the consequences of Quetelet's data collection process since the time of the *Correspondance*. De Decker, a Catholic philosopher and future prime minister of Belgium, claimed that "free will . . . disappears" in the "social milieu" of institutions, and that if one wants to be in a position to "rule society" they must give up their own "personal elements."[51] What this meant was that collective service, in either the state or science, required free will to be sacrificed, adding a literal sense to the abstract "law of large numbers." As a statesman and philosopher, de Decker cleverly inverted the question, asking not what data collection meant for those counted but what it meant for those *doing the counting*.[52]

De Decker, perhaps unwittingly, answered the objection to the mathematical sciences of man lodged by his fellow Catholics Bonald and Maistre. The counterrevolutionaries had seen state administrators as mere organs of a larger—and less material—social intelligence ensured by the church and crown, but de Decker saw little harm in giving up some individual autonomy to participate in a collective state, with government taking over the role of ensuring collective intelligence. Like Condorcet, de Decker imagined that scientific knowledge and political progress went hand in hand, and Quetelet's social physics had always been presented as a tool for governments. While past critics of reductive sciences of man had worried that mechanical analogies might turn men into machines or might limit human action and thought to a few simple rules, Quetelet's science of man seemed more a threat to the intelligence, or "human level," of the social scientist himself. As Jonathan Crary noted in his study of nineteenth-century observation, "an observer is one who sees within a prescribed set of possibilities, one embedded in a system on conventions and limitations."[53] Crary also pointed out that the Latin root *observare* means "to conform one's actions" or "to comply with." It is easy to see why Quetelet had seen his networks of observers as synonymous with the "average men" at the center of gravity in society. When it came to quantification in particular, administrators and researchers may indeed have some of their identity "stripped" away, and sociologists of quantification have noted that "conventions are at the heart of processes of quantification."[54] For Quetelet and de Decker, and for the legions of natural and social science administrators who emerged in the coming years, choosing to limit one's freedom of action in the service of administering state and science was a kind of bureaucratic noble bargain. However, as will be seen in the next section, not all applications of "social physics" would be content to determine the lives of the observers alone.

Crystals of Complicated Form: Jevons, the Logical Machine, and the Irrational Amusements of the People

In 1989, the economic historian Phillip Mirowski published *More Heat than Light*, a book that sought to demonstrate that the near entirety of modern economics—arguably the most successful social science of the twentieth century—was based upon "the reduction of all phenomena" to a handful of metaphors drawn from the world of late nineteenth-century physics.[55] He claimed, "The progenitors of neoclassical economics theory boldly copied the reigning physical theories of the 1870s," setting up the dominant paradigm of professional economics for well over a century. To make the point clear, Mirowski's subtitle argued for "economics as social physics." Quetelet had of course named his own science *physique sociale*, and the French social theorist Auguste Comte had wanted to use the same title for his science before adopting the neologism "sociologie," but as Mirowski pointed out, "physics" changed drastically in the decades following the sciences of man developed by Quetelet and Comte.[56] In physical theory itself, there was in fact a "gulf" between the "proliferation of fluids and ethers and forces and unconnected mathematical models" of the 1840s and the "consolidated discipline of the 1860s."[57] No longer just a helpful metaphor for applying the tools of the sciences to study society, "social physics" in the nineteenth century ossified into a research program that bound its practitioners to a restricted set of assumptions and ideals.

While the transformation of social physics into neoclassical economics involved a number of figures, the clearest connection to Quetelet's program was William Stanley Jevons, a political economist and logician who offered the most systematic English-language reduction of economics to physics in the nineteenth century. Jevons believed that Quetelet's social physics had been the "beginning of social science," but his metaphor went beyond general statistics and averages to get at the actual structure of the human mind.[58] Unlike other major social scientists of the nineteenth century who sought to study society as an independent entity—Comte, Mill, Durkheim, etc.—Jevons was content to treat human thought and action as simply another form of matter.[59] Although Jevons wrote several practical works of political economy and science—including influential studies of British coal reserves and the formation of clouds—the focus here is limited to his construction of a logical machine and his desire to show that there could be a unified methodology in the natural and social sciences, each of which chipped away at any distinction between mind and matter. In helping to develop the mathematical foundations of a new kind of political economy, he also helped create a new and idealized "man" known as *homo economicus*. Yet Jevons's social scientific thought was not always so neat, as the messy business of human thought and behavior often led him to frustration that people were not acting in the way his models said they would, and his ideas were far from the moral philosophy developed in Paris and Edinburgh that had originally given rise to political economy.[60] The social sciences of Condorcet and Quetelet had imagined some form of standardization for researchers, but Jevons's own attempt at social physics often required more direct intervention in human affairs.

Jevons stood apart from previous "social physicians" because of his insistence on establishing a foundational logic of scientific thinking. While Condorcet believed probability theory could allow the collective will to be known through numbers, and Quetelet believed the numbers might speak for themselves, Jevons devoted as much time to philosophy of science as he did to data collection and experiment. As detailed in Harro Maas's intellectual biography, Jevons dove headlong into some of the most profound debates over the philosophical underpinnings of a science of man.[61] In order to establish the potential for a true social science, Jevons in particular had to account for two of the most influential figures of Victorian England: John Stuart Mill (1806–1873) and William Whewell (1794–1866). As has been traced in several excellent histories of this era, Mill and Whewell themselves disagreed about how science "worked," and their disagreement set the terms of debate over the possibility of a science of man.[62] For Mill, there was a clear separation between the sciences of matter and those of people, with the truths of the latter requiring introspection and reflection rather than the tools of the natural sciences. There was simply too much complexity and, importantly, diversity in human affairs to apply any unifying methodology in the physical and social sciences.

For Whewell, the gap was large but not insurmountable; he noted that in "Mental and Social Science, we are much less likely than in Physical science to obtain new truths by any process which can distinctly called inductive."[63] The reason was twofold. The first problem echoed Mill, in that human actions appeared infinitely more diverse than those in nature and therefore showed fewer patterns. To use modern terminology, there were simply too many variables to isolate. The second problem was that experimentation, especially in moral sciences like history and philology, was nearly impossible. Yet Whewell still believed knowledge could be gained through a unified process of careful data accumulation and analysis, a process the historian Susan Faye Cannon deemed "Humboltian Science" after the approach of the German naturalist Alexander von Humboldt (1769–1859).[64] As Maas explained, however, Jevons rejected *both* Whewell's and Mill's separations, arguing instead that "mechanical explanations replaced the trust that was put in introspection."[65] While Mill in particular promoted philosophies of science that attempted to wall off the mind from reductive mechanical analogies, and Whewell had suggested a more careful form of analysis, Jevons simply saw no reason for *any* distinction between methodological programs for the sciences of man and those of matter.

In the work to impose mechanical analogies onto the working of the human mind, none of Jevons's projects looms larger for a history of artificial intelligence than his "Logical Machine," a fully realized apparatus made of blocks and gears that Jevons constructed in an attempt to show how the mind could make logical conclusions through strictly mechanical means. In a talk from 1870 explaining how such a machine might work, Jevons began with a few simple propositions, most importantly that "mechanical assistance" was inseparable from "calculation" and "mental operations." For example, the Greek word for "calculation" was based on using pebbles to count, and the abacus had a long history in several countries; even the human hand was a technology for aiding calculation. More radically, Jevons then asserted that nearly "every step accomplished in the progress of the arts and sciences has produced some mechanical device for facilitating calculation."[66] If machines could obviously *assist* with mental operations, why then could they not simply conduct these operations themselves? Tracing a long history of both imagined and actual "calculating machines"—from Pascal's Arithmetic Machine to Charles Babbage's Analytical Machine—Jevons declared that such machines were themselves proof of the fact that human thought was mechanical. As he put it: "Mind thus seems able to impress some of its highest attributes upon matter, and to create its own rival."[67]

Yet the mathematical calculation captured in technologies of the past was only one aspect of the mind, and Jevons argued that past machines had been limited because they depended primarily upon deductive logic using existing mathematical rules.[68] As Jevons had argued elsewhere, simple deduction or mathematical rule-following was an unsuitable methodology for any true science, including a science of man.[69] For Jevons, the height of a thinking "instrument" would be to perform logical *induction*, which allowed truths to be gathered from experience through inference. None other than Francis Bacon had described a "metaphorical" instrument that might perform such operations, but Jevons lamented that there had not been "even a single attempt to devise or construct a machine which should perform the operations of logical inference." Therefore, Jevons built a machine that he believed was capable of pure reason—in his words, "an analytical engine of very simple character, which performs a complete analysis of any logical problem impressed upon it. By merely writing down the premises or data of an argument on a key board representing the terms, conjunctions, copula, and stops of a sentence, the machine is caused to make such a comparison of those premises that it becomes capable of returning an answer which may be logically deduced from them." For Jevons, this meant that the "actual process of logical deduction is thus reduced to a purely mechanical form."[70] In practice, this allowed a user to press a series of keys that could "represent" a statement. Instead of words themselves,

Jevons used a complex set of notations. When a user pressed the keys, a series of rods would "remove" the illogical answers from the face of the machine, showing what was logically possible and impossible from the given input.

Jevons's device is a staple of histories of the "quest" to build an AI, where it is often referred to as a "logical piano," but few of these histories also discuss Jevons's role as a social scientist. While modern triumphalist histories often display the Logical Machine as one more step toward the perfection of a thinking machine, Jevons's intent was not to make a smart machine but to simplify thinking to the level of the machine in order to better understand the mind. As historians have noted of the proliferation of so-called thinking machines of the era, including the even more famous Analytical Machine of Charles Babbage (1791–1871), the effect was not to demonstrate remarkable advances in technology, given that the machines themselves had few if any advantages over human thought. Rather, the intent of the machine was to demystify and simplify aspects of the mind.[71] Maas has argued that it would be "absurd to suggest that Jevons believed the human mind consisted of a set of wooden levers," but the absurdity was only in the choice of materials to construct the mind.[72] Jevons did not believe there were wooden levers in the brain, but he had expressed earlier interest in the reductive theories of Richard Jennings (1814–1891), who sought to find the laws of political economy in "brain fibers." Jevons's own references to the "plastic fibers of the human brain" also indicated that he likely did believe that human thought, deliberation, and ideas were products of the gears of the brain.[73] Indeed, in his *Principles of Science* (1874), Jevons made the mechanization of thought a prerequisite for a proper understanding of scientific methodology. As he noted in the first of many reductive analogies, "Consciousness would almost seem to coexist in the break between one state of mind and the next, just as an induced current of electricity arises from the beginning of the ending of the primary current."[74] Even the many varieties of consciousness could be treated equally, as Jevons explained that "various logical processes"—including deduction, induction, generalization, analogy, classification, and quantitative reasoning—were just the "same principle operating in a more or less disguised form."[75]

After completing a tour de force of the various methods of thought, all reducible to what he called the "Principle of Substitution," Jevons remarkably concluded *Principles of Science* with a seemingly ambivalent stance on the possibility of using the methodology of science to gain certain knowledge of the natural and social world. Noting that "science can extend only so far as the power of accurate classification extends," Jevons pointed out that nature itself is "often devoid of strong lines of demarcation" and that it could not be supposed that "every group of natural objects will be found capable of rigorous classification."[76] And in a statement that might seem to rule out any potential science of man, Jevons concluded, "In spite of all the boasted powers of science, we cannot really apply scientific method to our own minds and characters, which are more important to us than all the stars and nebulae."[77] Moving on from the "so-called physical sciences to those which attempt to investigate mental and social phenomena," Jevons warned that "the same general conditions will apply." In a statement that might undercut Jevons's status as founding father of neoclassical economics, he even noted that the "laws of supply and demand have a complexity surpassing our powers of mathematical treatment." Rather, "all the functions involved are so complicated in character that there is not much fear of scientific method making rapid progress in this direction."[78] As for a "science of history," Jevons called it "an absurd notion."[79] Astonishingly, Jevons even echoed the concerns of the harshest critics of the sciences of man, noting the "too exclusive study of particular branches of the physical science seems to generate an over-confident and dogmatic spirit."[80]

How then to reconcile the preceding seven hundred pages of text—where Jevons explained the mechanical rules by which human minds operated—with such skepticism about applying these laws toward a science of society? If individual minds operated according to mechanical principles in a form similar to his Logical Machine, why would social science *not* be possible? In many ways, Jevons's skepticism anticipated Alan Turing's "paradox" from the introduction to this book, where the great British mathematician questioned the possibility of an inductive social science while still believing in the possibility of a thinking machine. In one of his more notorious phrases, Jevons even wondered if the "tender mechanism of the brain" could be "reduced to the expenditure of a determined weight of nitrogen and phosphorous," with the result that we may no "longer hold that mind is distinct from matter." If this were true, human minds would become "merely crystals . . . of complicated form."[81]

While the kind of social physics Jevons practiced drew on analogies from the more prestigious world of physics, the process was not so simple as applying the basic tools of the natural sciences to the study of people, the mantra for generations of social scientists covered in the first third of this book. The mind might be reducible to the laws of physics and mechanics, but the laws of physics and mechanics were not quite as simple as they were in previous eras. For Jevons, this meant that there might be more of a possibility to *alter* the subjects of these laws. In fact, Jevons often blamed people themselves for their failure to act in ways predicted by his science of man. Sandra Peart noted, "Jevons was confident that he knew how his subjects should act; if they failed to fulfill his conditions for equilibrium spending, he was ready and willing to recommend policies to correct the so-called improvidence and immorality of the lower classes."[82] For example, Jevons repeatedly emphasized that people, in particular those in poverty, consistently made "wrong choices" and that improvements would not be possible without changing the mindset of an entire class of workers. In a work entitled "Amusements of the People," Jevons complained that the entertainments in England were "woefully backward" and that "pure and rational recreation for the poorer classes can hardly be said to exist at all."[83] Instead, Jevons declared that social reformers needed to adopt a wide range of reforms "conjunctively" and "in simultaneous play" to "correct the bad habits of a population."[84] As so often happened, when the sciences of man failed to produce accurate accounts and predictions about human thought and action, the recourse remained the same: alter the condition of the human level to better fit the model.

Conclusion

Unlike many of the other attempted sciences of man seen so far, Jevons's work was far from scandalous. While critiqued by Mill and others, he was nonetheless at the forefront of one of the most influential ideas of the past hundred years: the "Marginal Revolution," the theory at the heart of modern economics which claimed that human beings acted rationally in order to allocate scarce resources. As historians have termed it, Jevons had helped give birth to a new kind of man, *homo economicus*, whose actions could be reduced to a handful of influences that could be expressed mathematically similar to physics.[85] So triumphant was social physics that by the late nineteenth century, one survey of theoretical physics could even write, "It may be truly stated that human intelligence begins and ends with [the conservation of energy]."[86] Put most starkly here, even "intelligence" became a simple output of physical theory. Remarkably, eighty years after Condorcet had sketched out a hope for a "social mathematics" while in hiding from the repressions of the Reign of Terror, and just a few decades removed from Quetelet's strained calls for data collection on social action, Jevons had embedded the idea of social physics into the heart of one of the most powerful social sciences of the age.

In different ways, the social mathematics and social physics of Condorcet and Quetelet had imagined individual variation as lawlike deviations from a norm, and they had offered various proposals to account for why human beings often acted in very *unlawful* ways. For Condorcet, the solution had been education combined with complex voting procedures to better reveal the will of collective individuals. While Quetelet believed that deviations from his "average man" were "monstrous," they were also of little importance, as the summing and averaging of all individual traits naturally canceled out those who deviated from the expected average. For Jevons, however, who devoted much of his last few decades of work to trying to understand how to "reform" the British public, the "perturbations" of individual human beings became more pronounced and problematic. Though Quetelet mostly lamented that crime, poverty, and deviance existed year after year, Jevons offered far more potential solutions, devoting dozens of tracts to "improvements" in the lives of everyday people.

Jevons's confidence in the predictive powers of economics demonstrates that the sciences of man changed significantly in the course of the nineteenth century, and not just at the level of theory. The belief that the methodologies of the natural sciences could be applied to humanity went from a niche and dangerous idea to a core foundation of economics, the most influential social science of the twentieth century. Jevons with his Logical Machine had even managed to create an *actual* "machine man" that seemed far more materialistic than anything imagined by La Mettrie. Rather than cause a scandal, Jevons received praise. What made the actual machines of Jevons and Babbage more palatable than the banned works of d'Holbach, Helvétius, and La Mettrie in the previous century? In part, it was because the separate sphere of human activity had been simplified through continued mathematical reduction. The social mathematics of Condorcet and social physics of Quetelet had shown that even supposedly independent features of human behavior could be expressed in mathematical form.

Perhaps more importantly than abstract debates about free will and materialism, however, actual human beings were being confronted more often with their own actions as products of physical and natural laws. If people imagined themselves as products of natural laws that could be uncovered through scientific methodology, they may just begin to behave in new ways. As the prophets had warned, the sciences of man would then be not *revealing* any great truths about humanity but actively *creating* them. Indeed, Jevons had imagined such a process in his *Principles of Science*. Expressing concern that atheism and materialism were the natural result of *teaching* a scientific methodology, he suggested that the social sciences should at least act "as if" people still believed they had free will.

Our own hopes and wishes and determinations are the most undoubted phenomena within the sphere of consciousness. If men do act, feel, and be as if they were not merely the brief products of a casual conjunction of atoms, but the instruments of a far-reaching purpose, are we to record all other phenomena and pass over these? We investigate the instincts of the ant and the bee and the beaver, and discover that they are led by an inscrutable agency to work to a distant purpose. Let us be faithful to our scientific method, and investigate also those instincts of the human mind by which man is led to work as if the approval of a Higher Being were the aim.[87]

While Jevons was praised by clergymen for this statement, it was not quite the powerful defense against materialism and atheism he had imagined.[88] Aside from equating a belief in a "Higher Being" with the instincts of an ant, Jevons also left unsaid what might happen if people *stopped* acting "as if" they were "instruments of a far-reaching purpose." What would a science of man study if people did come to believe they were "products of a casual conjunction of atoms," Quetelet's determinist critics had argued? Even more, what would the new social mathematicians and social physicians come to think about their own lives as they studied people "as if" they were in fact atoms, with no distinctions between methodologies of natural or social science? Last, as the conclusions from these sciences were spread among governments, institutions, and the public alike, would people themselves continue to live "as if" their lives had a higher purpose? Physical matter cared little about what theories said about them, but Jevons recognized that social science might have the possibility to alter, rather than reveal, the nature of human thought and behavior.

In ways that clearly marked a departure for the social sciences from the natural sciences, the process of reducing human actions to scientific methodologies required great intervention in the lives of *actual* people. Although Jevons had tried to argue for uncertainty in both social and natural science, it was hard to imagine that astronomers would need to intervene in the paths of celestial objects for their theories to function properly. As critics had worried, the spread of scientific accounts of behavior were not merely neutral theories but instead had the possibility to create the very type of "people" they imagined. With the Industrial Revolution in England bringing people and machines into greater contact, the kind of people who were being subjected to social scientific study underwent significant change as analytical foci.

Political economy, which Jevons had helped transform into the social science of economics, was more than simple theorizing based on observational data. As seen in the next chapter, the biggest changes in the "human level" came not only through abstract ideas of thinkers like Condorcet, Quetelet, and Jevons but also through the very real and material changes occurring in the lives of men and women tending to their machines.

5

Tuning *the* Mind

The Failure of the "March of Intelligence" in
Nineteenth-Century British Political Economy

Morals and metaphysics, politics and political economy,
the way to make the most of all the modification of smoke;
steam; gas, and paper currency; you have all these to
learn from us. . . . We are the modern Athenians.
—Mr. Mac Quedy in Thomas Love Peacock,
Crotchet Castle (1831)

LIKE MUCH OF THE WRITING OF THE GREAT SATIRIST THOMAS LOVE PEACOCK
(1785–1866), the fictional Mr. Mac Quedy's proclamation that the Scottish
were "modern Athenians" was intended as a howler.[1] As a member of an
eclectic cast of social thinkers, men of science, socialists, landed aristocrats,
romantics, fops, and reactionaries that made up Peacock's *Crotchet Castle* (1831),
Mac Quedy was a none-too-subtle caricature of the popular Scottish political
scientist J.R. McCulloch (1789–1864), a journalist and part-time university
professor who had done more to influence popular perceptions of political
economy than just about anyone prior to John Stuart Mill.[2] For Peacock, the
lawlike certainties of political economy made it a science of man particularly
ripe for mockery, as throughout the story Mac Quedy's grand pronounce-
ments are undercut by the Reverend Doctor Folliott, a likely stand-in for the
author. When, for example, Mac Quedy insists that political economy is the
"science of sciences," Folliott responds by calling it a "hyperbarbarous technology."[3] As for the glories of science and industry that led Mac Quedy to call
the Scottish "modern Athenians," Folliott jokes, while the two survey a lavish

spread of prawns, anchovies, trout, and herring, that the only advancement in Scotland is that the "country is pre-eminent in the glory of fish for breakfast."[4]

Although *Crotchet Castle* skewered a great range of British ideas of the age—from the zeal of utopian reformers like Robert Owen and Edwin Chadwick to the poetic musings of Samuel Coleridge to the predictable reactions of Thomas Carlyle—its main target was the *popularization* of political economy, or what became known in England and Scotland as the "march of National intelligence." Political economy had originated in the British Isles as an explicit "science of man" and had particularly deep roots in Scotland, where David Hume, Francis Hutcheson, Adam Ferguson, and Adam Smith made some of the earliest attempts to capture the "laws" of human action through applying scientific principles to moral philosophy.[5] Yet Scottish moral philosophers of the eighteenth century did not pay much attention to popularizing their ideas. After all, a science of man based on physics or astronomy should operate *independently* of what the subjects of that science believed. In the 1830s, however, popular political economists sought to redefine social and economic laws as the collective output of individuals who bore personal responsibility for their own thoughts and behavior. William Stanley Jevons had contrasted his idealized vision of the human level with the seeming irrationality of the British public, but Jevons was only one of many in a long line of British political scientists frustrated with the failures of what the Reverend Doctor Folliott mocked as the "march of the mind." When, for example, in *Crotchet Castle* the political economist Mac Quedy claimed that "education" will "make all the difference" in the march of the mind, Folliott responded with less certainty for the sciences of man, noting that "the mass of mankind" were only "blockheads of different degrees."[6]

By focusing on educating individual minds—rather than simply studying human behavior using the tools of the natural sciences—political economy as a science of man in the nineteenth century therefore took a very different route from the French geometric spirit, Quetelet's social physics, and the political economy of the Scottish Enlightenment. An example of this shift in economic thinking can be seen in the work of the Scottish historian Adam Ferguson (1723–1816), who claimed in *An Essay on the History of Civil Society* (1767), one of the earliest English-language attempts at a science of man, that "Mankind are to be taken in groupes; as they have always subsisted. . . . Every experiment relative to this subject should be made with respect to entire societies, not with single men."[7] Like Quetelet's vision that individual will disappeared in the milieu of large numbers or Bonald's assertion that social intelligence was irreducible to the level of the individual, early political economists like Smith and David Ricardo (1772–1823) similarly focused on how "iron laws" and impersonal forces affected "groupes."[8] Yet, by the middle of the nineteenth century, popularizing political economists had redefined political economy as a science dedicated to practical interventions in individual intelligence. By the late nineteenth century, Jevons could even claim that political economy was a field that "presumes to investigate the condition of a mind."[9]

In part, the shift in the site of analysis from groups to individual thoughts and actions emerged because of the increasing interaction of British workers with Mac Quedy's vaunted "modification of smoke," "steam," and "gas," which forced political economists to reconcile the obvious hardships of industrial life with the seemingly harmonious laws of social science.[10] Was a "march of intelligence" even possible alongside the notoriously dreary factory work that attended economic progress? Though the field of political economy had begun with attempts by Smith, Ferguson, and Hume to find orderly and general laws of social behavior, by the middle of the nineteenth century political economy seemed unable to account for the stunning rise in poverty, soaring machine accidents, labor unrest, and the simple destruction of wide swaths of British social, economic, and cultural life.[11] While the deleterious physical effects of humanity's first extended interaction with machines in England and Scotland are well known, less attention has been given to how industrial work with machines prompted political economists to look inside the mental world of the worker.[12] Whereas a myopic technological determinism might imagine machines alone as the culprits in the simplification of human intelligence, this chapter argues that the shift to individual minds in popular political economy was a crucial tool in limiting the potential and scope of human thought and action. As these popularizers argued in their stories and pamphlets, if human intelligence could not accommodate the predictions and mechanical laws of the "science of sciences," the fault lay in the complexities of human intelligence itself.

When it became clear that political economy could not accurately account for the reality of its subjects, changing the minds of those subjects therefore became more essential for it to succeed as a science of man. To trace this transformation, the chapter begins, naturally, with Smith, whose ideas on political economy came to be paradigmatic throughout the first half of nineteenth-century Great Britain. Though rarely discussed as a rigid science of man in the same way as d'Holbach's *Système de la nature* or Quetelet's social physics, Smith's earliest methodological writings indicate that political economy emerged out of a similar effort to create a harmonious science based on the tools and methodologies of the natural sciences. While seemingly convincing on paper, in the decades following the Napoleonic Wars factory life in England grew especially grim, and worker revolts from 1815 through the 1830s indicated that the laws of political economy did not seem natural for those whom Friedrich Engels (1820–1895) deemed the "working class."[13] Faced with the possible failure of political economy as a science of man, British reformers therefore attempted to counter the effects of industrialization through popular fiction and didactic pamphlets, including Harriet Martineau's *Popular Illustrations of Political Economy* and Charles Knight's *Workingman's Companion* series. By 1845, when Knight and Martineau published *Mind amongst the Spindles*—short stories written by the girls and women employed in a Lowell, Massachusetts, factory—the failures of the sciences of man had been reframed as a failure of working-class attitude and intelligence. Smith, for example, believed that inherent individual intelligence mattered little for political economy, but the curated stories of *Mind amongst the Spindles* placed responsibility for desultory factory mental life squarely on the inability of workers to attune their minds to the reality of laws of social science. The "iron laws" of political economy could obviously not change, and so the minds of the workers would need to be changed to accommodate these laws. By the middle of the nineteenth century, altering the intelligence of the industrial laborer at her machine had therefore become a crucial site for ensuring the status of political economy as a science of man.

The Imaginary Machine of Adam Smith's Political Economy

The intellectual roots of many social sciences like sociology and psychology comprise a tangled and knotted history, but few economists would disagree that Adam Smith (1723–1790) was the primary founder of their field. As a recent popular biography—appropriately titled *Adam Smith: Father of Economics*—details, Smith thoroughly shaped the development of classical economics and still dominates today in modern surveys, academic citations, and general knowledge of the field.[14] For many in the nineteenth century too, Smith's influence was all-encompassing. In the words of the political economist J.R. McCulloch, Smith's *Wealth of Nations* (1776) had "done for Political Economy what the Principia of Newton did for physics."[15] Scholarly debate exists over whether Smith's science of man had truly appropriated Newton in the way the French Enlightenment had imagined—through the strict application of tools in the natural sciences to human affairs—but many have argued persuasively for what Margaret Schabas has called Smith's "debts to nature."[16] Though not nearly as reductive in tone or methodology as d'Holbach or Helvétius, Smith nevertheless imagined political economy as a science based in the techniques successfully deployed in the natural sciences.

Despite naturalist language being mostly absent in Smith's two most famous works—*Wealth of Nations* (1776) and *Theory of Moral Sentiments* (1756)—his "debts" to physics, astronomy, mathematics, and mechanics appear most clearly in his earliest work, a set of posthumously published texts known as the *Essays on Philosophical Subjects* (1795).[17] While Smith had asked his assistants to burn nearly all his papers in a fireplace just steps from his deathbed, he insisted to his friend Hume that the *Essays* be published, preserving the sketches of what has been called a possible "universal intellectual history."[18] The most detailed survey of the *Essays*, which varied in style and tone, concluded that "Smith maintained an interest in the natural sciences and appears to have read widely on the subject throughout his life."[19] In addition to the *Essays*, which range from astronomy to ancient physics to philology, Smith's interest in the natural sciences can also be seen in his vast library and early education under Francis Hutcheson (1684–1746), who had himself created a proposed science of man based on mathematical principles.[20] Though in the text itself Smith obscured almost all potential influences for *Wealth of Nations*, its "debt to nature" becomes more clear in the context of these earlier essays.

The lack of bibliographic information from Smith's life makes it difficult to know when he wrote the *Essays on Philosophical Subjects*, but their structure indicates that they were the beginning of a larger methodological work based on Newtonian ideas.[21] The first and longest essay, *History of Astronomy*, for example, began not with the earliest discoveries in astronomy or principles but with three rather curious extended sections on "Wonder," "Surprise," and "The Origins of Philosophy."[22] For Smith, it was the "wonder" and "surprise" of natural objects that gave birth to the sciences in the first place, and the opening paragraphs of *History of Astronomy* suggested that knowledge and scientific investigations were an almost mechanical and automatic reaction to external stimuli. The feeling of surprise, for example, was an "irresistible force" most notable for the "violent and sudden change produced upon the mind." Smith noted that the power of surprise was so overwhelming at first that "in its full force [it] takes at once entire and complete possession of the soul." The observer, lacking almost any active role in the process, was rather a passive receptor as the "heart springs to joy."[23] Yet just as quickly as the heart or mind was overtaken by surprise, it tried to establish some form of "equilibrium." The mechanical nature of wonder and surprise echoed Descartes, and it was therefore inevitable that wonder was "the first principle which promot[ed] mankind to study philosophy."[24]

In addition to the mechanical means by which Smith described both the origins of philosophy and the mental process, the opening of *History of Astronomy* also suggested that the resolution of wonder itself could bring about an "equilibrium" for the mind. Smith claimed "wonder" was an extraordinary prod to scientific thought, but it was also a feeling that faded with true understanding. Like Pascal, who worried about the "boredom" that would result from a true understanding of nature, Smith asked, "Who wonders at the machinery of the opera house who has once been admitted behind the scenes?"[25] Similarly, in natural investigations Smith claimed that "confusion" and "giddiness" were gradually erased as the mind made order out of chaos. While the opera house may present a delightful form of wonder, Smith also invoked the confusion of those who enter "the work-houses of . . . dyers, brewers, and distillers" and can make little sense of the machinery. An inexperienced observer who saw these small workhouses would feel "confused, then giddy, and at last distracted," as the equilibrium of the observer's mind would be disturbed by a seemingly unnatural sequence of events. Therefore, for Smith, echoing d'Alembert's embrace of "linkage," scientific investigations were merely intentional efforts to achieve what came naturally to the mind of the workman at home in his workhouse: the ability to "join together . . . a chain of invisible objects."[26]

Smith's belief that scientific investigation could resolve confusion and restore mental equilibrium framed his entire history of astronomical investigation. As he claimed at the outset, "the repose and tranquility of the imagination is the ultimate end of philosophy."[27] Smith objected especially to those astronomical theories that "tended to embarrass and confound the imagination," even when they might be closer to the truth. For example, although the Greek idea of "concentric Spheres" was "rude and inartificial," Smith nevertheless praised it as "capable of connecting together, in the imagination, the grandest and most seemingly disjointed appearances in the heavens."[28] Like Condillac, Smith also believed that systems, like machines, performed best the less complicated they appeared.[29] For example, Smith claimed that Hipparchus's "imaginary machine" of the solar system worked far better than Aristotle's "Fifty-Six Spheres" because it had fewer moving parts.[30] Likewise, Copernicus's banishment of the "embarrassment of epicycles" allowed for "a more simple machine" that "connected together, by fewer movements, the complex appearance of the heavens." Copernicus's simplified system managed to be a "coherent," "simple," "intelligible," and "beautiful" machine.[31] Indeed, in its enthusiasm for aesthetic beauty over the messy business of machines, Smith's *History of Astronomy* was representative of what Franz van Lunteren describes as the "gradual exchang[e]" of the "Cartesian clockwork vision" of astronomy for "a view in which planets moved harmoniously throughout space."[32]

For those who know the Smith of *Theory of Moral Sentiments*, it may not be such a surprise to see the consistent mention of imagination and feelings in the work of one of the founders of the "dismal science." Yet the *History of Astronomy* shows just how closely Smith linked "systematic" thought in the natural sciences to mental equilibrium, as he shared a belief with Condorcet and Jevons that scientific thinking entailed mental progress. For Smith, the various models and accounts of the planetary movements were not objective certainties but rather workable "systems of thought" that alleviated the "embarrassments" racking the imagination. And systems, as Smith wrote in a much-quoted excerpt, "in many respects resemble machines. . . . A system is an imaginary machine invented to connect together in the fancy those different movements and effects which are already in reality perfected." Smith's reliance on machine metaphors has largely gone unnoticed, but he also referenced dozens of actual machines in his writing, including "fire-engines," "forges," "water mills," the "condensing engine," and James Watt's "copying machines."[33] As much as he was inspired by the great systems of the heavens, he was equally impressed by the manner by which the gadgets around him contributed to mental equilibrium.

Smith's methodological approach in *History of Astronomy* can help to better understand how *Wealth of Nations* could emerge as a foundational science of man. His political economy was not necessarily the final word on human action but simply the most workable system of the moment. In *Wealth of Nations*, Smith was in fact far from the later "iron laws" of Ricardo, and his invisible hand was accompanied by a great belief in plasticity of human ability. As he claimed in *Wealth of Nations*, a "progressive state" of intellectual activity was "cheerful and hearty" in comparison to the "stationary state," which was "dull."[34] Like Pascal, Diderot, and d'Alembert, Smith believed that *thinking* about the social and natural world was better than having one simple set of rules. In contrast to later social sciences, he was also open to the possibility that people and groups may change over time. Even individual "genius" was far from static: "The difference of natural talents in different men is, in reality, much less than we are aware of; and the very different genius which appears to distinguish men of different professions, when grown up to maturity, is not upon many occasions so much the cause, as the effect of division of labor. The difference between the most dissimilar characters, between a philosopher and a common street porter, for example, seems to arise not so much from nature, as from habit, custom, and education" (*WN*, 23). By believing in no "difference of natural talent" and that "genius" could be produced through "habit, custom, and education," Smith echoed Helvétius and the French sciences of man as well as the understanding of much of the Scottish Enlightenment on the importance of work ethic over natural intelligence. As his friend Hume had declared, "how nearly equal all men are . . . even in their mental powers and faculties."[35]

Smith's belief that any person could become a "porter" or "philosopher" might, however, suggest another "Adam Smith Problem," as it seems hard to square with one of the key "laws" from *Wealth of Nations*.[36] Smith famously argued that division of labor with machines, rather than broad education and training, was the key to economic success, and the productive value of division of labor in fact *required* specialization. For example, in his analysis that begins Book I, Smith praised the physical "dexterity" that comes through specialized labor with machines at a pin factory, which he imagined could even be good for the mind. As he claimed, the "consequence of the division of labour" meant that "men are much more likely to discover easier and readier methods of attaining any object, when the whole attention of their minds is directed toward that single object" (*WN*, 17). In other words, Smith argued that the kind of peace of mind found in the "work-houses" could be found by specialized laborers in a factory. Furthermore, Smith's belief in education was explicitly limited to the young, pre-specialized worker, meaning that a child then might become a porter or a philosopher, but an adult could not easily switch from one form of labor to another. In his words, "the most essential parts of education . . . can be acquired at so early a period of life, that the greater part even of those who are to be bred to the lowest occupations, have time to acquire them before they can be employed in those operations" (*WN*, 432).

By combining the idea of a plastic mind with the division of labor, Smith imagined something like intelligence ("genius") as emerging through the "education" and "customs" of a child at a particular task, an intelligence that could then be extended throughout the child's lifetime. What was less noted was that this kind of intelligence was completely different from the "equilibrium" of the mind felt by the worker at his brewery or distillery. The specialized worker may master one job, but as Smith had argued in *History of Astronomy*, true mental equilibrium resided in the ability to comprehend an *entire system*, not just one part. Even worse, in Book V, Smith provided what might seem to be a significant disturbance in the otherwise harmonious world of division of labor in *Wealth of Nations*. In a notorious passage worth quoting at length, he claimed,

> The man whose whole life is spent in performing a few simple opera-
> tions, of which the effects too are, perhaps, always the same, or very
> nearly the same, has no occasion to exert his understanding, or to
> exercise his invention in finding out expedients for removing difficul-
> ties which never occur. He naturally loses, therefore, the habit of such
> exertions, and generally becomes as stupid and ignorant as it is possible
> for a human creature to become. The torpor of his mind renders him,
> not only incapable of relishing or bearing a part in any rational conver-
> sation, but of conceiving any generous, noble, or tender sentiment, and
> consequently of forming any just judgment concerning many even of the
> ordinary duties of private life. (*WN*, 429)

Though a foundational figure of the Enlightenment, Smith's warnings about specialization leading to a "torpor of the mind" anticipated the concerns from the counterrevolutionaries in chapter 3, who had warned that schoolchildren drilled in only mathematics and the exact sciences were reduced to mere geometric reasoners. Smith's worry that industrial labor would make people as "stupid and ignorant as it is possible for a human creature to become" also hinted at the challenges of later popularizers of political economy who attempted to "elevate" the mind of the worker at her machine.

As many frustrated students of Smith know, he never reconciled his belief in the "natural" advantage of division of labor with machines with his concern that it drastically reduced intelligence.[37] Rather than a simple contradiction or one infelicitous paragraph, Smith's acknowledgment of the deleterious effects of specialized labor in fact is deeply ironic given the importance of labor to "equilibrium" of the mind. As Smith demonstrated in both *History of Astronomy* and *Wealth of Nations*, there was a perfectly good solution to the problem of stunted intelligence: preindustrial labor. Smith took some aspects of political economy, like trade, to be original "principles" of human existence, but he specifically noted that division of labor was *not* a transhistorical law like gravity, and that it had only evolved under specific social and historical conditions.

In *Wealth of Nations* Smith wrote, "The nature of agriculture . . . does not admit of so many subdivisions of labor." Furthermore, while in industrial societies "the spinner is almost always a distinct person from the weaver," in agricultural societies, "the ploughman, the harrower, the sower of the seed, and the reaper of the corn are often the same" (*WN*, 13). In *History of Astronomy*, Smith had also championed the easy mind of the distiller who could, prior to specialization, "join together . . . a chain of invisible objects." Especially given his belief in the plasticity of individual intelligence, it would seem almost cruel that political economy championed a system of specialization that knowingly degraded human intelligence to a level of "torpor" and "ignorance."

A Beauty of a Different Kind: Harriet Martineau and the *Illustrations of Political Economy*

To add to the possible despair, Smith also knew there was little workers could do to combat the division of labor with machines. Though he never lived to see the Luddites, Captain Swing riots, and the vast antipathy of workers toward new forms of machine labor, he nevertheless predicted the course of British revolts and repression that would occur over the next century. Referring to collective groups of workmen as "combinations," Smith explained the futility of any efforts to contradict the laws of political economy.

> They are desperate, and act with the folly and extravagance of desperate men, who must either starve, or frighten their masters into an immediate compliance with their demands. The masters upon these occasions are just as clamorous upon the other side, and never cease to call aloud for the assistance of the civil magistrate, and the rigorous execution of those laws which have been enacted with as much severity against the combinations of servants, labourers, and journeymen. The workmen, accordingly, very seldom derive any advantage from the violence of those tumultuous combinations, which . . . generally end in nothing, but the punishment or ruin of the ringleaders. (*WN*, 66)

Between 1810 and 1830, the plight of "desperate men" confirmed Smith's belief that a violent response to division of labor with machinery would be futile. While the European countryside had a long tradition of machine breaking in response to lost wages, Smith predicted correctly how newly aligned industrial capital would call on state "magistrates," leading to the "rigorous execution" of laws intended to stop the revolts. As detailed in the popular account *The History of England during the Thirty Years' Peace, 1816–1846* (1850), the "lawless course of drunkenness and plunder" could only be met with the "sure retribution of offended laws."[38] As Smith foretold, "Masters" did indeed "clamour" for the magistrates, especially during the Luddite revolts from 1811 to 1816, when 12,000 soldiers were used against the machine breakers. As Eric Hobsbawm noted, this force amounted to more soldiers than the Duke of Wellington took into the Peninsular War against Napoleon in 1808.[39] Smith was also right that that one of the chief results of "combinations" was the "ruin of the ringleaders," as tens of thousands of British laborers were sentenced to prison or removed from their families and exiled to Australia for crimes against private property.[40] As the authors reported in their history of thirty years of "peace," in just one of hundreds of uprisings, the "ruin" occurred when thirty-four men were sentenced to death for robbery and burglary.[41] It seemed, then, that Smith was correct that the will of workers would need to bend toward the laws of political economy.

Given the failures of earlier theorists to account for machine breaking, it fell to other, lesser-known, reformers in the 1820s and 1830s to investigate how political economy could explain—and perhaps counter—the deleterious effects of division of labor with machines.[42] Jevons would complain about irrational Victorian minds, but the clearest and earliest connection between machine breaking and worker intelligence in the context of political economy emerged in the popular fictions of Harriet Martineau (1802–1876), author of *Thirty Years' Peace*.[43] As Martineau noted of the intelligence of the machine breakers, "There never before was such an organized system of havoc resorted to by men who were at once grossly ignorant and pre-eminently crafty."[44] A member of the Unitarian church who was deeply sympathetic to the plight of the poor, Martineau had begun her reforming career writing for the Houlston publishing house, which relied on cheap and disposable books for tales of moral virtue, and where impoverished and marginalized characters overcame their difficulties through individual grit and effort.[45] Linda Peterson has shown how Martineau eventually objected to the "Houlston tracts" because of their focus on "individual experience and personal testimony" rather than "the context of larger economic and political forces."[46] A key founder of sociology, Martineau believed in a more "scientific perspective" on the prospects for the poor and later authored the first English translation of an abridged version of Auguste Comte's *Cours de Philosophie Positive* in 1853. As Peterson claimed of the Houlston era, Martineau began to move "away from individual morality, religious piety and economic resignation . . . toward economic knowledge and political consciousness."[47] While Houlston reformers had aimed for the soul of the dispossessed, Martineau's science of man went straight for the mind.[48]

Although published during her time at Houlston, Martineau's short story *The Rioters* (1827) is one of the best examples of early fictional efforts to explain how the minds of workers at the machines could be influenced by social science.[49] Called the "first important social novel" of the decade, Martineau based *The Rioters* on newspaper reports, believing that "the subject of machine-breaking was a good one" to fictionalize.[50] Told from the perspective of an unnamed industrialist arriving by train in Manchester, *The Rioters* begins with the observation that the city appeared in "a great state of confusion, with an unusual appearance about the place." After the industrialist inquires about the commotion, a waiter responds straightforwardly: "The rioters, Sir, have come into town." Thousands of displaced weavers had marched into town and "have broken every loom to pieces." Their subsequent exchange echoes Adam Smith's language on the hopelessness of worker resistance. After the industrialist asks, "What good do they expect to do by violence?," the waiter responds that they are not acting rationally but that "they are starving and desperate, and ready to turn their hand to any mischief."[51]

While Martineau allows empathy for those that have "sunk lower and lower" because of the "distresses of the time," the story makes clear that the causes of the riots are rooted in worker ignorance rather than the laws of economics.[52] After ruminating on the plight of the Brett family, the industrialist decides that the best solution is a lecture on political economy, and it is here that the story becomes, in the words of Maxine Berg, a "nauseating didactic" tale.[53] Taking up most of the story, the conversation involves the industrialist parrying nearly every argument Mr. Brett is able to make against the machines.[54] The Brett patriarch begins by making his case that the "machinery" had led to the "falling off" in trade and that "it was plain enough, that if wood and iron did the weaving, the weavers must starve." Mr. Brett also faults the masters and machines for mass production, which led to the "market being overstocked," noting that "if they had kept to hand-loom weaving, they would have no more goods than are wanted."[55] When the industrialist tries to interject, in sound economic language, that the expansion of trade would benefit the Bretts in the long run, the latter responds, "That will be too late, Sir, to do any good. What use will the revival of trade be, if we all starve in the mean time?" For Brett, this meant that machine breaking was the only option, and he notes simply that "it would be all the better if all the power looms were done away with" so they could establish "hand-loom weaving again."[56] After hearing the complaints, the industrialist returns to the principles of division of labor, the law of wages, and the productive benefits of machines for the whole, claiming that "if the poor people who have been rioting here understood the truth a little better, they would be sorry for what they have done." Convinced that the unrest of machine breaking is "a matter which you do not understand," the industrialist leaves the Bretts to counsel the rioters at the factory.[57]

While the story's didactic approach may appear somewhat forced and obvious in retrospect, Martineau actually saw *The Rioters* as too opaque, as the early 1830s brought about even more spates of factory arson and machine breaking, most notably in the Captain Swing riots in the English countryside.[58] Perhaps believing that the lessons of political economy had been lost in the drama of the Bretts' plight, Martineau returned to fiction in 1832 with *Illustrations of Political Economy*, an inexpensive collection of short stories that sought to highlight the key scientific principles of political economy. Sounding like the Adam Smith who had judged astronomical systems throughout history based on their abilities to soothe the "embarrassments of imagination," Martineau too sought "peace" as an epistemological criterion. And as many scholars have shown, Martineau's *Illustrations* had the explicit goal of using the scientific certainty of political economy to "calm" the social and political tensions throughout England in the first third of the nineteenth century.[59] As she described in her autobiography, upon learning "the conception of general law . . . my laboring brain and beating heart grew quiet, and something more like peace than I had ever known settled down upon my anxious mind."[60] Though the lessons may have appeared condescending—one account calls them "tranquilizing tales"—it seems that Martineau was interested in bringing a similar peace of mind to anxious industrial laborers, and the *Illustrations* sold extraordinarily well, with over 10,000 weekly copies and close to 150,000 total sales.[61]

Taken as a whole, the *Illustrations* makes clear that Martineau wished to move from the consolations of fiction to the necessary indoctrination of scientific principles, as it seemed unlikely that machine use or the laws of social science would change. As she noted in the introduction to the series, in a lightly veiled critique of her earlier work and the morality tales of the Houlston stories, "It is many years since we grew sick of works that pretended to be stories, and turn out to be catechisms of some kind of knowledge which we had much rather become acquainted with in undisguised form."[62] Therefore, the tales in *Illustrations* were bracketed by a list of "principles" that could guide the reader through each story. In these principles, Martineau echoed the classic formulation of Smith, that "labour is economized through division of labour" because "men do the most quickly work which they stick to." So harmonious was this law that it could only be obstructed by human minds. In her words, "As the materials of nature appear to be inexhaustible, and as the supply of labour is continually progressive, no other limits can be assigned to the operations of labour than those of human intelligence. And where are the limits of human intelligence?"[63]

While *The Rioters* placed the lessons of political economy directly in the mouth of the industrialist, the *Illustrations* story "The Hill and the Valley" took a more oblique approach. Instead of the misery of the Bretts, the story instead begins with an idyllic description of the life of John Armstrong, a man described as basking in the pleasures of the unindustrious life. Armstrong's wealth is inherited, and he spends his days wandering about the untrammeled nature of his large estate, fishing and playing the flute with a neighbor. In what initially seems like the beginning of a classic technophobic story, Armstrong's routine is disturbed by the arrival of a brash industrialist named Wallace, who challenges Armstrong's georgic paradise, pointing out that "your way of life would not suit persons . . . who wish to rise in the world."[64] After Armstrong replies that he is happy to gaze out across his pristine land and has no need for industry, Wallace responds that he is building an iron foundry next to his property and that Armstrong would soon "see hundreds of human beings thriving where there are now only woodcocks and trout."[65]

Armstrong is given plenty of space to elaborate on the problems of industrialization and, to a modern eye concerned with ecological destruction, it can even seem like he has the better of the argument: his rural bliss will surely be destroyed by the iron foundry. Yet Martineau offers a subtle twist. After Wallace returns with his wife, Armstrong gives a long speech, which the narrator describes as "eloquent upon the inelegance of smoke, and the rows of houses, and the appearances which attend an iron-work." When Armstrong appeals to Wallace's wife, however, to know "whether laying waste of the works of nature was not melancholy," Mrs. Wallace "could not agree that it was." In her words, "it was true that a grove was a finer object at this distance than a cinder ridge, and that a mountain-stream was more picturesque than a column of smoke: but there was a beauty of a different kind which belongs to such establishments. . . . There was . . . the beauty of the machinery." [66]After Mrs. Wallace surprises Armstrong with an appeal to "a column of smoke" as a "beauty of a different kind," her husband returns to arguments from political economy, noting the importance of machinery in the division of labor and warning Armstrong that he is an anachronism, uniting "the functions of Capitalist and Labourer" in one person. Calling Armstrong an "uncommon case" in his self-sufficiency, Wallace charged that if everyone adopted Armstrong's simple ways of living, "life would have become barbarous in comparison." As anyone could see, "those who object to machinery do not perceive its nature and effects."[67]

After this surprising turn, Armstrong fades into the background, ceding the stage to the inevitable iron foundry revolt that functions as the climax of the story. Unlike *The Rioters*, which spends little time on *why* machines were introduced, in "The Hill and the Valley" Martineau explains that a fall in iron demand necessitated the introduction of machine work and a decrease in wages. Although the workers initially "understand" the idea that falling prices meant falling wages, the narrator laments that some of the "least industrious" had been fired and that there remained accounts of uneducated workers "swearing at the machinery."[68] Additionally, workers who saw the overlap between falling wages and machinery led many to remember "how prosperous they had all been once when less machinery was in use."[69] Though Martineau allowed that there were some workers who were "too wise" to blame the machine and who recognized that "machinery was the consequence and not the cause," the frustration of the workers is again framed as understandable but counterproductive. As the narrator asks: "What could have been done besides?"[70]

As discontent gives way to violence, Martineau places more blame on worker ignorance, faulting human intelligence rather than the laws of social science. When an accident occurs that leads to the death of a young boy at the hands of one of the new machines, the narrator claims that the death was tragic but exaggerated, and that it had little to do with machines, an "unnatural accident" outside of the laws of political economy and machine work. As the narrator notes, no one would fault a "furnace" if a worker had fallen in and died or the "hearth" if someone had burned their hand, "but the new invention was now to blame for everything." Similar to *The Rioters*, Martineau gives the workers their say before explaining how counterproductive it all is, as one worker notes that the management will simply say the boy made a mistake and that he had been better off in the factory than in "begging for bread" on the streets. Unmoved by this logic, a "fiery orator" lifts the boy and carries him to his mother, who gives full voice to the irrational anger of the workers: "'O, my boy, my boy,' cried a dreadful voice at this moment. 'I will see my boy, I will see who murdered him, I will have revenge on whoever murdered him! Oh, you are cruel to keep me away! I will have revenge on ye all!'"[71]

In descriptions of the riots, Martineau dwells on the specifics of machine breaking and violence more than in *The Rioters*, and she clearly understood the importance of tension in the storytelling process. It would, after all, make for quite a boring tale if all the workers believed in the laws of social science from the outset, and Martineau offers a rich description of the riot: "The crowd came pouring over the opposite ridge, pell-mell, brandishing clubs and shouting as if every one of them was drunk. In front was a horrid figure. It was the mother."[72] When the mob reaches the factory, Martineau reminds the readers of the futility of their actions, as she imagines "the least ignorant" workers realizing that "destruction would ensure the total ruin of the iron works." Perhaps most surprisingly, an industrious worker named Paul jumps into the fray in order to *defend* the machines, picking up an "iron bar" and threatening to do violence to any man who harms a machine. As always, the violence is finally quelled by the arrival of soldiers and the arrest of all involved. Paul is arrested along with the rest but goes away smiling, knowing that he will be released. Upon receiving his freedom, Paul is praised by Martineau's narrator for defending the machines and helping the soldiers track down the guilty parties.[73]

As in *The Rioters*, the final word goes to the industrialist, as Wallace returns to his smoldering factory to give a speech to the machine breakers. Following the principles of political economy, he explains how machines could have kept the factory two-thirds full but that because of human ignorance it now had to be permanently closed. While the workers gather around Wallace, he explains for over ten pages how capital and machinery work together to benefit the worker, and he chides them that "the revenge for which you have snatched . . . is as foolish as it is wicked."[74] Unlike the industrialist from *The Rioters*, Wallace reminds them of the inevitable result suggested by Smith—that while the masters will suffer, they can simply build a new factory and "suffer least." Wallace then hires Paul to handle the last accounts of the factory and leaves town with his wife to build a new foundry with a new set of investors.

Before wrapping up the story, Martineau makes one final criticism: not of the workers themselves but of those who should know better. On their way out, they encounter "Old Armstrong," whose prominence in the story has faded since his daily strolls in the valley. While Armstrong will still not admit that industrial life is superior to his old ways, he does acknowledge that there is no chance of returning the factory grounds back to the unspoiled valley that existed before Wallace's arrival. Faced with the choice of seeing a full factory or its empty remains, Armstrong grudgingly admits he prefers the former to the factory husk that now occupies his once-vibrant valley: "There is no comparison between a settlement where art and industry thrive, and a greater number of human beings share its propensity every year, and a scene like that, where there is everything to put one in the mind of man but man himself."[75] Although Armstrong's conversion is not complete, Martineau must have imagined that readers would return to the principles at the end of the story, reasoning that the scientific laws of political economy could bring order to the human mind, if not to the factory floor itself.

From Education to Accommodation: Charles Knight, the March of National Intelligence, and the "Mind amongst the Spindles"

Harriet Martineau was not alone among late-Georgian reformers in trying to "calm" worker unrest and reform minds through education, although she may be unique in believing that theories in the sciences of man alone would change the distressed minds of displaced factory workers. Far more common in the 1830s was the approach of her friend and occasional collaborator Charles Knight (1791–1873), author of a series of cheap pamphlets on machinery and publisher of *The Penny Encyclopedia* and *The Penny Magazine*.[76] More than Martineau, Knight was a true believer in the "march of the mind," the loose idea mocked by *Crotchet Castle* that worker and social unrest could be soothed through broad public education in the arts and sciences. Knight had coined the phrase "march of national intelligence," and he published the earliest works of the Society for the Diffusion of Useful Knowledge, the group founded by Henry Brougham (1778–1868) that Peacock parodied as the "Steam Intellect Society." Though he was not nearly as well known as Brougham, J.R. McCulloch, or Martineau, Knight's pamphlets and publishing efforts capture an important evolution in British thinking about industrial labor, as he moved away from the idea that education in political economy itself would quell worker revolts. Regrettably, Knight concluded, given the harsh laws of political economy, the best education could offer was a brief respite from an otherwise dreary life.

For a man who often touted the benefits of specialization, Knight led an eclectic life. As his lone biographer described, Knight's success was due to a "lack of specialization," and he was fortunate to begin his writing and publishing career "before the division of labor had been widely adopted."[77] Knight's breadth of contributions to the debates over machine breaking, political economy, and worker education have largely gone unnoticed, in part because it is difficult to attach any clear and consistent position to his vast output. As an indication of just how varied Knight's ideas were, it has been claimed that one of his books was authored by Brougham, that Knight had been the author of a book by Charles Babbage, and that Knight was Charles Dickens's model for Thomas Gradgrind in *Hard Times*.[78] So persistent was the last rumor that Knight himself wrote to his friend Dickens to ask if he was indeed the inspiration for the cruel and calculating patriarch of Coketown.[79] Though Dickens assured Knight that he was not, it is impressive nonetheless that Knight could be mistaken for the amiable, aloof, and optimistic Brougham, the perpetually disgruntled mathematician Babbage, and the hard-hearted Gradgrind.

Knight did however share with Brougham, Babbage, and Gradgrind the unflinching belief in the science and laws of political economy, in particular the importance of specialized work with machines. Like Martineau, Knight was initially inspired to write *Results of Machinery* (1831) by a spate of machine breaking in 1830, the turbulent decades after the Napoleonic Wars convincing him that the harmonious laws of political economy would not obtain on their own.[80] Knight initially believed that machine breaking was counterproductive and that the riots would stop as soon as the working public could understand how harmful they were. Published as part of a cheap series of pamphlets called "The Workingman's Companion," *Results of Machinery* spoke directly to factory workers, claiming, "We address you as capable of reasoning." For evidence that reasoning would work, Knight quoted the 1827 House of Commons committee on "distressed labourers," where a number of weavers testified that "they know perfectly well that machinery must go on, that it will go on, and that it is impossible to stop." As he wrote, echoing the industrialist in *The Rioters*, the people "are wickedly and ignorantly destroying the property of the farmer and the manufacturer in the belief that machinery can be stopped or put down."[81] As evidence of their ignorance, Knight lamented that machine breakers had become "embarrassed" at not knowing which machines to break and that "there are fearful mistakes at the bottom of their furious hostility to machines."[82] Though subsequent histories have argued that the machine breakers knew exactly what they were doing, *Results of Machinery* was intended as a direct communication to workers to prevent any further "mistakes."[83]

Like Peacock's Mac Quedy, Adam Smith, and Martineau, Knight originally thought the benefits of division of labor and machines would trickle
down to workers. Knight claimed of industrialization, "There are great
temporary inconveniences in the introduction of a new machine" and "the
state of change is doubtless a state of suffering."[84] Putting himself in league
with the workers, Knight admitted that the wealthy can withstand these
changes and that comfort is for "the rich, not us." In *Rights of Industry*, Knight
even hearkened back to Smith's concern that extensive machinery and strict
division of labor deadened the mind of the industrial worker, noting that
"it was debasing to the human intellect and morals to make for ever the
eye of the needle, or raise a nap upon woollen cloth."[85] Yet Knight quickly
waved away the objection that specialization at machines entailed mental
drudgery: "The division of labour will not press severely on any man," because people were free to take any job they wished, and so could move from
job to job without "debasing" their intelligence. As he wrote in *Results of
Machinery*, it would be best therefore to approach a "state of society where
the laborers would be many and lightly tasked . . . giving a direction by its
intelligence to the mere physical power which it had conquered." Anticipating a society so "advanced" that all work required a "certain level of intelligence," Knight believed that "all that was purely mechanical" would be left
to the machines.[86]

As seen in the 1830s and 1840s, however, the idea that intelligent men
and women might one day be able to leave the machines behind became
untenable. While the social physics of Quetelet had imagined that only social
scientific *workers* would have to sacrifice their free will, the laws of political
economy seemed to stretch much further into the minds of the masses. For
example, in her review of hundreds of working-class biographies from the
time, Regenia Gagnier reported that in the 1840s the workers began to see
themselves as "atoms of masses" that had become "a piece of machinery,
dehumanized and disregarded."[87] As described by Jonathan Glickstein in his
history of mechanization in England and America, Knight and Martineau's
hopes for education had been overrun by a concern that "the mastery of
specialized tasks by unskilled laborers . . . did appear to promote a narrow
intelligence."[88] Glickstein included an 1852 passage from a clergyman in
Massachusetts that perhaps best captured just how incompatible industrial
work was with Knight's hopes in the 1830s for a "march of national intelligence." This was not Smith's speculation about mental "torpor" and
"ignorance" from 1776 but rather the accumulated empirical evidence from

generations of factory work. As the clergyman claimed, in factories, "the minute subdivision of labor has a direct tendency to dwarf the intellect. A man is less of a man of intelligence [and] skill . . . when he merely watches a spindle or mends a web, than if he took the wool or cotton home, and brought the finished cloth to market." In a note that recalled Bonald's warnings from the beginning of the century, he added, "Improvements in machinery tend to make the operative less and less an intelligent agent, and more and more a machine."[89] Like Diderot, who imagined that specialized intellectual labor led to the "stripping away" of individual personality, specialized physical labor also brought about a "less and less . . . intelligent agent." Although Smith had once praised the "dexterity" of specialized work, and J.R. McCulloch had argued that the division of labor created "mental excitement," it seemed to most that the lessons of the sciences of man did not elevate the mind in the way reformers since Condorcet had hoped. Rather, adapting oneself to the laws of social science led to a precipitous decline in mental life.[90]

It is perhaps for this reason that Knight, along with his friend Martineau, seemed to change tactics. Instead of offering lessons in popular political economy that might cure the ailing intelligence of workers at their machines, they offered more palliative—though equally didactic—approaches, suggesting that *self-alterations* in the human mind would help bring human experience into better harmony with the laws of science. In 1845, the two collaborated on *Mind amongst the Spindles*, a collection of fiction and poetry from girls and women who worked in factories in Lowell, Massachusetts. In these stories, originally published in America as *The Lowell Offering*, Knight and Martineau presented examples of how the mind might be changed, even simplified, to accommodate the threat that division of labor with machines made to intelligence. In an acknowledgment that the societal "march of the mind" might tend to lower intelligence, Knight quoted the original American editor of *The Lowell Offering*, who claimed that "a day of constant manual employment, must in some sense measure unfit the individual for the full development of mental power."[91] Knight himself wrote, challenging a century of social thought from Condillac and Smith through the *sciences sociale* of Condorcet and Quetelet, "the hard imperious laws which regulate the production of wealth" were not always in agreement with "the increase of human happiness." Yet rather than acknowledge a deficiency in the laws themselves, Knight insisted that the stories of the Lowell girls and women proved that it was up to the individual intelligence of each worker to rise above the din of the machines.

As he explained in the preface, those who wrote and read "lifted themselves up into a higher region than is attainable by those . . . whose minds are not filled with images of what is beautiful and true." Summing up the lesson for the "immense body of our factory operators," he hoped they would learn that "their strength, as well as their happiness, lies in the cultivation of their minds" (*MAS*, xiii).

Although Martineau had once written unflinchingly of the suffering of displaced factory workers, she agreed with Knight that the Lowell stories proved that factory life could be endured with the right mental adjustments. It was her idea to publish the stories in England, and her initial letter to Knight was included in the introduction to the American edition of *Mind amongst the Spindles*. Martineau had traveled to America in the 1830s, taking a customary tour of New England factories along with her friend Ralph Waldo Emerson (1803–1882), the writer and essayist who claimed that he would not be "deceived into admiring the routine of handicrafts and mechanics, how splendid soever the result."[92] Though Emerson complained that the "splendid results" of machinery resulted from the "labors of stupid men," Martineau's descriptions of workers looked nothing like the lives she had imagined in *The Rioters* and "The Hill and the Valley."[93] Rather than the miseries of the Brett family, Martineau claimed that while the girls worked 70-hour weeks, "I saw no sign of weariness among them." Martineau also asserted that she could find "no complaints peculiar to mill life" and that the girls were "not wan and depressed under their labours." For Martineau, the factory girls of Lowell had avoided physical and mental pain because "their minds are kept fresh," and she attributed the happiness to "the invigorating effects of MIND in the life of their labours" (*MAS*, xvi; emphasis in the original). In the same way that laws of political economy had brought peace to her own mind, Martineau concluded that "the institution of factory labour has brought ease of heart to many" (*MAS*, xvii).[94] Instead of having to learn the laws of political economy—as in *The Rioters*—workers could better their lives at the machines simply by changing their expectations.

Like *The Rioters* and Knight's first pamphlets, many of the stories were quite clear about the hardships of factory life. In "The Spirit of Discontent," for example, a character named Ellen complains that she detests the work because she does not wish to be "obliged to rise so early in the morning, nor be dragged about the ringing of the bell, nor confined in a close noisy room from morning till night." Explaining to her roommate why she wants to leave, Ellen claims that it was "just as if we were living machines" (*MAS*,

36–37). In "Prejudice Against Labour," a sympathetic character named Mrs. S. worries that factory workers are "quite as ignorant as southern slaves" and doubts any chance of mental elevation: "What chance can they have for improvement . . . they are driven like little slaves to and from their work for fourteen hours each day, and dare not disobey the calls of the factory bell" (*MAS*, 70). In a later story, "Susan Miller," a deacon warns young Susan not to travel to the Lowell factories because of "how hard it will seem to be boxed up fourteen hours a day among a parcel of clattering looms, or whirling spindles, whose constant din is of itself enough to drive a girl out of her wits" (*MAS*, 86). Though the stories are not uniform, nearly all include a character who declares that factory life is hard, particularly upon the mind, and that it should be avoided at all costs.

These criticisms of factory work echoed British working-class biographies that Knight and Martineau would have been familiar with, such as William Dodd's *A Narrative of the Experience and Suffering of William Dodd, A Factory Cripple* (1841), a work so detailed in its anatomy of factory life that it helped influence the 1844 Factories Bills, which addressed the danger of working with machines.[95] While Dodd shocked Victorian audiences with gruesome tales of physical machine accidents, he also noted that "the boys and girls who were employed in the factories . . . are generally weak, stunted . . . and ignorant in mind. . . . Their whole faculties have been absorbed in the daily routine of factory life." As for the parents, "they make . . . but 'sorry' heads of families, and their children, as a matter of course, are compelled, by dire necessity, to pass through the same dull, tedious, miserable state of existence."[96]

In the stories written by the Lowell girls and women, however, the "dull, tedious, miserable state of existence" of factory life could be overcome through self-alterations of the mind. Ellen may have thought she was a "living machine," but her roommate reminds her that she may be expecting too much, chiding her, "If we expect to find all sunshine and flowers in any station in life, we must surely be disappointed" (*MAS*, 36). Rather than being consoled, Ellen learns that life is inherently difficult and that a bee "gathers honey even in a poisonous flower" (*MAS*, 37). In "Susan Miller," the titular character describes a "dull pain in her head and a sharp pain in her ankles" but immediately rebukes herself that "it would not always seem so." After noting a "tiresome monotony" in her co-workers, Susan recognizes that if she simply accepted her lot, each day would seem "shorter and pleasanter than the last" (*MAS*, 90). When the deacon who had warned Susan of the harsh life of factories reports that he has heard of "girls who have had their hands torn off by the machinery," Susan calmly dismisses the concern, noting, "If I am careful I need not fear any injury" (*MAS*, 86). In "The Sugar-Making Excursion," a group of girls escape factory life by literally making lemonade out of lemons (*MAS*, 61).

Unlike the education of previous stories and pamphlets, *Mind amongst the Spindles* suggested that political economy could succeed largely through reducing mental life to pleasant expectations. For example, in one of the longer stories, "Leisure Hours of the Mill Girls," it is made explicit that emancipation from drudgery can be found through simplifying the mind. During a day of rest, the experienced "mill girl" Isabel explains to a new worker named Ann why some of the girls seem so much happier than others. After noting that, in spite of "inconveniences, factory girls are as happy as any class of females," she warns Ann about the causes of their co-worker's misery: "Yes, Ann, it is strange that not everyone prefers happiness. Indeed, it is quite probable that every one does not prefer it. But some mistake the modes of acquiring it through want of judgement. Others are too indolent to employ the means necessary to its attainment, and appear to expect it to flow in to them, without taking any pains to prepare a channel. Others, like our friend Alice, have constitutional infirmities, which entail upon them a deal of suffering, that to us, *of different mental organization*, appears wholly unnecessary" (*MAS*, 132; italics added). When Ann asks if it were possible for the gloomy Alice to change her attitude, Isabel responds in language that seems to indicate that no amount of education can help some girls: "You know the minds of different persons are like instruments of different tones. The same touch thrills gaily on one, mournful on others." Ann suggests that perhaps the simile is wrong, that "different minds may be compared to the same instrument *in* and *out* of tune" (*MAS*, 132). Whatever the case, both girls return to their leisure time, laughing at the "gloominess" of Alice and lamenting her inability to make herself happy.

While well written and, at times, deeply moving, the short stories collected and published by Martineau and Knight in *Mind amongst the Spindles* represent a very different approach to the mind of the worker than *The Rioters, Illustrations*, or the "Workingman's Companion" series. Gone completely were consoling lessons about political economy, appeals to collective sacrifice, or even the consumer benefits of machines. As Edwin Chadwick's *Report on the Sanitary Condition of the Labouring Population of Great Britain* (1842) and Engels's *Conditions of the Working Class* (1845) had made clear, the much-lauded material benefits of specialized labor with machines had not yet arrived. Perhaps in an attempt to counter the crushing details from British working-class autobiographies, *Mind amongst the Spindles* offered personal stories rather than abstractions about political economy. While this could be in part explained by the gender and age of the Lowell girls, or even the unique American location, Knight's introduction and Martineau's letter made it clear that the lessons of the stories were for *all* machine workers in England.[97] The iron laws of political economy were not just descriptive and abstract laws that were discovered through the objective analysis of collective behavior, but were also a fixed pitch and arrangement to which the human mind must be attuned.

Conclusion

"While the hands perform low, menial service, the soul untrammeled is away and reveling amidst its own creation of beauty" (*MAS*, 117). In *Mind amongst the Spindles*, the success of political economy as a science of man did not rest in correctly predicting or evaluating the collective state of human thought and action, as it had since Adam Smith and the Scottish Enlightenment. Neither did it lie in simply educating workers on how political economy worked, as it had for Harriet Martineau and Charles Knight in their first efforts at popularization. Rather, for the mill girls of Lowell, the sciences of man required the self-discipline of individual workers to give up mental life nearly completely. As seen in the quote above, drawn from the aptly titled story "Joan of Arc," division of labor—the key to political economy as science since Adam Smith—required a kind of mental martyrdom composed of mindless work and spiritual transcendence to actually operate. After all, although Smith had tried to set forth the principles of political economy as a science of man in 1776, the proliferation of machine breaking, arson, and uprisings that occurred over the next three generations seemed to make clear that *Wealth of Nations* was not going to bring the order to human affairs that Newton had brought to the heavens. The "little machine" of systemic economic thought had broken down as worker uprisings led reformers—some cynical, some genuine—to try to explain what went wrong to the workers themselves.

The story of nineteenth-century workers and their machines often did mirror Smith's first discussion of the division of labor in *Wealth of Nations*, where factory life *itself* imposes a kind of order (or "torpor") on the mind, yet this chapter has argued that there was more to the interaction of man and machines than a tale of technological determinism where machines radically reshape the human mind. Indeed, if factory life did in fact reduce the mind to a state of torpor by itself—as so many claimed—it would be hard to account for the many intelligent and coherent responses of workers, and there would have been no need for the many popular pamphlets and stories trying to convince factory workers that they were wrong. Instead, as this chapter has argued, the simplification of the human mind also required a more thorough diffusion of the ideas of political economy. Out of a number of political and military responses to machine breaking, this chapter has focused on some of the most popular and widespread accounts of the sciences of man of the age. Although Martineau's *Illustrations*, Knight's "Workingman's Companion" tracts, and their collaboration *Mind amongst the Spindles* differed in tone and content, they shared two key ideas common to reductive social scientific thought: the laws of social science were as fixed as those of the heavens, and it was the mind of the industrial laborer that would need to conform to better accord with such laws.

The idea of investigating the worker's mind and intelligence to justify a science of man was mocked, like so many other things, by Thomas Love Peacock and his mouthpiece the Reverend Doctor Folliott. Well before Marx pointed out the material basis of capitalist ideology, in *Crotchet Castle* Peacock claimed that the "science of sciences," political economy, did no more than reflect the particular prejudices and interests of a handful of industrialists, capitalists, and reformers in London and Scotland. When Mr. Skionar, a stand-in for the poet Samuel Taylor Coleridge (1772–1834), claimed that it was beyond human ability to truly know the "nature of things," deferring judgment to a "transcendental intelligence," this was not an invocation of the sublime but a sign that the idea of any kind of guiding intelligence in Great Britain could only be seen as a joke. Aside from Mac Quedy, Folliott, and Skionar, the colorful cast also included Mr. Chainmail, a hidebound reactionary in the mode of Maistre who wanted to return society to the Middle Ages; Mr. Toogood, an expositor of "quadrangular paradises" based on the socialist reformer Robert Owen (1771–1858); and Mr. Firedamp, a character of unknown provenance who believed that the presence of water was the cause of all humankind's ills. For Peacock, all of these sciences of man were equally absurd, and all ultimately required outside interventions, often violent, to make the messy world of human behavior accord with such laws. As a sign of the collective ineptitude of the sciences of man, *Crotchet Castle* ends with a retreat to "Chainmail Hall," where machine breakers, armed with a "sixpenny treatise on mechanics," set to attacking the group.[98] After Folliott laments that the "march of the mind . . . has marched into my rick-yard, and sets my stacks on fire, with chemical materials, most scientifically advanced," the group abandons their speech making and various sciences. Instead, in confirmation of Smith's prediction, the gathering takes up Mr. Chainmail's collection of medieval weapons to violently suppress the "the march of the mind."[99] While this chapter has focused on nonviolent coercion found in popular educational guides, the conclusion of *Crotchet Castle* is a darkly humorous reminder of what often happened when blustering social scientists did not see their theories obtain in the real world.

6

The Descent *of* Man
(*and* Intelligence)

Organic Social Science from Spencer to Galton

No training or education can create [intelligence].... You
must breed it.
  —Karl Pearson, "On the Laws of Human Inheritance" (1904)

IN 1851, ONE OF THE MILLIONS OF VISITORS TO THE STORIED GREAT
Exhibition in London found fault with the purely mechanical vision of human
life and nature, declaring, "A man can put together a machine; but he cannot
make a machine develop itself." Noticing a mechanical piano in the "Musical
Instrument" gallery, he went on to claim that "the ingenious artizan, able as
some have been . . . to produce a mechanical pianoforte-player, may in some
sort conceive how . . . a complete man might be artificially produced; but he
is unable to conceive how such a complex organism gradually arises out of a
minute germ." For those who know their eminent Victorians, the enthusiasm
for the organism over the machine (and the tortuous phrasing) could only have
come from Herbert Spencer (1820–1893).[1] Although Spencer would devote
tens of thousands of pages to explaining how organic growth structured all
life in the universe—from the unicellular organism to the human mind to
the cosmos itself—he was not beyond marveling at the ability of machines.
In the second edition of his celebrated *Principles of Psychology* (1872), Spencer
even compared the structure of the brain to the *piano-mécanique* he saw at the
exhibition. Echoing the idea from *Mind amongst the Spindles* that minds could be
"in and out of tune"—and anticipating the Logical Piano of William Stanley

Jevons—Spencer noted that we might "consider the cerebrum and cerebellum as like vast magazines of tune-boards." Despite imagining that one day "kindred appliances of a higher order" could approach intelligence, Spencer still saw the mechanical alone as "an illustration that falls short in many ways."[2] For Spencer, mechanical intelligence was not enough.[3]

A small moment in the history of social thought and artificial intelligence, Spencer's comments at the musical instruments section of the Great Exhibition indicated a larger development in how the sciences of man viewed intelligence and even life itself. While the Great Exhibition marked the triumph of classical political economy, the 1850s also saw the instauration of a competing methodology for a science of man, one based on biological analogies of evolution rather than mechanical or physical metaphors.[4] As will be seen, the sciences of man that grew out of this belief—including Spencerian "sociology," social Darwinism, and eugenics—often challenged the belief that the individual mind was malleable, plastic, and subject to improvement, potentially dashing the hopes of the sciences of man from Helvétius and Condorcet through Martineau and Jevons. And as seen in the previous chapter, popularizers of political economy had grounded the success of their science of man in the ability to *change the minds* of those workers subject to economic laws through education. Yet the new biological social sciences seemed to leave little room for improvements in intelligence based on instruction or appeals to individual minds. By the time of the statistician and eugenicist Karl Pearson, who combined organic ideas with the mass collection of data from social physics, such hopes for progressive improvement in intelligence required selective "breeding" rather than education.[5]

While the publication of Charles Darwin's *On the Origin of Species* (1859) is the landmark event that usually marks the beginning of organic thinking in the sciences of man, the 1850s were a particularly turbulent era of ideas even before the *Origin*. As recent scholarship has shown, a number of the most important developments in the growth of the organic sciences of man, including what became known as social Darwinism, preceded Darwin's theories.[6] Even the Great Exhibition itself, usually described as a monument to classical political economy, or the apotheosis of divine Newtonian world order, became a harbinger of later nineteenth-century thought because of its subtle and not-so-subtle nods to racial and ethnic hierarchies.[7] To examine how these biological sciences of man emerged out of political economy as a competing account of human thought and action, the chapter begins with an analysis of two of Herbert Spencer's works from the early 1850s that critiqued classical political economy prior to Darwin: *Social Statics* and "A Theory of

Population." While Spencer believed that organic and biological sciences had superseded the creaky mechanisms of previous sciences of man, his initial idea of evolution *maintained* the optimistic and progressive spirit of the Enlightenment toward the human mind, and at times even matched the enthusiasm of Condorcet. In the wake of the theories of natural selection offered by Darwin and Alfred Russel Wallace, however, "Spencerism" became decidedly more conservative about the ability for human progress, with human "intelligence" potentially chained to its lowly origins in the physiology of animals.

The middle of the chapter therefore examines Spencer's influence on Wallace's "The Origin of Human Races" and Darwin's *The Descent of Man*, two organic sciences of man that profoundly disagreed about the nature and potential of human intelligence to, in Wallace's words, "escape" from the laws of natural selection. While both men had taken a similar methodological approach toward theorizing natural selection through modification as the key to evolution, they later disagreed whether human beings could eventually govern such a process. Whereas Wallace imagined the potential of human intelligence to *direct* the evolution of not only humans but the entire planet, Darwin was characteristically cautious. In *The Descent of Man*, in which he finally offered his own thoughts on how evolution through natural selection might influence human moral and intellectual traits, he claimed that intelligence was like any other trait in any other species: helpful to be sure, but still subject to natural selection.

The interdependent ideas merged in the work of Francis Galton, whose grand project to breed a better human race combined the progressive ideas of early Spencer, classical political economy, and Darwin's belief that intelligence was forever shaped by natural selection. As a science of man that would inspire a host of practical implementations—from ideas of "racial hygiene" in England and Germany to forced sterilization in the United States—Galton's vision of eugenics nevertheless seemed to be the only way for an organic science of man to improve collective human intelligence *and* be fully Darwinian. Once the province of divine understanding, in the work of Galton's science of man, intelligence could only thrive by being brought down to something that could be bred as easily as prized livestock.

The Fungus and the Oak Tree: Herbert Spencer's Progressive Organicism

In spite of efforts by many historians to correct the record, the social science developed by Herbert Spencer is still often reduced to the idea of "social

Darwinism," a theory that supposedly took Darwin's ideas and "applied" them to society. As a popular sociology textbook declared, in boldface and italics, "Spencer became a proponent of a doctrine known as **social Darwinism**, which *applied to society Charles Darwin's notion of survival of the fittest, in which those species of animals best adapted to the environment survived and prospered, whereas those poorly adapted die out.*" It went on to claim that "Spencer reasoned that people who could not successfully compete in modern society were poorly adapted to their environment and were therefore inferior."[8] While few historians would accept such an incoherent account, which inverts the influential arrow (Darwin in fact borrowed the phrase "survival of the fittest" from Spencer), more subtle accounts of Spencer remain influenced by Richard Hofstadter's highly popular *Social Darwinism in American Thought, 1860–1915* (1945), a book that essentially held Spencer responsible for the evils of rapacious American capitalism but said little about his early immersion and influence in European ideas of evolution.[9] As Robert Richards pointed out over thirty years ago, Ernst Mayr's authoritative history of evolutionary theory, *Growth of Biological Thought* (1982), devoted only three paragraphs to Spencer.[10] Even in a wide-ranging and impressive collection of sixty-three essays on Darwin and evolution, Spencer is only referenced a handful of times, with his most substantial appearance in the essay "Social Darwinism."[11] Yet today historians mostly agree that both Spencer's and Darwin's ideas emerged out of a similar milieu of evolutionary thought. Rather than one idea "applied" to another, they were mutually reinforcing.[12]

Spencer's early work on social "evolution" was actually quite progressive and much closer to the optimistic visions of the eighteenth-century sciences of man than might be imagined based on his dour reputation.[13] As Paul Elliot has chronicled, Spencer likely first encountered evolutionary thought in his hometown of Derby, where his father directed the Derby Philosophical Society, a group founded on the principles of progressive enlightenment and providential evolution propounded by Erasmus Darwin (Charles's grandfather).[14] One detailed account noted of Spencer's early immersion in Enlightenment ideas, "Spencer's notion of evolution is thus markedly pre-Darwinian" and was a "theory of progress" rather than "adaptive change."[15] And, as the philosopher of science Michael Ruse has declared, "Evolution is a child of the Enlightenment."[16] Somewhat remarkably, Spencer's earliest works could even appear similar to Condorcet's *Sketch* in their vision of historically produced change, a far cry from the Spencer who would later be endorsed by social Darwinists, conservatives, and Libertarians.

In *Social Statics* (1851), for example, Spencer called for a progressive and organic social science to triumph over the mechanical analogies of political economy. Written and published the same year he visited the Great Exhibition, just months after he had moved to London to work as a journalist with *The Economist*, the work was as dismissive of mechanistic social thought as it was of the *piano-mécanique*. Spencer believed that the ideas that had influenced the Crystal Palace—political economy and utilitarianism—were limited by their mechanist approach and that there was another "possibility of a strictly scientific morality."[17] In *Social Statics*, Spencer was unwavering in his critique of previous sciences of man that sought to "calculate" expected behaviors. Although targeting Jeremy Bentham's utilitarian philosophy, his criticism could have applied equally to social physics, the French sciences of man, and almost all classical political economy from Smith through Jevons. As Spencer claimed, governments that relied on calculating science had erred in trying to "form just estimates . . . of the probabilities of future events."[18] Instead of the predictable forces of Helvétius, d'Holbach, and Quetelet, Spencer imagined the futility of "grasping at once the multiplied phenomena of this ever-agitated, ever-changing sea of life, to derive from them knowledge of their governing principles." Spencer believed these theories of human action were in fact anti-Newtonian, joking that much of the quantitative sciences of man looked like what physics would have been had Newton studied the skies "without any previous investigation of the properties of terrestrial matter." Trying to infer laws of humanity through merely counting human actions, as Bentham and Quetelet had done, was as fruitless as if Newton had only "set himself to tabulate the accumulated mass of observations, and to educe from them the fundamental laws of planetary and stellar equilibrium."[19] Spencer believed any science of man that treated human beings as an inanimate mass of atoms was a failure, and the only way to determine a "true theory of society" was through "inquiring into the nature of its component individuals." In a direct challenge to classical political economy and social physics alike, which had tried to study human action in groups, Spencer invoked what he called the "moral sense," which "warns us against adopting any fundamental doctrine which cannot be expressed without presupposing a state of aggregation."[20]

Spencer believed such a "moral sense" revealed in humanity an intelligence that ordered the development of all things, and it was here where he explained how evolutionary ideas might upend the mechanical sciences of man. Even though people and civilizations varied throughout history, humanity was becoming increasingly aware of a law that governed society and all creation, a law that Spencer took as "sure, inflexible, ever active, and having no exceptions."[21] The law was what Spencer would later call the "survival of the fittest," a phrase that would subsequently take on so much cultural and social baggage that it can be difficult to see in its original formulation.[22] In *Social Statics*, however, he referred only to the "universal law of physical modification." Arguing that no amount of society, culture, or education could change the basic fact that organisms adjusted to their natural environment, Spencer presented his "universal law" as a simple deductive truth for biological development.[23] As he claimed of the necessity of such a law,

> no other arrangement of things can be imagined. For we must adopt one of three propositions. We must either affirm that the human being is wholly unaltered by the influences that are brought to bear upon him . . . or that he perpetually tends to become more and more *un*fitted to those circumstances; or that he tends to become fitted to them. If the first is true, then all schemes of education, of government, of social reform . . . are utterly useless. If the second is true, then the way to make a man virtuous is to accustom him to vicious practices, and *vice versa*. Both of which propositions being absurd, we are compelled to admit the remaining one.[24]

Ironically, strict social Darwinists and eugenicists would later take up Spencer's first "absurd proposition"—believing that social amelioration was pointless for those born with "inferior" traits—but the early Spencer was far from such a belief. Not a lawlike system to weed out the weak, Spencer's initial articulation of his ideas made fitness an internal and rational response by an individual's moral sense to the world around him rather than a simple fortuitous draw in the genetic lottery.

It would take Charles Darwin over a decade between the *Origin* and *The Descent of Man* to apply the biological theory of evolution to the moral and intellectual world, but Spencer made the leap from the natural to the social sciences on the same page. As he claimed of his early iteration of the principle of fitness, "the universal law of physical modification is the law of mental modification also."[25] This basic analogy—that the "fitted" nature of things like intelligence and morality were equivalent to the fitness of complex structures in nature—became the cornerstone of much of Spencer's sociology, providing the foundation for the ten volumes of metaphysics and quasi-theology that made up his magnum opus: *The Synthetic Philosophy* (1896). In *Social Statics*, for example, Spencer saw each adaptation in nature and society as the removal of evil because, in his words, "all evil results from the non-adaptation of constitution to conditions."[26] Specifically contrasting his vision of the universe with a "lawless world" having "no order" and "no certainty," he believed his "science of social life . . . assumes perfection in the elements in which it deals." Describing a world that sounds very different from what is normally evoked by "survival of the fittest," Spencer assured his readers, "Progress, therefore, is not an accident, but a necessity. Instead of civilization being artificial, it is a part of nature." Summing up the enthusiasm of the early Spencer, Paul Elliot has claimed, somewhat more vividly, that "even the most drug-mellowed hippie of the 1960s or 1970s could scarcely have framed a more optimistic view of eventual social harmony."[27]

The optimism continued in "A Theory of Population" (1852), a short journal article that attempted to demonstrate that while all of nature operated according to a "law of universal modification," there was still a divide between living and non-living matter. Remarkably, in just thirty-eight pages, Spencer both invented a new theory of life and applied it across the entire spectrum of living beings. Although he admitted that it would be arbitrary and "not universally applicable," Spencer proposed "*co-ordination of action*" as a new definition of life. In this view, inanimate objects like crystals developed only because of a single action of "accretion." In contrast, even the "lowest organic process" of the unicellular organism coordinated actions of "accretion" and "disintegration." For Spencer, "it is in the balance of these two actions that the life consists. It is not in the assimilation alone; for the crystal assimilates: neither is it in the oxidation alone; for oxidation is common to inorganic matter: but it is in the joint maintenance of these—the co-ordination of them. So long as the two go on together, life continues: suspend either of them, and the result is—death."[28] For Spencer, this was why human beings were not, as Jevons later speculated, mere "crystals of complicated form."

With this new principle of "co-ordination" in hand, Spencer embarked on a breakneck tour of how various forms of life developed, from single-celled organisms to the most complex human societies, or what he referred to as everything from "monad to man."[29] Like many Victorians, Spencer assumed the natural world was subject to the "pressure" of population growth, taking his cue from the work of the economist Thomas Malthus (1766–1834), who believed that increasing population would always undercut human material progress. For Spencer, there were essentially two strategies for survival in a Malthusian world: either organisms could reproduce at a staggering rate or they could develop more complex "co-ordinations." In an example of the former strategy, Spencer noted that the lowly fungus survived through the production of an enormous number of spores, most of which never managed to reproduce. Even though it was a poorly coordinated effort, leading to the vast destruction of most of the spores, the sheer number of opportunities for reproduction ensured its continued success. For an example of "co-ordination," Spencer suggested an oak tree, a far more advanced organism with a number of specialized and hierarchical functions, which survived despite producing fewer acorns in its lifetime than the number of spores a fungus produced "in a single night." To make up for the lack of potential "offspring," therefore, an oak tree needed to develop specialized parts of the organism. In short, "division of labor" was one possible successful coordination of actions in response to the pressures of population.[30]

Although high levels of both reproduction (the fungus) and complexity (the oak tree) had proven successful, each approach required a different development in the coordinating "elements." As Spencer explained, the less coordinated actions were, the less important each individual part became—for example, in the replaceable segments of a worm compared to the body parts of a more advanced animal.[31] Similarly, each part of a fungus could exist and develop on its own in a way a branch from an oak tree could not. As evidence for the striking independence of the individual elements at the lower end of the complexity scale in nature, Spencer quoted from William Carpenter's description of the *Haematococcus binalis*, a "reddish slime" that essentially existed as a series of distinct, independent, and completely superfluous "individuals" held together by a mucus. Carpenter wrote in *Principles of General and Comparative Physiology* (1838), "Every component cell of the aggregate mass that springs from a single germ, being capable of existing independently of the rest, may be regarded as a distinct individual."[32] In a formulation that might trouble Spencer's later Libertarian admirers, this meant that independent "individuals" were the hallmark of the *least* advanced organisms. As organisms (or civilizations) progressed, however, their individual members lost the ability to rapidly reproduce and fend for themselves, becoming *less* independent and more reliant on the organism as a whole. In language that would seem more meant for societies than red slime, Spencer concluded that "progress" in nature entailed a "necessary antagonism between the development of the individual and the increase of the race."[33]

"A Theory of Population" proposed "co-ordination of action" as the key to organic life, but this did not necessarily mean *human* societies could choose between the strategy of the fungus or the oak tree. In his concluding remarks, Spencer proposed that humanity, alone among organisms, may have developed an escape from the Malthusian trap of limited resources. After noting that it was time to apply the "law which we have thus traced throughout the animal kingdom . . . to mankind," Spencer said humanity was in a "transition" period where it seemed possible that fertility might finally be brought under control. As Spencer surveyed the means through which fertility and complexity might be balanced, he examined a number of human faculties that had contributed to such a transition. Abilities like "strength," "swiftness," and "mechanical ability" had certainly led to better means of coordination, but humanity could hardly be said to excel in any one of these traits. Instead, Spencer identified "intelligence" and "morality" as the coordinated actions that necessarily developed in response to further pressures of population. The result was a kind of paradise of complexity, where Malthusian concerns led not to endless poverty and starvation but rather to greater and greater intelligence: "[The] inevitable redundancy of numbers—this constant increase of people beyond the means of subsistence—involving as it does an increasing stimulus to better the modes of producing food and other necessities—involves also an increasing demand for skill, intelligence, and self-control." As Spencer wrote, therefore, "every improvement is at once the product of a higher forms of humanity, and demands that the higher form of humanity carry it into practice."[34] Here, then, was the progressive evolutionary theory of Erasmus Darwin, not Charles. By imagining an "inevitable" growth of "intelligence" and "self-control," Spencer's first attempt at a science of man rescued humanity from the lower organisms and inorganic matter. Rather than a complex society requiring its members submit to a brutal process of weeding out the weak, the coordinated actions of intelligence and morality might provide an escape from the jungle.

Crucially, however, Spencer's escape did not apply to *all* humanity, and one might be able to guess from his later admirers which civilizations he believed were fungi and which were the oak trees. For anyone who might doubt that these sentiments could manifest themselves in scientific racism or think that eugenicists needed Darwin's *Origin* to imagine the possibilities of ranking civilizations on an evolutionary scale, Spencer asserted at the end of "A Theory of Population" that sciences like phrenology had already done much of the work to distinguish between the coordinating abilities of various peoples. After a dubious tour through the evolution of distinct species, Spencer suggested that the growth of intelligence and moral sense could be confirmed by an increase in the capacity of the nervous system. Relying on skull measurements from the biologist Richard Owens (1804–1892) that showed an increase in "mean capacities of the crania," Spencer declared it an "ascertained fact" that the nervous centers had increased since the origins of civilization.[35] As he claimed, in civilizations where there had been a "further enlargement of the nervous center," there had been a concomitant "further decline in fertility." It could only have occurred because more complex civilizations had more complex desires than "primitive groups." For example, while the "Pacific Islander" had "all his wants provided by Nature," Spencer argued that the "Englishman, generation after generation," had to bring "ever-increasing knowledge and skill" to the "satisfaction of his wants." Such was the "discipline" of nature, and in the final pages Spencer made it clear that this "discipline" was the key to all progress in humanity. In one of the most notable tautologies in the history of ideas, he wrote: "All mankind . . . subject themselves more or less to the discipline described . . . but, in the nature of things, only those who *do* advance under it eventually survive." While many traits might be important to survival, Spencer highlighted a concern for those "families and races" that did "not stimulate . . . greater mental activity." In a chilling comment for any organic science of man, Spencer concluded that such "families and races" would therefore be "on the high road to extinction and must ultimately be replaced by those who the pressure does so stimulate."[36]

Spencer never actually used the phrase "survival of the fittest" in this passage, but the anticipation of a more limited and less progressive organic science of man might be clear.[37] While Spencer would spend a lifetime working out the particulars of how mental, biological, social, and moral thought all developed along similarly evolutionary lines, he notably dropped from *Social Statics* much of the progressive vision of how evolution could lead to a stable equilibrium.[38] Many have argued that Darwin's *Origin* forced Spencer to give up his progressive ideas, but in his revisionist intellectual biography, Mark Francis argues that a mental breakdown in 1856 caused such a shift, claiming that until then Spencer's "ideas had been bathed in the harmonious light given off by the Crystal Palace."[39] Elliot may have overstated the case in claiming that Spencer was more optimistic than "drug-mellowed hippies," but there was certainly a kind of grandeur in Spencer's initial formulation of life, at least for those civilizations with sufficiently large enough skulls. As Francis explained, by the 1860s, Spencer had dropped the utopian vision, as the "great organic-inorganic divide that had run through Spencer's early scientific writings had altogether disappeared."[40] The great triumph of his earliest work—to try to mark out what elements of nature constituted life—was subsumed by a reductive and nearly theological science he called "sociology."

Later Spencerism can therefore be accused of being doubly damaging to the idea of progress for human intelligence in the organic sciences of man. It not only challenged the harmonious mechanism of Enlightenment social sciences but also undercut the glory of the organic metaphor that animated his earliest work. He had been influential in arguing for the organic over the mechanical, and separated the mind from inorganic matter, but somewhere in his decades of philosophy the historical improvement and growth in human intelligence was lost, as was the idea that coordinated actions were at the heart of advanced civilizations. The reactionary social Darwinists in Hofstadter's America never seemed to have read the early Spencer. Although he believed *Social Statics* and "A Theory of Population" provided an organic, rather than mechanical, model for society, Spencer did not imagine at the outset that it would become less progressive. In trying to replace the "calculating" and mechanical sciences of man with an organic idea, however, he severed the link between the progress of science and ascent of human intelligence that Condorcet had forged. Instead, by the end of the nineteenth century, Spencerism would be transformed into a science of man that managed to repackage a host of discarded or discredited assumptions from political economy, the "calculating" social physics of Quetelet, and a simplified application of the laws of evolution through natural selection. The resulting science was called eugenics—or the science of good genes—and it rightly concludes this survey of nineteenth-century attempts at a science of man.

The Great Escape: Intelligence in Wallace and Darwin

The leap from Spencer's vision of competition creating smarter "families and races" to the idea of selectively breeding human populations might seem a natural development, especially considering that the full title of "A Theory of Population" indicated it was a search for "the General Law of Animal Fertility." Spencer had claimed that progress occurred only through eliminating "unfit" members, and if all "evil" emerged from maladapted organisms, then why would "unintelligent" people or races not also be on the "high road to extinction"? Yet Spencer's theories alone could not have led to eugenical or social Darwinist solutions for one reason: Spencer always maintained, even in the face of mounting criticism, that higher traits and successful adaptations could be passed down through instruction and education. In its simplest form, he remained a committed Lamarckian, believing in the original ideas of the French naturalist Jean-Baptiste Lamarck (1744–1829) that acquired abilities and faculties could be inherited.[41] For Spencer, Lamarckian transmission of inherited characteristics meant people and groups could progress through education rather than being subject to a natural elite (social Darwinism) or evolving through "selective breeding" of desired traits (eugenics). Even in his later work, Spencer never adopted any of the key features of what came to be called eugenics. Tellingly, while Francis Galton coined the term as early as 1883, Spencer lived another twenty years and never seems to have used the word in the 1,500 pages of his *Autobiography*, published in 1904.[42]

Neither can eugenics be seen as a simple "applied" Darwinism. As noted, most of the key claims that emerged in social sciences like social Darwinism and eugenics—the superiority of Europeans, the particular strength of British character, the elevation of individual intelligence as the key to progress—could be found well before Darwin in the organization of the 1851 Great Exhibition and the ideas of Spencer. Indeed, as Karl Marx (1818–1883) pointed out when the *Origin* was published, Darwin's work read much like a naturalization of British economic and political life, sarcastically noting that it was "remarkable how Darwin rediscovers, among the beasts and plants, the society of England with its division of labour, competition, opening up of new markets, 'inventions' and Malthusian 'struggle for existence.'"[43] Rather, Galton's influential science of man developed out of a series of paradoxes in the wake of Spencer's vast popularity and the dual discovery in 1858—by Wallace and Darwin—of natural selection and descent through modification. As traced below, eugenics in part emerged out of a debate between Darwin and Wallace over two related aspects: the progressive possibility of an evolutionary science of man and the ability of "intelligence" to transcend natural selection. While Condorcet's social science was able to imagine the endless progress of the human mind for everyone, and political economists held out hope for educating the masses, the stubborn insistence of Darwin that, contra Lamarck, intellectual traits were subject to natural selection alone left few options for such improvement. Wallace would offer a near-religious alternative that allowed for intelligence to transcend its origins in natural selection, but by the end of the nineteenth century, it seemed any science of man that wanted to improve the collective intelligence of society *and* embrace descent through natural selection would have to do so through breeding out less intelligent members.

In contrast to *On the Origin of Species*, which received considerable criticism for seeming to deny humanity a special place in nature, the work of the naturalist Alfred Russel Wallace (1823–1913) showed that one could simultaneously believe in both natural selection *and* humanity's exalted intellectual status. Wallace, whose letter from the island of Ternate near Indonesia famously motivated Darwin to finally publish his ideas, had come to nearly the same conclusion as to how different species emerged.[44] While Darwin and Wallace had similar experiences in formulating the idea of descent through modification and natural selection—both were devoted naturalists, and both had come upon their ideas while observing the fauna of island archipelagos—they differed substantially in their later application of their theories to humanity. For example, in "The Origin of Human Races and the Antiquity of Man Deduced from the Theory of Natural Selection" (1864), Wallace offered perhaps the most optimistic suggestion—at least from a European perspective—for an organic science of man in the wake of the *Origin*. Whereas Darwin made little distinction in the *Origin* between how natural selection might operate on animals and modern people, Wallace believed that modern "man" had reached a point where natural selection essentially ceased to matter. He asked, "Is there anything in human nature that takes him out of the category of those organic existences, over whose successive mutations it has had such a powerful sway?"[45] What Wallace wanted to know was simple: Was there something that had developed in human beings that allowed them to escape the "jungle" and to exit the unvarying and, apparently, non-purposeful world of natural selection?

His answer was an unequivocal yes: intelligence was the escape. Wallace's 1864 paper drew on much of the hopeful beliefs of Spencer's "A Theory of Population," where increasing complexity in organisms required that each individual member of the organism become more dependent on others. Wallace described animals in the way Spencer described "primitive" civilizations or fungi, where there was no division of labor and where "each must fulfill *all* the conditions of existence." In the cruel logic of the jungle, in these groups the "weaker" members of society were left behind in order to benefit the group. In Wallace's words, before the introduction of division of labor and specialization in modern man, "natural selection keeps all up to a pretty uniform standard."[46] In "man," however, Wallace believed the case was "different" because he was "social and sympathetic." In this world, "some division of labour takes place; the swiftest hunt, the less active fish, or gather fruits; food is to some extent divided. The action of natural selection is therefore checked; the weaker, the dwarfish, those of less active limbs, or less piercing eyesight, do not suggest the extreme penalty which falls upon animals so defective."[47] In other words, the survival of the fittest did not apply to members of advanced societies. Social and sympathetic feelings, or what Wallace called the "intelligence of foresight," meant that, remarkably, "man would cease to be influenced by natural selection in his physical structure and form."[48] Like Spencer, Wallace theorized that once higher faculties emerged, the only form of competition would be in the mind. As he put it, "there came into existence a being in whom that subtle force we term *mind* became of greater importance than his mere bodily structure."[49] With the emergence of this mind, more-advanced groups need not wait for necessary physical adaptations to randomly emerge and spread throughout society through biological inheritance.

Had Wallace stopped here, he would not have been far from a simplistic science of man that saw intelligence as the latest adaptive tool in nature's toolbox, a means—like a thicker coat or more powerful claws—by which a species can better propagate its own kind. This was generally how Darwin himself viewed intelligence: an exceptional biological adaptation, but an adaptation nonetheless. Yet for Wallace, intelligence and emotion meant that humanity had "escaped the influence of those laws which have produced unceasing change in the animal world." Humanity became "a being apart" that was not tied to the "power which leads to the rigid destruction of all animals who cannot in every respect help themselves."[50] Not only could humanity escape the law of natural selection, but they could also bring the rest of nature along with them, as Wallace imagined how a more developed humanity would "take away some of that power" of natural selection, leaving humanity in control of a vast world of domesticated plants and animals. Rather than a world of endless and savage competition, Wallace's take on Spencer resulted in a permanent *pax humana*, with intelligent human beings directing evolution for the good of all living things.

Wallace's vision of a more perfect nature directed by human intelligence eventually became a form of natural theology, and he has sometimes been the preferred naturalist for those who object to Darwin's materialist account.[51] Yet as grand a view as Wallace seemed to present—where human intelligence and altruism led to perfect harmony—it should be noted that such a vision had frightening consequences for those peoples and civilizations who had not yet advanced to Wallace's perceived level of intelligence. His grand progressive ideas were published, after all, in an essay intended to show that the human "races" were in fact separate species. Wallace believed that the evolutionary adaptations of division of labor, intelligence, and moral feeling must have emerged so early in human history that it put "civilized" races on a separate evolutionary track from "primitive" ones. Referencing the subtitle of *On the Origin of Species* to argue for the natural superiority of some races over others, Wallace claimed, "It is the same great law of '*the preservation of favoured races in the struggle for life*' which leads to the inevitable extinction of all those low and mentally undeveloped populations with which Europeans come in contact."[52] While one might think that the "escape" from natural selection would alleviate future conflict among civilizations, Wallace's teleological view meant that in the future such "extinction" would continue, as humanity's "mental constitution may continue to advance and improve till the world is again inhabited by a single homogeneous race."[53]

Wallace's "homogeneous race," which he referenced numerous times, was not the human race in 1864 but some intelligent race yet to come. In a disturbing paragraph to read from the perspective of the twenty-first century, Wallace explained how his particular science of man led to great progress through massive "extinctions." He wrote, "If my conclusions are just, it must inevitably follow that the higher—the more intellectual and moral—must displace the lower and more degraded races; and the power of 'natural selection,' still acting on his mental organization, must ever lead to the more perfect adaptation of man's higher faculties to the conditions of surrounding nature."[54] In the last paragraph of his paper, Wallace even added the final Spencerian kicker, arguing that a world of mentally superior creatures would lead to "perfect freedom of action" without the need for "restrictive laws" or "compulsory government," resulting in "as bright a paradise as ever haunted the dreams of seer or poet."[55] While the early Spencer imagined this "paradise" as emerging through the gradual education and the inheritance of acquired characteristics, Wallace's full embrace of natural selection as the *only* means of intellectual improvement left the extinction of inferior races as the sole option for human progress.

Wallace's belief that natural selection offered a "bright paradise" for a "homogeneous race" was not the only possible view of evolution in human society. In fact, the prospects for an evolutionary science of man by that *other* theorist behind natural selection would seem far gloomier. Although Charles Darwin (1809–1882) famously declared that there "was a grandeur" in his view of life presented in the *Origin*, many responded in horror to the idea that the human species emerged through the same natural process as all other organic life. While Darwin in the *Origin* said little about how the ideas of natural selection might apply to human societies, or even how it may have influenced the development of human intellectual and moral faculties, legions of theorists like Wallace and Galton emerged in the coming decades to explain what Darwin's theory of evolution through natural selection meant for a science of man. Darwin himself only used the loaded and progressive word "evolution" once in the first edition, claiming in a letter to Spencer in 1858 that "I treat the subject simply as a naturalist."[56] And for the next decade, Darwin largely remained silent on what evolution through natural selection meant for human mental and social life, leaving the speculation to others. As Robert Richards noted in his detailed study of Darwin's notebooks and journals from the time, there was "no evidence that he planned an extensive treatment of human intelligence and morality" before 1869.[57] Yet by 1871, in response to the growing interest in Spencer and Wallace's progressive sciences of man, Darwin declared in his characteristically understated way that it was "time to put together my notes, so as to see how far the general conclusions arrived at in my former work were applicable to man."[58]

Given the seemingly bleak title of *The Descent of Man* (1871), Darwin's most extended statement on a science of man is full of incomplete sketches, inconsistencies, and provocative ideas, and he admitted that it included "hardly any original facts in regard to man."[59] While Darwin made no great claims for consistency or unity of thought in the *Descent*, he was still at his most confounding in alternating between praising mankind for its separation from the animal kingdom and undercutting this very separation through deflationary analogies. For example, at times he seemed to be trying to defend the *Origin* against those who thought he had reduced humanity to an animal. Calling mankind "the most dominant animal that has ever appeared on earth," Darwin agreed with Wallace and Spencer that what gave humans "their immense superiority" were "intellectual faculties" and "social habits."[60] He also agreed that such faculties were the result of natural selection, as scarce resources in the "struggle for existence" meant that mankind's intellectual powers and ability to care for one another emerged as powerfully effective adaptive mechanisms.[61] Darwin too believed there was "an enormous" gap between the "mental powers" of what he called the "lowest savage" and the highest ape. Yet in language that challenged Wallace's vision of separate mental and moral life among races, Darwin noted that the people from the island of Tierra del Fuego who traveled with him on the *Beagle*, whom he called "the lowest barbarians," still "resembled us in disposition and in most of our mental powers."[62] And throughout the *Descent*, Darwin made clear that intellectual and social faculties were responsible for the wide separation of man from other animals, the kind of argument that might seem to assuage the concerns of those who believed natural selection through evolution debased the human species. In this, he was no different from Spencer and Wallace, believing that human intelligence was special, and he referenced both authors often in support of the idea that intelligence was an extraordinary adaptive tool.

However, as digressive and conflicted as the *Descent* was, one thing was clear: Wallace's proposed "escape" from natural selection through intelligence was off the table. In fact, Darwin's argument for the similarity of mental ability in all humans was part of a larger vision that placed collective human intelligence as just one among many traits humans shared with other animals. While Darwin put Englishmen and Fuegians on the same scale, the *Descent* was also full of an astonishing number of humbling arguments that challenged human exceptionalism. After praising human intelligence, for example, Darwin followed with the comment that there was "no fundamental difference between men and higher mammals in their mental faculties." Rather, as he noted of physical structure, even the staunchest critics had to admit that "every chief fissure and fold in the brain of man has its analogy in that of the orangutan." Even more disconcerting to some, there was "no dispute as to the resemblance in fundamental characters between the ape's brain and man's."[63] In other parts of the book, Darwin went well beyond comparisons to human's closest ancestors, claiming that "the brain of an ant is one of the most marvelous atoms of matter in the world, perhaps more so than the brain of a man."[64] Darwin also quoted a story on a popular expression in South America that argued that mules, among all animals, "reason best." For Darwin, this story of the rational mule, "dictated by long experience, combines the system of animated machines, better perhaps than all the arguments of speculative philosophy." Even at the level of morality, Darwin claimed "dogs possess something very like a conscience."[65] All of these examples, and many more, served to highlight the key contribution of the *Descent* to any proposed organic science of man: his infamous claim that in all its faculties—even intelligence—humanity bears the "indelible stamp of lowly origin."[66]

The *Descent* proved to be a challenge to the more progressive evolutionism developed by Spencer and Wallace. Though Spencer was mentioned often, Darwin almost always brought him up as a means to draw distinctions between his own work as a "simple naturalist" and what he called Spencer's "master art of wriggling" and "wonderfully clever" ideas.[67] Even in a famous passage of the *Descent* where Darwin appeared to argue for harsh social Darwinism or eugenics, the full context shows that he was not prepared to accept a cruel science of man that left each member of an organism to fend for itself, even if it led to some greater good. Despite later critics often using the following passage as a cudgel to condemn Darwin as a eugenicist, it is worth quoting in full as it represents almost the entirety of Darwin's intentional contributions to anything that might look like a Spencerian science of man or eugenics.

> With savages, the weak in body or mind are soon eliminated. . . . We civilized men, on the other hand, do our utmost to check the process of elimination; we build asylums for the imbecile, the maimed, and the sick; we institute poor-laws; and our medical men exert their utmost skill to save the life of everyone to the last moment. . . . Thus the weak members of civilized societies propagate their kind. No one who has attended to the breeding of domesticated animals will doubt that this must be highly injurious to the race of man. It is surprising how soon a want of care, or care wrongly directed, leads to the degradation of a domestic race; but excepting in the case of man himself, hardly any one is so ignorant as to allow his worst animals to breed.[68]

Darwin might seem then to agree with Malthus, Spencer, social Darwinists, and Galton that humanity hurts its own "race" through caring for the weak, and there certainly was a connection between those who adopted Darwin's science and later projects to declare certain people "unfit" for life.[69] And in polemics that blame evolutionary ideas for creating the benighted modern world, the "application" of Darwinian struggle to society begins with Darwin himself.[70] Yet Darwin followed these lines with something he probably felt was almost too obvious to mention, that the "aid which we feel impelled to give to the helpless" was *itself* part of our evolved "social instincts." For Darwin, helping others was as much a part of the human evolutionary advantage as having a large brain or the ability to use tools.[71] Though Darwin admitted this sympathy may be "incidental" and "more tender and widely diffused" than it needed to be, civilized societies, *as a whole*, benefited from these social instincts. In language that would seem to directly contradict Spencer's utopian plans for *Social Statics*, as well as later eugenical plans to destroy the lives of the weak for the betterment of all, he wrote that humans could not "check our sympathy, even at the urging of hard reason, without deterioration in the noblest part of our nature."[72]

While interpreting Darwin's views is always challenging, it is even more difficult to appreciate his thoughts on the role of natural selection for intelligence. As he wrote to his friend George Romanes, Darwin was not entirely sure himself: "When I think of how it has bothered me to know what I mean by 'intelligence' I am sorry for you in your great work on the mind of animals."[73] For the purposes here, however, Darwin seems clear on two points regarding the possibility of intelligence as an "escape" from natural selection. The first was that civilizations do progress and decline, and this was due in part to the cultivation of "higher intellects" among their populations. Such a claim might be enough for the eugenicist or social Darwinist who believed that the role of a social scientist should be to discover the means through which this intelligence could then be propagated and encouraged through selective breeding. However, in the *Descent* Darwin was equally clear that, *once they emerge*, the relative levels of intelligence in society had very little to do with natural selection. Even if they did, superior moral instincts, themselves the product of evolution, would prevent humanity from "applying" any lessons or directing this evolution. Though Darwin often resorted to metaphors of domestication, he never came close to suggesting in any of his works that society could somehow breed their best selves or that humanity might enter a new kind of accelerating and progressive world with sufficient intelligence. Instead, as the philosopher Daniel Milo noted in his study of Darwin's unique and non-reductive methodology, "The equating of fitness with superiority and the pursuit of perfection" was the product of other biologists who saw "progress" as "natural."[74] For Darwin, who did not share the progressive evolution of his grandfather Erasmus, Spencer, or Wallace, there was no danger that he would dream up something like eugenics or social Darwinism on his own.

You Must Breed It: Galton's Progressive Route for an Organic Social Science

Given Darwin's challenge to Wallace and Spencer, it might seem like the organic sciences of man had few options for improving human intelligence. Yet there was a biological science of man that managed to be progressive, Darwinian, and reductive: the eugenics of Darwin's cousin Francis Galton (1822–1911). The only catch was that it required removing the exalted status given to intelligence by Wallace and Spencer. Galton's belief that all human traits were inherited, and that the "human stock" could be improved through techniques borrowed from animal and plant breeders, allowed him to develop a theory of progressive human development that did not require intelligence to "escape" from the law of descent through natural selection. Rather than intelligence as outside of natural selection, Galton on the contrary argued that *because* intelligence was on the same evolutionary plane as animal traits, it could be improved in similar ways.[75] In fact, the novelty of eugenics as a social science was that it fully maintained Darwin's belief in the "indelible stamp of lowly origin" while offering a progressive vision of improvement. In doing so, however, Galton needed to reduce intelligence to yet another biological trait that could be measured.

Galton's eugenics first broke from the organic social sciences of Spencer and Wallace by denying that human intelligence was on a separate scale from animal traits. Darwin had largely argued the same thing in the *Descent*, but Galton's language moved far beyond the cautious studies of his cousin. For Galton, the overlap in human and animal intelligence was so obvious that in the "Race Improvement" chapter of *Memories of My Life* (1908), Galton included the subheading of "Breeding Dogs for Intelligence" along with "Eugenics," "Eugenics Laboratory," and "Duty Towards Race Improvement."[76] While Darwin had been hesitant to offer blatantly offensive conflations such as dog breeding and human intelligence, and had rarely spoke of "race improvement," Galton saw domesticated animals (and even plants) as perfectly fine models for social development. At his famous 1904 announcement of the eugenics program at the Sociological Society at London University, entitled "Eugenics: Its Definition, Scope and Aims," Galton thoroughly blurred the line between human and animal in ways which Darwin had avoided. After listing the attributes of the "best specimens" of humanity—"health, energy, ability, manliness, and courteous disposition"—Galton asked his audience to "recollect that the natural difference between dogs are highly marked in all these respects."[77] Where and how Galton saw "manly" dogs was unclear, but it is notable that canines appear in his work far more often than apes or any of the other animals Darwin suggested as humanity's closest kin. When he gave the Huxley Lecture to the Anthropological Society the same year, Galton began his explanation of "Hereditary Variety" by noting that human "faculties" differ as much as "those of domesticated animals, such as dogs and horses," noting that the "variety of aptitudes . . . in dogs is truly remarkable."[78] One might object to such a vision of humanity, but such an attack could not come from the Darwinian flank.

Galton's vision of the biological nature of intelligence was closer to the ideas of his cousin, but Darwin was loath to accept the idea that individual intelligence was the result heredity alone. According to Robert Richards in his history of evolutionary ideas of the mind, Darwin "was acutely sensitive of the social consequences of equating men with animals and therefore mind with brain."[79] Darwin too saw his own success in the terms presented by the Great

Exhibition, which argued that British superiority was built on individual industry rather than good fortune or genes. As Janet Browne has demonstrated, he also believed the *Origin* itself was the result of his great *labor* rather than inherited natural *genius*, and the earliest biographies of Darwin stressed that he was "a hard worker" and a "busy, productive, self-made man" rather than a natural genius on the level of Newton.[80] Galton, whose temperament and concerns were drastically different from his cousin's, had no such scruples about devaluing individual mental effort and assigning genius to the fortunes of the well-bred. As he argued in the preface to the second edition of *Hereditary Genius* (1892), Victorian values that privileged hard work and effort were no longer necessary, as "what was true for the year 1869 does not continue to be true for 1892."[81]

Galton's belief that intelligence was the result of inheritance, not hard work, may have marked the greatest break between eugenics and previous British social sciences. For Jevons, Knight, and Martineau, it meant that their endless efforts to educate the British populace about the laws of political economy were in vain.[82] Economists David Levy and Sandra Peart have shown how the idea that success and intelligence were *earned* was intimately tied to "classical economics," which from "Smith to Mill . . . rejected racial explanations of observed behavior."[83] Though Galton's work hewed close to Darwinian ideas, it must have been difficult for the industrious Darwin to see Galton reduce human traits like intelligence to good breeding, as Darwin's own ideas were very much indebted to the ideas of Adam Smith and the Scottish moral philosophers.[84] Darwin had found *Hereditary Genius* "interesting" and "original" but wrote to his cousin, "I have always maintained that, excepting fools, men did not differ much in intellect, only in zeal and hard work," pointing out that "I still think this is an eminently important difference." Although in the same letter he claimed to be a "convert," nothing in the *Descent* suggested Galton had significantly influenced his cousin's views.[85]

The shift from the Newtonian order of classical economics to the biological order of eugenics did not emerge in an intellectual vacuum, however, and Galton's science of man (like Darwin's) often seemed to reflect the hierarchies and tensions inherent in the British Empire. For example, when Galton gave the Herbert Spencer Lecture on "Probability, the Foundation of Eugenics," he made sure to open with the claim that "we are born to act, and not to wait for help like able-bodied idlers, whining for doles."[86] Galton, who inherited his wealth and spent parts of his early years idling in Africa, also saw eugenics as the perfect vehicle to justify British rule throughout the world. In the speech on "The Possible Improvement of the Human Breed," Galton assured his audience of the importance of the imperial project, that for "no nation is a high human breed more necessary than to our own, for we plant our stock all over the world and lay the foundation of the dispositions and capacities of future million of the human race." In case the point was made too subtly, in "Eugenics, Its Definitions, Aims, and Scopes," Galton claimed his science was necessary in order to "fulfill our vast imperial opportunities."[87] And in contrast to a century of social thought from Helvétius, d'Holbach, and Smith to Condorcet, Mill, and Quetelet, which took the essential equality of individuals as a given, Galton declared, "It is in the most unqualified manner that I object to pretentions of natural equality."[88] As he claimed of "intelligence" in his revised *Hereditary Genius*, "the natural ability of which this book mainly treats, is such as a modern European possesses in a much greater share than men of lower races."[89]

Had Galton's ideas on intelligence emerged only in his theories, he likely would have remained a curious outlier, perhaps widely read like Spencer but today largely forgotten. However, as Wallace and Spencer faded from view, Galton's name became increasingly popular in the scientific world, and his first work, *Hereditary Genius*, was praised more in the scientific press than anywhere else.[90] Indeed, Galton's methodology had a practical influence far in excess of almost any other social science of the age, as his vision of "savage numbers" contributed to some of the greatest horrors of the twentieth century.[91] While the story of how eugenics contributed to deeming certain lives "unworthy" is well known, less noticed is that the practical and conceptual *tools* Galton developed in search of his theory of inherent difference between and within human races have endured in the social sciences, embedding bias in numbers for generations.[92] A partial list of methodologies used in the quantitative sciences of man that Galton developed would include the use of psychological questionnaires, detailed enumerations of physical and mental characteristics, the anthropological lab, fingerprint analysis, photographic arrays, twin studies, the statistical theories of correlation and regression, and the very dichotomy (and phrase) "nature vs. nurture."[93] Rather than revising his ideas, Galton spent fifty years trying to prove the same point in different ways through methodological innovation: most traits and abilities are largely hereditary. As has recently been argued, in his quest for "extreme determinism" and "hard hereditarianism," Galton developed so many practical tools because statistical methodology *was* his theory.[94] As the historian of science Ted Porter noted, like Quetelet's contributions, such techniques and ideas then "moved up" into biology and genetics, showing that now discredited ideas from the sciences of man could find a permanent home in the natural sciences.[95]

While Galton's science of man pointed toward a future of self-reporting and massive data collection on the physical, moral, and mental behavior of human beings, it remained at his death, in some senses, a quaint operation. Galton was one of the last "gentleman" scientists, relying on his inheritance to fund his far-flung operations, but near the end of his life he made two great steps toward embedding his ideas in institutions. The first was the creation of the anthropometrical lab, which debuted at the International Health Exhibition in London in 1884.[96] To come full circle from the Great Exhibition, the 1884 exhibition was held in part with the remaining funds from the 1851 event. Such a setup was perfect for Galton, who created a laboratory to measure body size and strength in 9,337 visitors, all of whom paid for the opportunity. Though he noted that a few "were apparently altogether not sober," it was an "astonishing success," and Galton imagined an "ideally perfect laboratory . . . admit[ting] a stream of persons passing continuously through it" who could perform the tests on themselves.[97] Though Galton only tested physical characteristics and reflexes, one of the visitors to his lab, James McKeen Cattell, would go on to develop the first seeds of intelligence testing in America.[98]

The statistical tools developed by Galton were embraced throughout the world in the social sciences, and in England Galton's legacy was ensured through the donation of a large portion of his inheritance to the founding of a Eugenical Laboratory, a Galton Professorship in eugenics at the University of London, and the first prominent journal of a statistical science of man: *Biometrika*. While Galton originated many of the tools of the social sciences, the institutionalization would be led by his friend Karl Pearson (1857–1936), who, as the first Galton Professor and editor of *Biometrika*, shaped nearly all quantitative social sciences in the first half of the twentieth century. It was Pearson, not Galton, who claimed that "you must breed" intelligence. Education was simply not up to the task:

> We are ceasing as a nation to breed intelligence as we did fifty to a hundred years ago. The mentally better stock in the nation is not reproducing itself at the same rate as it did of old; the less able and the less energetic are more fertile than the better stocks. No scheme of wider or more thorough education will bring up, in the scale of intelligence, hereditary

weakness to the level of hereditary strength. The only remedy, if one be possible at all, is to alter the relative fertility of the good and the bad stocks in the community.[99]

The sciences of man had been trying for over a century to offer a "scheme of wider or more thorough education" to "bring up" intelligence, but by Pearson's time, quantitative and evolutionary theories denied that progress might come through the reform of institutions. While those progressive social scientists who wanted to avoid "alter[ing] the relatively good and bad stocks in the community" might have wished to fall back on Darwin—who insisted that intelligence was merely at the mercy of natural selection—the "simple naturalist" was far from a font for progressive hopes. After all, a book entitled *The Descent of Man* was not exactly what one turned to for optimism. By the time of Pearson, the great hope for an organic science of man that could benefit human mental life through education and other ameliorative efforts seemed at an end.

Conclusion

In his 1907 Herbert Spencer Lecture, delivered at Oxford, Galton summed up how much had changed in the quantitative sciences of man because of the Darwinian challenge to the Newtonian metaphor. As he wrote, "Variability . . . was disregarded by the old method of statistics, that concerned themselves with averages. . . . A population was treated by the old methods as a structureless atom, but the newer methods will treat it as a compound unity."[100] Francis Galton was not alone in redirecting the sciences of man and the geometric spirit away from the harmonious and abstract laws of Newton toward the enumeration of biological traits, nor was he unique in exploring the importance of intelligence for human evolution. Nor, as we have seen, can *Origin of Species* be credited as the sole reason for the shift in social scientific thought from the laws of physics to the laws of biology. The Great Exhibition, which many believed could have been a monument to the Newtonian laws of political science, instead heralded a competitive science that separated races and peoples by national characters. Herbert Spencer too had imagined the social world as analogous to the biological struggle for existence, adding the key component that "intelligence" was the only means to set humanity free. More than any other thinker of the time, he dramatically challenged the harmony and equilibrium of classical political economy, replacing it instead with a progressive vision that benefited the "fittest." Whereas Alfred Russel

Wallace tried to accommodate Spencer's ideas with the clear facts of natural selection, Darwin's great contribution to the sciences of man was to provide an extraordinary amount of evidence to prove common origins among species and a common origin of all humanity. By the time of the *Descent*, Darwin had provided such a staggeringly high number of facts about animal life that most biologists had to concede the common origins of species and peoples, with the attendant conclusion that human intelligence, while impressive, was merely on the same scale with the animal world. Though a blow to Spencer and Wallace, Galton was all too happy to see human intelligence as yet another trait of the animal kingdom, subject to the laws of heredity and good breeding. Galton's science of man departed from the crude mechanisms of earlier ideas—replacing the "structureless atom" with a "compound unit"—but he had helped create a new form of reductive social science.

As opposed to the abstract progressivism of Spencer and Wallace, Galton's creation of a biological and statistical science of man required an extraordinary level of participation by individuals in data collection and experimentation, as he moved from historical surveys of great men to increasingly direct data collection at his anthropometry lab. He claimed, "The Enlightenment of the individual was a preamble to eugenics."[101] Earlier statisticians like Quetelet had been explicit that social physics was meant for governments and not individuals, but Galton believed eugenical research could be deployed by the individual. The truths of eugenics, unlike social physics—and certainly unlike the laws of physics or evolution through natural selection—*required* that people believe. Like Knight and Martineau before him, Galton made his science of man dependent on individuals recognizing the theory. Going well beyond even the hopes for popular political economy in the 1830s, Galton argued that eugenics "must be introduced into the national conscience, like a new religion," and he imagined "eugenics being a religious dogma among mankind."[102] To this end he spent his last few decades involved in a massive project of encouraging the people of England and around the world to self-report nearly every aspect of their physical and mental life, thus allowing the eugenicist to determine their fitness for future reproduction.[103]

Eugenics became the point at which two great trends, covered in the previous two chapters of this book, coincided. The first was the increasing attention to data collection and statistical manipulation first found in Condorcet's social mathematics and Quetelet's social physics. Quetelet had stunning success in creating interest in his various counting schemes, but as seen in chapter 4, he lacked much of a theoretical framework to organize the data into a full science of society. Like the strictest inductivist, he may have imagined the numbers would reveal themselves. Galton demonstrated, however, how numbers wedded to an organic account could provide a more comprehensive theoretical framework for human data collection. The second significant trend was that intelligence was further downgraded. Once synonymous with divine understanding, from Newton through Adam Smith, by the time of Galton intelligence was inseparable from height, weight, and hair color as a simple heritable animal trait. As seen in chapter 5, even at their most "nauseatingly didactic" and condescending core, the popularizers of political economy had imagined the human mind capable of alteration through education. Political economy had increased focus on the individual mind, but even thinkers like Mill, Spencer, and Wallace imagined human intelligence as outside the normal course of biology. Political economy had increasingly shifted its focus from groups to individual intelligence, but Galton's ideas inhibited the reformers because of his claim that each individual's intelligence was locked in at birth. Even genius, where Galton began his studies, could be bred like a prized animal. Through combining the statistical methodology of social physics with the practical intervention of political economy into the minds of its subjects, eugenics managed therefore to make intelligence the centerpiece of one of the most influential sciences of man in world history, even as it became just another biological trait. The young Spencer who visited the Great Exhibition may have bristled at the idea of the mechanical sciences of man reducing the human mind to a piano, but one has to wonder if he had more in mind for his organic social science than simplifying intelligence to a point where it was level to the faculties of a dog.

PART III

Social Science *in* America

7

The Sacrifice *and* Rebirth *of* "Man"

Practicing Sociology and Economics in America

On the altar of her method of study, sociology sacrifices man.
—Ludwig Gumplowicz, *The Outlines of Sociology* (1899)

BY THE END OF THE NINETEENTH CENTURY, AS THE VARIOUS SCIENCES OF MAN had begun to coalesce into the disciplines of the social sciences, the economist and social theorist Thorstein Veblen (1857–1929) had noticed that the era had also given birth to a new kind of "human material" to investigate.[1] As he claimed, in language worth quoting at length:

> In all the received formulations of economic theory . . . the human material with which the inquiry is concerned is conceived . . . in terms of a passive and substantially inert and immutably given human nature. . . . [The] conception of man is that of a lightning calculator of pleasures and pains, who oscillates like a homogeneous globule of desire and happiness. . . . He is an isolated, definitive human datum, in stable equilibrium except for the buffets of the impinging forces that displace him in one direction or another. Self-poised in elemental space, he spins symmetrically about his own spiritual axis until the parallelogram of forces bears down upon him, whereupon he follows the line of the resultant.

While Veblen was only referencing economic theory, the "passive" and "inert" form of "human datum" he described appeared in a host of nineteenth-century sciences covered in the previous three chapters. Indeed,

Veblen's description of a person in "elemental" geometric space strangely recalled Adolphe Quetelet's unpublished notebooks, even down to the reference to the "forces" of a "parallelogram," where a person is drawn and "squashed" by the impersonal forces of bureaucratic work. And Veblen's claim that "man" had been reduced to a "calculator of pleasures and pains" was a direct attack on the economics that William Stanley Jevons created out of social physics. In the popular political science tracts of the nineteenth century, too, people were depicted as victims of the impersonal laws guiding factory work, only a proper understanding of which might make them tolerable. And while Veblen had suggested economics as an "evolutionary science," organic accounts of human intelligence hardly offered much relief from "impinging forces" outside of one's control.

As will be seen over the final three chapters, the move away from the traditional sciences of man based on Enlightenment optimism and progress to the new disciplines of sociology, economics, and psychology hardly offered better prospects for conceptions of the "human material." To substitute the language of AI experts in the introduction to this book, Veblen worried at the end of the nineteenth century that the "human level" was being reduced to a narrow and simplistic idea completely unrelated to how actual humans thought and behaved in the world. As he noted, in contrast to the "isolated" person reduced to simple pleasures and pains, the true "individual . . . enters into each successive action as a whole . . . on the basis of . . . economic, aesthetic, sexual, humanitarian, [and] devotional interests."[2] Human acts were, in other words, irreducible, and Veblen, like Diderot, d'Alembert, the prophets of the past, and Spencer before him, proposed that a proper science of man would study the entire chain of cultural and social influences.[3] His "human material" was not a stable target consisting of predictable laws that could be expressed in mathematical formula but a dynamic and shifting "whole," which represented the constellation of all the historical "institutions" a person had encountered in their life.

Veblen's methodological plan for what became "institutional economics" did not, in the long run, succeed. Rather, economics instead became a "mathematical science" in the middle of the twentieth century, in part through differentiating itself from other forms of social science.[4] Indeed, in historical retrospect, Veblen's ideas seem closer to the "habitus," "thick description," and "milieu" of anthropologists and sociologists or to the "interpretive philosophy" of the 1970s, and his proposal for economics as an "evolutionary science" hardly seems connected to the world of marginal utility, supply-and-demand curves, and the rational *homo economicus* that have dominated economics departments for generations. Yet the professional contexts and guiding ideas of academic economics and sociology were not so far apart at the beginning of the twentieth century. In fact, both sociologists and economists shared similar reforming interests, as they tied their ideas and research to specific plans to ameliorate the difficult lives of average citizens caught up in "modern" civilization. Rather than the harsh reforms of Victorian political economists, or the laissez-faire approach of social Darwinists, many American social scientists at the onset of the twentieth century argued that the "human material" of thought and behavior was not the simplistic product of a series of natural and determined laws but instead the dynamic consequence of people in their historical, cultural, and socioeconomic contexts.

Indeed, by the middle third of the twentieth century, both economics and sociology seemed to have sacrificed the "man" of previous centuries, jettisoning many of the moral and ethical ideas of reform that had accompanied earlier sciences of man. To examine the sacrifice and rebirth of "man" in sociology and economics, this chapter shifts the book's geographic focus again, this time to America, where institutions of higher learning transformed diffuse fields into practical disciplines through the application of methodological tools borrowed from the natural sciences.[5] While both disciplines initially attracted social reformers driven by practical and ethical motivations to improve people's lives, by the beginning of World War II the professionalization of economics and sociology would leave many of the most committed reformers outside academia. Indeed, by the 1950s, moral philosophy, which had animated the idea of a science of man for nearly every figure covered in this study, was cast as the antithesis to proper social scientific work.

Because the formation of American sociology and economics as disciplines are large stories that have been told before, the focus here will be limited to the shifting methodologies and professional standards. For sociology, the story will be told through the experience of a little-known sociologist named Earle Eubank, whose Zelig-like tour of international sociological institutions and personal odyssey in the first decades of the twentieth century provide a guide to the most substantial development in the discipline. Eubank, the son of missionaries, witnessed firsthand how sociology abandoned reform efforts in favor of esoteric theories that pushed individual experience to the margins. While his story offers only a narrow window onto the larger developments of sociology, his detailed personal notes provide rare insight into how discipline formation could transform the lives of social scientific researchers themselves.

Just at the moment when social scientists were expected to become inter-changeable and narrowly defined researchers, Eubank's life and experience are a reminder of the rich complexity of "whole" human lives.

For economics, the story will begin with Veblen and a group of anti-re-ductive thinkers known as "institutionalists," examining how the contest over mathematical and statistical methodologies played out over several de-cades, with particularly attention given to the work of Wesley C. Mitchell, a longtime economics professor at Columbia University who helped create the New School for Social Research and a number of statistical organizations. As economics became more theoretical and mathematical, however, the actual minds and behaviors of people became secondary. Especially in the work of the Chicago economist Frank L. Knight, moral thinking became the enemy of social science, as economics was divorced from almost any "applied" theory. In the move from social amelioration toward abstract theory and mathema-tization, which reached its logical end in the work of Milton Friedman, the real thoughts and actions of human beings were ignored completely. In the disciplines of sociology and economics, the "human level" was either reduced to the point of abstract simplicity or "sacrificed" altogether in the name of proper scientific methodology. At the same time, the broad human material of social scientists continued to be narrowed, as new methodological imperatives meant that researchers too were expected to conform to a circumscribed range of motivations. The human material, which Veblen had imagined as the locus of a complex historical and cultural process requiring rich description and analysis, was reduced in many corners of academic social science to a simplistic set of assumptions or, more remarkably, banished altogether.

The Dead Science: Earle Eubank and the Construction of a Discipline

One of the more bizarre items among the papers of the American sociologist Earle Eubank (1887–1945) is a clipping of a "Ripley's Believe It or Not" car-toon from a Cincinnati newspaper. Towering over an image of Babe Ruth and a drawing of two Civil War bullets meeting in midair, the frightening image of Auguste Comte (1798–1857) looms above the following caption: "Carrying in his crazed brain the seeds of wisdom!" With darkened eyes and a furrowed brow, Comte appears insane, and the note mentions that the "raving maniac" wrote his *Cours de philosophie positive* from memory while in the Esquirol Insane Asylum.[6] Believe it or not? For historians of the social sciences, it is easy to believe, as Comte's emergence as the founder of a science of man is indeed one of the more amazing stories in nineteenth-century ideas.[7] Comte endures in all histories of social thought because of his creation of the French word *sociologie* and because his conception of "positivism" was one of the first systematic efforts to develop a scientific methodology based on facts and experiment alone. In founding positivism as a methodological approach to science, Comte's "ravings" in the *Cours* offered to many the first true science of man that was independent of the geometric and reductive systems of the French sciences of man and social physics.[8]

Comte was certainly an important founder for the field, but Eubank's interest was more personal: the image was based upon a reproduction he had made of a painting in France. As part of a plan to write a two-volume history of sociology, Eubank had traveled throughout Europe in the 1930s to interview and document the people he believed to be the "Makers of Sociology," and the comic is just a small piece of the much stranger story of Eubank's travels through international sociology.[9] A notable moment in 1930s transnational social science, Eubank's image of Comte, and even his larger project to write a history of the "Makers of Sociology," trails a number of events in historio-graphical prominence, falling somewhat far back behind Talcott Parsons's synthesis of European ideas or the influx of Central and Eastern European refugees into American sociology departments.[10] Part of the reason may be that Eubank—despite being one of the earliest graduates of the University of Chicago's influential Department of Sociology and Anthropology, longtime chair of the Sociology Department at the University of Cincinnati, and author of dozens of articles in prominent sociological journals—did not leave much of a scholarly legacy. His most successful work, a textbook called *The Concepts of Sociology* (1932), was used as an example of the "failed" American theoret-ical approach prior to Parsons, and his massive work to compile a history of sociological "makers" was never finished.[11] As a contributor to the internal development of American social sciences, Earle Eubank has likely received the deserved amount of attention.

However, Eubank's efforts to make sense of the field of sociology in the 1930s offer a unique and valuable perspective on the moment in the first decades of the twentieth century when the broad and diffuse idea of a science of man transformed into a professional discipline. The idea that there was a distinct set of knowledge known as the "social sciences" emerged around the same time as American sociology departments, and few were better witness to the consequences of this transformation than Eubank.[12] Sociology as practiced in the first decades of the twentieth century not only threatened to sacrifice "man" entirely as an abstract category, but in many cases it denied the humanitarian and egalitarian impulses that had inspired the first Enlightenment *sciences humaines* and many of the first social scientists. Eubank, like many of the first professional sociologists, began his studies as a means of social work but ended up encountering abstruse and racist theories from Europe. Though Eubank did not precipitate these changes and soon found himself in a profession he did not recognize, his attempts to navigate the demands of a discipline reveal the narrowing possibilities for social science.

Eubank began his project to write the history of sociology on October 25, 1935, when he sent his outline to E.A. Ross, editor of a series on sociology for the publisher D. Appleton–Century. There, he offered a twofold rationale. First, he wanted to record "the essential information concerning the life and writings of the leading persons who have made fundamental contributions to sociological theory." Second, and most importantly, he wanted "to observe the steps by which our present discipline has been built up as 'a slowly evolving aggregate.'"[13] The quotations marks were Eubank's and, from his perspective, it was quite possible to see a "slowly evolving aggregate" for the discipline of sociology in 1935. Eubank had received his sociology PhD (summa cum laude) at the University of Chicago in 1916, just a few decades after the Colby College sociologist Albion Small (1854–1926) was enlisted to start the department, the first devoted to sociology at an American university.[14] The son of missionaries, Eubank had spent part of his early life in Nigeria and the Philippines in service to the poor and had seen sociology as a natural progression in alleviating the hardships of the modern industrial world. He first learned of the discipline through reading a popular textbook, *Practical Sociology*, and chose the field only because an advisor had claimed it was the best path to become a social worker. In 1927, he called the years between 1908 and 1914 a time when he practiced "sociology as a reform and humanitarian movement."[15] As someone who had witnessed the earliest stages of the formation of a discipline, he reiterated to Ross that his book was "not to be a series of detached chapters, but a unified synthesis."[16]

In historical retrospect, however, it is hard to see October 1935 as a moment of slow aggregation or synthesis in the history of American sociology. Just two months after Eubank mailed his proposal, the field suffered a significant fracture at the American Sociological Society (ASA) meeting in New York City, where a major "rebellion" arose against the University of Chicago's domination of the field and recent marginalization of social reformers.[17] The immediate cause of the split had been a suggestion that a new journal replace the *American Journal of Sociology* (the official organ of the ASA and housed at Chicago), but it also reflected a long-simmering debate between academic sociologists over what it meant for sociology to be "scientific." Though sociology had once been a popular field, in the years surrounding the rebellion, ASA membership had fallen precipitously, losing over a third of its members between 1930 and 1937.[18] In terms of public influence, the field was also far removed from its success in the first decades of the century, when Eubank was among hundreds of social reformers who saw the discipline as a path to helping Americans navigate the harsh realities of modern industrial life. Eubank would have been well aware of these challenges, as he was friends with many of the "rebels" and a long-standing member of the ASA. In her detailed anatomy of the event, historian and sociologist Patricia Lengermann even identified Eubank as one of the "rebel fringe" sympathetic to the idea that the "Chicago school" had overridden the field.[19] Seen from this perspective, Eubank's proposed "aggregate" and "synthesis" seemed to be an attempt to imagine into existence a common history in order to defeat the factionalism he witnessed in his own professional circumstances.

While Eubank's hope for disciplinary unity is easy to understand, more surprising may be his desire to break with the university from which he graduated in 1916. Indeed, in the same chart that named Eubank a "rebel," Lengermann listed "Chicago graduates of the 1920s and 1930s" as "loyalists."[20] What changed at Chicago in just five years after Eubank graduated that caused him to rebel? In short, it was William F. Ogburn (1886–1959), the department chair at Chicago from 1919 to 1927. In a presidential address to the ASA in 1929 that synthesized the ideas he had spread at Chicago, Ogburn called on sociologists to abandon the kind of social work that had attracted people like Eubank to the field. In a bid at making the field properly scientific, Ogburn claimed, "It will be necessary to crush out emotion and to discipline the mind . . . to taboo ethics and values . . . and . . . we shall have to spend most of our time doing hard, dull, tedious, and routine tasks."[21] In a direct rebuke to the "social gospel" and reforming efforts that animated the earliest work at Chicago, he claimed in his presidential address, "Sociology as a science is not interested in making the world a better place." Ogburn was insistent that the field could not become immersed in "ethics, religion, commerce, education, journalism, literature and propaganda," but should be limited to "discovering new knowledge."[22] While such a project might indeed be beneficial to establishing scientific bona fides, Ogburn did not present a particularly romantic vision of the sociological research of the future, criticizing the very kind of personalized work Eubank wanted to accomplish in his "Makers of Sociology" book. Eubank had wanted great personal stories for his book, but Ogburn declared,

> There will be no virtue in a merely stimulating article. The sine qua non of scientific publication will be verification and evidence. Verification in this future state of scientific sociology will amount almost to a fetish. There will inevitably be a great many unimportant and uninteresting things verified. Thus science will utilize the dull and uninteresting person, just as logic utilizes the paranoiac, as social philosophy utilizes the fanatic, and as intellectualism utilizes the daydreamer. For science will rest on a base of a great deal of long, careful, painstaking work. And many stupid persons can be careful, patient, [and] methodical.[23]

Eubank's proposed book of interviews seemed to fall under the category of "merely stimulating," and it is easy to see why many sociologists chafed under Ogburn's plan to replace engaged social reformers with "stupid persons" to conduct the "long, careful, painstaking work" of science. The plan to verify a "great many unimportant and uninteresting things" seemed to make almost a mockery of the positivist approach, and the call for "the dull and uninteresting person" echoed the concerns of Bonald and Maistre that education in the geometric spirit alone stunted human intelligence. Condorcet had imagined social science as understandable to all, and Quetelet and his friend de Decker had insisted the social physicist needed to give up some personality in order to study and govern society, but Ogburn's verification "fetish" took on new meaning in a professional discipline. As the historian Robert Bannister and others have noted, Ogburn was at the center of the age of "scientism," where "value-free" data collection and experimentation were imagined as the path toward "objectivity."[24]

Eubank was not alone in his concern over Ogburn's explicit calls to repudiate Chicago's legacy of social work and theory in favor of strict empiricism. In 1972, a sociology professor named Kenneth Barnhart noted in a collection of interviews with Chicago graduates that "one of the criticisms that I had of modern sociology . . . is that they get so wrapped up in gabbedy-goop . . . that the average student can't see any relationship [to] . . . anything that is practical."[25] Echoing Eubank's experience, Barnhart claimed that at Chicago he "had dumped overboard the theology I had grown up with."[26] The rebellion with which Eubank sympathized also began with the complaints of his friend

L.L. Bernard (1881–1951), another Chicago student who had become disillusioned by the direction of the sociology department at his alma mater. In 1928, Bernard received a letter from Chase Going Woodhouse (1890–1984), an economist at Smith College who would go on to play a leading role in the suffragette movement and serve in the US Congress. Aware that Bernard was leading the insurrection against Ogburn's school, she warned that at Smith, the introductory course in sociology had been "entirely revised" in 1919, with "all work on charities and corrections taken out . . . and put into a special course on mal-adjustment." In part, this was because of the work of Frank Hankins, a quantitative sociologist and future president of the ASA discussed in chapter 8, whose own approach to sociology echoed Ogburn's.[27] Woodhouse noted that a course based on Hankins's textbook, *Introduction to Sociology*, "began with a study of biological evolution, went on to Anthropological Data, and then a study of social institutions." She lamented, "No applied Sociology has been taught" because it had been moved to the school of social work.[28]

In 1927, Bernard wrote to his former Chicago professor Charles Ellwood (1873–1946), the inspirational leader of the rebellion, telling him, "It is not the same group who were in charge of the department when you knew it years ago."[29] In far stronger language than Woodhouse, Ellwood declared full opposition to Ogburn's scientism, repudiating the kind of geometric spirit that had captivated reductionist social scientists for centuries: "The erection of physical science methods into a dogma in the social sciences is a betrayal of the social sciences."[30] In an article responding to Ogburn's speech, Ellwood, a former president of the ASA and another Chicago graduate, declared, "Sociology is in danger of becoming a dead science . . . due in large part to the invasion of the spirit and method of the so-called natural sciences." Ellwood's own speech as president just five years prior to Ogburn's had been an impassioned plea against "intolerance" rather than a call for the "fetish" of "verification." As David LoConto has written, Ellwood felt that Ogburn's address had been a "personal attack" and a "death blow" to the sociology he had known.[31] In giving a "valedictory" speech upon his retirement, Ellwood offered a vicious attack on scientism in the field, a denunciation that may have contributed to his virtual anonymity in the history of the discipline today, a marginalization that continues in spite of the fact that he was the most translated American sociologist in the first half of the twentieth century.[32]

Like Ellwood, for Eubank the attack on the "ameliorative" element was personal. The son of missionaries, Eubank had started his career in sociology under the tutelage of Charles Henderson (1845–1915), a former minister and proud champion of engaged social reform whom Eubank called his "lode-star."[33] Henderson's work consisted largely of accumulating statistics about social hardships in Chicago, and his research and teaching were much closer to what is now considered social work than to social science.[34] Though Henderson was part of the original Department of Sociology and Anthropology at Chicago, he had been a minister for twenty years before joining the university. While at Chicago, Henderson had worked with Jane Addams (1860–1935) and members of Hull House, attempting to lessen the effects of Chicago's rapid urbanization and industrialization.[35] Henderson also taught courses at the Social Science Center for Practical Training in Philanthropic and Social Work, co-teaching classes on poverty and dependence with Graham Taylor (1851–1938), the leader of the "social gospel" movement who also held a temporary appointment at Chicago.[36] Taylor himself had consistently trained ministers alongside sociologists and hardly saw the difference between the two, claiming that the "acme of sociology is to develop the life of the individual out of mere self-conscious experience into a personality that shares the life of the whole brotherhood of man." This of course could not be further from Ogburn's call "to crush out emotion and to discipline the mind." Indeed, Eubank's advisor Henderson, along with Addams and Taylor, were precisely the people Ogburn was trying to remove from the world of academic sociology.[37]

As Ogburn knew well, the social work and social gospel of Henderson, Addams, and Taylor were not incidental to the founding of sociology but in fact crucial sources for students like Eubank. As Anthony Oberschall claimed in one of the first surveys of the origins of the discipline, "without the active social support of various reform groups, it is very doubtful that sociology as an autonomous academic discipline would have been established."[38] In a field that managed to combine a critique of industrial society with a progressive interest in social well-being, the reformist aspects of sociology made it the perfect science for someone like Eubank, and a majority of the ASA presidents in the first few decades were either sons of ministers or ministers themselves.[39] Historians have argued that the American Protestant tradition of reform partly explained why, in spite of producing almost no lasting sociological theories in the first three decades of the twentieth century, American schools produced the most sociologists. According to Roberto Sala, this was in contrast to Germany, where theory dominated but where there was not a single sociology professor until 1920.[40] In fact, early sociology was so dependent on social work that Addams—perhaps the best-known reformer of the age—turned down several opportunities to join Small's department at Chicago, choosing to dedicate her time to Hull House rather than teaching and research. As James Chriss points out, for Addams, being labeled an academic "sociologist" likely would have been a step down from her pragmatic work to alleviate poverty.[41]

Yet in spite of his early interest in social reform and Henderson's work with the poor, Eubank finished his studies under Albion Small, the department's founder and a leading theoretician who left a strong imprint on Eubank's much-maligned textbook.[42] Although Eubank claimed he had been attracted to theory in sociology, he left social work only after Henderson's death, claiming that "social problems . . . were my first love."[43] Small had been almost Comtean in his attempt to organize the principles of sociology and spent most of his life writing endless Spencerian histories of different theories, never engaging in any of the practical work of "verification" proposed by Ogburn. Neither did Small participate in the kind of richly detailed and personal industrial studies conducted by Robert Park (1864–1944), the dazzling star of the department who had briefly served as Eubank's interim advisor after Henderson. Park, who was at the heart of the rebel group that rejected Ogburn's programmatic reforms, held deeply "antiscientistic tendencies," which led him to be "openly disdainful of statistical social science."[44] While Oberschall claimed that in the early years, "almost all academic sociologists were textbook writers, not researchers or writers of scholarly publication," many of the textbooks were attempts at theoretical "synthesis" in the vein of the massive projects completed by Spencer and Comte.[45] In moving from Henderson and Park to Small, Eubank therefore observed firsthand how fragmented sociology in American universities had been between the theoretical and practical poles. Like many in the early years of the Chicago school, he entered as a reformer and left a theoretician.[46]

Therefore, by the time Eubank was set to travel to Europe to write his book on the history of sociology, the two traditions he had learned at Chicago—the social gospel movement and Spencerian organicism—were on the wane.[47] Instead, Eubank encountered a rogue's gallery of system builders, organicists, and Comtean cultists. As detailed in his voluminous notes and interviews, Eubank traveled to the home of Comte, going down a rabbit hole of the "leader's" so-called Temple of Humanity. He also reviewed dozens of pages of the French *Revue Internationale de Sociologie* (*RIS*), even though the French sociologists he met—mostly schooled under Émile Durkheim—imagined the *RIS* as something of a joke. In Germany, Eubank did not fare much better, given that few German sociologists felt free to discuss their ideas openly as the increasing intolerance of National Socialism toward systematic sociology had allowed for only organic and racially based eugenical systems. Although he reported having great dinners and long conversations, his confused notes seemed very far away from the committed social worker of two decades prior.

While Eubank's "Makers" project left behind dozens of boxes of notes on the history of sociology, when he returned home to Cincinnati, he was unable to finish the project. In part, this was because of dedicated efforts to save "Tosha," a Czech sociologist named Anton Obrdlík whom he had met at the end of his travels. In contrast to sociology in America and Western Europe, the "practical sociology" of Czechoslovakia focused on the ameliorative and personal efforts, and Eubank recounted an easy rapport with many of the most important names in the field. Indeed, when he returned to America, Eubank almost immediately dropped his project when he learned that Czech and German sociologists were seeking positions in the United States to avoid Nazi persecution. After a long and costly attempt to bring Tosha and his family to America, Eubank finally prevailed upon his friend Kenneth I. Brown, the president of Hiram College, to hire Obrdlík. By January 1940, the Obrdlíks had made it to New York, with Eubank learning that Tosha had claimed that "the Nazi's allow[ed] me to leave with exactly $12."[48] Obrdlík had also learned that the Eubank family had paid for his travel, to which Eubank replied, "It has been one of the great pleasures of our lives to borrow money to do this."[49]

Eubank died five years after helping rescue Tosha, and his history of sociology was never completed. Yet it seems that his turn toward practical work to help individuals brought comfort late in life. As he claimed later of his decision to enter sociology: "Social work as a profession had not yet come into my line of vision, but if it had done so, I should have definitely said then, as I did later on, that social work was the field which I desired to enter."[50] Eubank set off in the summer of 1934 in order to explain the origins of a discipline that, at least in America, represented the strongest link between the twentieth-century social sciences and the classical nineteenth-century sciences of man. Economics, psychology, anthropology, and other twentieth-century disciplines all drew from the same well of nineteenth-century thought, but sociology in particular had kept alive the vision of French and Scottish Enlightenment authors that the social sciences could be as rigorous and all-encompassing as any natural science. Yet by the time of Ogburn's presidential address, sociology no longer had room for the broad reforms and humanist motivations of Enlightenment social science. There was grand rhetoric about progress, but there was no evidence that anyone was attempting to improve human happiness, as everyone from d'Holbach to Smith had wanted. Compared to the verification "fetish," the reform efforts of Quetelet, Knight, Martineau, and even Galton appeared far closer to the social work that had inspired Eubank to be a sociologist. But as Rousseau, Diderot, d'Alembert, and the counterrevolutionaries might have warned, a narrow methodological plan led only to the "dull and uninteresting persons" of twentieth-century social science work.

The "Thin and Formal Character" of Economics: From Social Physics to Positive Science

When Earle Eubank and other social reformers joined American sociology departments in the first decades of the twentieth century, they were not alone in seeing the social sciences as places to do practical work for the moral betterment of the nation. For these social scientists, to study human life scientifically meant to investigate moral and social conditions as lived by actual human beings. As Marion Fourcade pointed out in her history of the transformation of economics in American universities, these schools in the early nineteenth century had been built on reformist ethics, with the "vast majority of faculty . . . involved in preaching and missionary work."[51] Political economy was taught as a "branch" of moral philosophy, and the "social sciences . . . were thought to embody the highest moral purpose on which new academic institutions claimed to be built."[52] To give just one example of this kind of social science, at the University of Wisconsin Richard T. Ely (1854–1953) promoted the idea of economics as a science based on the possibility of intervention and reform. Drawing on the work of the German Historical School, which stressed that all social relations were products of an organic unfolding of history, Ely and others imagined economic "institutions" as grounded in culture and history, and that understanding the history of these relationships was the goal of social scientific investigation.[53] Ely, who had studied in Germany and been a member of Episcopalian Church, has been described as a "social gospeler, economist, and apostle of social reform," and his vision of economics as a social science was similar to the aims of early American sociology.[54] Like Eubank, L.L. Bernard, Charles Henderson, Robert Park, Jane Addams, and the first classes of sociologists at Chicago, Ely and later institutional economists offered the possibility to transform lives through detailed examinations of the actual state of life under the economic conditions of modern industrial capitalism.[55]

Fourcade and others have also shown that economics as a whole diverged from other sciences of man in the first decades of the twentieth century, as it "grew more distant from the rest of the social sciences" and "became increasingly reliant on mathematical formulation."[56] While sociologists like Ogburn banished morality from sociology in the name of experimentation and verification, economics banished moral philosophy in the name of mathematics. Both methodological challenges, however, assumed that moral and ethical thinking were opposed to "objective" science. Economics in the first half of the twentieth century saw a protracted struggle over the nature of what Veblen called the "human material" as more historical and descriptive economists critiqued the mathematical and mechanical drift of the field. While key institutionalists like Wesley Mitchell argued for rich statistical and descriptive accounts that allowed for more ameliorative interventions, economists like Frank Knight condemned the idea that economics as a social science should have any relation to practical interventions. By the time of the influential work of Knight's student Milton Friedman, actual human motivations had been subsumed in the quest to make economics a "pure science." Economics, which would become the most successful social science of the century, showed that the path to influence and success was not just to reduce actual human acts and thoughts to simplistic laws but simply to ignore the reality of human existence altogether.

The University of Chicago economist Thorstein Veblen recognized early on the connection between the new methodology of economics and the classical mechanism of the old sciences of man. As Veblen noted in "The Place of Science in Modern Civilization" (1906), what he called the "colorless mathematical formulation" relied on an "essentially metaphysical" belief in mechanical "causation."[57] For Veblen, this led to an "impersonal matter-of-fact" analysis of economic life, where the "machine process . . . displaced the workman as the archetype" for "scientific investigators."[58] While Veblen professed to have no opinion on the worth of these various archetypes, he did warn that such a form of analysis might not capture the wide variety of human life and thought throughout history. Reflecting the success of the kind of abstract "normal man" developed by Quetelet and Jevons, he asked: "How far is the scientific quest of matter-of-fact knowledge consonant with the intellectual aptitudes and propensities of the normal man?"[59] Veblen, who rejected the inherent superiority of modern Western life animating so many social theories of the age, saw the "impersonal" modern economic man as just one more cultural product in the history of humanity. Scientific knowledge was certainly dominant in many fields, but the "machine process" was merely taking its place in a long line of mythologies people had constructed to make meaning of their lives.

For Veblen, the great mistake in economics at the turn of the twentieth century was in treating "matter-of-fact knowledge" as some kind of transhistorical given or, worse, normative good. For example, Jevons had claimed in *Principles of Science* that "the first principles of political economy are so widely true and applicable that they may be considered universally true as regards human nature."[60] Such transhistorical and transcultural constants were tied to mathematics, as foundational figures like Jevons believed political economy had a "mathematical character," and he had even claimed that Adam Smith's "reasoning was really mathematical in character."[61] The "marginal revolution," which assumed human actions to be based on maximizing pleasures and minimizing pains, occurred in part because Jevons's work "radically changed the way man's mind was analyzed."[62] While the British economist Alan Marshall (1842–1924) had tried to temper the extremes of Jevon's "hedonistic calculus"—noting that economists should "burn the mathematics" after they confirmed their theories—the marginal revolution ensured that key aspects of social physics made their way into British and American economics classrooms. Statistical studies, lawlike patterns, mathematical laws, and simplified individuals would take the place of historical investigations into the lives of actual economic actors.

Veblen was not alone in criticizing the idea of "lightning calculators of pleasure and pain," or the "hedonistic preconception" that did not allow for growth or change in human nature. Just a decade after Veblen, fellow institutionalist Wesley C. Mitchell (1874–1948) warned in *The Rationality of Economic Activity* (1910) of the economist who "treats the concepts which modern men have gradually learned to use as if they were a matter of course . . . something generally human." In more transparent language than Veblen, Mitchell claimed that the rationality praised by many economists was "an acquired aptitude" and not a "solid foundation upon which elaborate theoretical structures may be erected without more ado."[63] As the economic historian Malcolm Rutherford noted, Mitchell exhibited "a concern with the reductive aspects of the psychological version of individualism often found in orthodox economics."[64] Or, in Mitchell's words, taking humans out of their institutional context was to make a "thin and formal character in comparison with the heir of all the ages."[65]

While Veblen had critiqued the entire approach of neoclassical economics, Mitchell did not disdain all "colorless mathematical formulations," comparing what he called the "statistical" approach favorably against the "mechanical." In his 1935 presidential address to the American Economic Association, cofounded by Richard Ely, Mitchell explained how statistics could be pursued as an independent process of acquiring economic knowledge. Echoing the call of d'Alembert for an "analyst" over a "geometer," and anticipating the approach of statisticians like John Tukey (discussed in chapter 9), in "Quantitative Analysis in Economic Theory" Mitchell explained how "quantitative workers" needed to go beyond being servants for lawlike theories of mathematics. In particular, he critiqued Jevons's "hedonistic calculus," for which he claimed there was "little likelihood" of "quantitative proof."[66] While Marshall had tried to replace a calculation of pleasure and pain with the "force motive" of money, Mitchell asked if there was "a better chance that we shall attain a statistical measurement of the force motives than that we shall measure pleasures and pains?" His answer: "I doubt it."[67] Noting the impossibility of statistical work to prove or disprove a theory on its own, Mitchell claimed that "interpretations are something which the theorist adds to the data, not something which he draws out of them." Adopting a bit of Veblen's rhetorical flourish, and foreshadowing critics of the high level of abstraction taken in later neoclassical work, Mitchell added, "In the present state of our knowledge of human nature," the interpretations of Jevons and Marshall "smack more of metaphysics than of science."[68]

Instead of high-minded theorists who might produce a *Principia* of economics, Mitchell suggested that "quantitative workers" in economics would have to produce thoughtful and carefully aggregated studies of life under various economic conditions. The big theories of classical economics—"reasoning on the basis of utilities and disutilities, or motives, or choices"—would, in Mitchell's view, "drop out of sight in the work of quantitative analysis."[69] Contrasting the coming age of data accumulation with the past age of genius, Mitchell predicted that "the literature which the quantitative workers are due to produce will be characterized not by general treatises, but by numberless papers and monographs." In this world, "knowledge will grow by accretion."[70] Mitchell's view borrowed from Quetelet's plan for social physics, which stressed collective effort over individual genius, but instead of the endless production of "average" researchers, Mitchell expected future quantitative workers to be judged according to their "creative capacity." Though they might produce "numberless monographs," they would not be Ogburn's verifiers or Quetelet's average men, as the reformist and practical spirit of Mitchell's methodological approach allowed for moral and ethical thinking to guide the research. Indeed, Mitchell's vision for economics as a social science managed to cleave apart the idea of the "mechanical type" and the quantitative approach that had united in social physics, casting almost all of Jevons's and Marshall's work in the disgraced mechanistic tradition of Smith and Ricardo while allowing for a return to something like "descriptive statistics." As he noted at the end of his presidential address, data collection and analysis would triumph, as "more abundant and more reliable data" would make "obsolete . . . the qualitative work of Dr. Marshall and others."[71]

Mitchell's plans to aggregate data, along with the work of later institutionalists like John R. Commons (1862–1945), led to a number of important institutions—including the National Bureau of Economic Research (NBER) and the Brookings Institute of Economics—and helped shape the research focus for the Federal Reserve.[72] At universities like Columbia, Wisconsin, and Texas, institutionalism held sway for decades, producing large and imposing documents of statistical accounting for academia and government offices. As Elizabeth Popp Berman claimed in her history of the rise of an "economic style" that opposed rich cultural and historical economics, the institutionalists had their most important success in the United States during the 1930s when the Roosevelt administration looked for practical solutions to the Depression. She noted that many economic policies from the New Deal through Lyndon Johnson's Great Society were "not grounded in the economic reasoning of efficiency and cost-effectiveness" but based on "universalism, equality and rights."[73] Although they might be considered "noneconomic values," this would have been a surprise to most economists and political economists up until the twentieth century; since the French and Scottish Enlightenment, the language of rights, laws, and morality *were* economic values.[74] As Berman states, however, the great shift in twentieth-century social science was to define the "economic style" as antithetical to talk of moral rights and laws.

Institutionalism triumphed in economics departments between the wars, but the battle for the "soul of economics" shifted after World War II, as a number of criticisms emerged of the kind of anti-theoretical proposals promoted by Commons, Mitchell, and the NBER.[75] Most influential was a 1947 review article entitled "Measurement without Theory," which attacked an NBER-produced volume coauthored by Mitchell. Written by Tjalling C. Koopman (1910–1985), the review seemed to approach the science of economics through the lens of the geometric spirit from two hundred years prior. After first grafting a reductive schema of Keplerian and Newtonian mechanics onto the NBER's presentation of business cycles—which Koopman admitted was not in the text itself—he then criticized the authors for failing to advance beyond the "Kepler stage" of empirical research to the level of Newton.[76]

Contrasting the ability to measure business cycles with "planetary motion," Koopman, who would win the Noble Prize for economics in 1975, used scare quotes to mock the "scientific 'strategy'" of the book, in which he complained that "measurement and observation preceded, and [was] largely independent of, any attempts toward the explanation of economic fluctuations."[77] Mitchell had argued that empiricism was more scientific than abstract theorizing and mathematical formulas, but Koopman's work helped reorient economics back toward theoretical hypothesis, embedding the phrase "measurement without theory" into a generation of textbook discussion of the institutionalists.[78]

As important as Koopman's work was in criticizing Mitchell's quantitative approach, less noticed was the concomitant attack on economics as a moral science launched by Frank Knight (1885–1972), an economist and social theorist who began his work as an institutionalist but became crucial to the later Chicago school of economics embodied by his successor and student, Milton Friedman.[79] While Knight left behind a host of formal contributions to economic theory, his most indelible legacy was to try to rid economics of "instrumentation," or the attempt to apply economic theory to real-world situations.[80] Though it has been claimed that "applied theory is largely missing in Knight's work," it might be more accurate to say that applying economics to ameliorative efforts was antithetical to his entire belief system.[81] Like many social scientists of the era, Knight was horrified by the "failure" of social science to predict or rein in the perceived massification and irrationality of the two World Wars.[82] Even before the First World War, Knight had criticized the interventionist aspects of economics. In "Social Science and the Political Trend" (1934), Knight claimed that the war and subsequent depression had led to the "discrediting and discarding of old economic and political points of view . . . and the rise of a new kind of leadership." For Knight, this included the "'New Deals' in Germany and the United States," which "abandoned" appeals to "logic" and "reason." Though conceding that Hitler was "perhaps more frank in his call to his people to 'think with their blood,' than Roosevelt," Knight nevertheless believed the interventions of institutionalists during the New Deal were "at bottom the same thing."[83]

The New Deal in America caused Knight to be concerned about the extension of social scientific ideas, but the "failures" of the Second World War led him to question the social sciences completely. In "Fact and Value in Social Science" (1947), Knight located the "root fallacy" in the belief that the social sciences imagined they "should be or can be a science in the same sense as the natural sciences." Social reformers who had tried to "parallel the modern achievements of the natural sciences" would find no Newton, as "no such revolutionary results were or are possible."[84] Knight was far from alone in this critique, but what separated him from other critics was that he located the largest failure of the social sciences in aping the "pragmatism" of the natural sciences, where sophisticated technologies had essentially proved correct the theories of physics, chemistry, and biology. Unlike Koopman's argument that institutionalism lacked a "Newtonian stage," Knight attacked the heart of the geometric spirit itself. In a broad critique of the entire history of attempts to extend the methodologies of the natural sciences to the study of humanity, he joked that the "best minds of our race" believe in the maxim "that since natural objects are not like men, men must be like natural objects."[85] While Knight agreed with Jevons on the importance of classification, he bluntly declared that "there simply is no real measurement of distinctly human or social behavior" and that "the data with which social sciences are concerned are themselves not objective in the physical meaning."[86] Challenging the entire project of statistical institutionalists, Knight claimed, "What is called measurement in the social sciences . . . is the averaging of estimates, and even the use of the term measurement is a misnomer."[87] Even biological approaches toward social science focused on "struggle and adaptation" were impossible, as he claimed it was "sheer dogmatism to assert" that human behavior could be reduced to the "purely physical."[88]

Knight might sound like a modern critic of reductive social science, but he was far more enamored with the possibility of economics as a "pure theory," which reduced economics to a study of abstract "efficiency" regardless of its connection to the real world. Although he did not use mathematical notation often, Knight cleared the ground of all interventionist, statistical, and moral approaches to social science for an even more audacious claim: the science of economics need not be concerned with actions at all. Because predictions could not be made based on empirical study, they should move to the realm of what has been called "pure science." Upending three centuries of belief that the study of humanity was more complex than nature, for Knight it was the *natural sciences* that were bogged down with reality, while "pure" economics could exist without reference to nature at all. Knight's own contributions to this kind of pure theory are somewhat muddled, confused, and paradoxical, but his efforts at the University of Chicago to place economics beyond even experiment and observation had profound influence.

Knight's work is largely forgotten today, but the ideas of his student and successor at Chicago, Milton Friedman (1912–2006), exert a far stronger hold on modern economic thought, especially at the level of public policy. Friedman has been the subject of praise and scorn as one of the chief contributors to the Chicago school of economics and because of his notorious work to influence foreign governments, though the focus here is limited to his methodology for economics as a profession.[89] Published in 1953, "The Methodology of Positive Economics" has been described as "the most cited, influential, and controversial piece of methodological writing in twentieth-century economics."[90] There Friedman admitted that dealing with the "interrelations of human beings" created a number of "special difficulties," and that in order to be effective at making predictions, economics needed to be more than "disguised mathematics."[91] Disagreeing with Knight, Friedman also recognized the perils of making a "pure science" like mathematics, which was nothing more than a "tautological" exercise without the ability to make predictions in the real world.[92] While Knight believed that economics might be above *all* confirmatory processes, Friedman claimed only that the assumptions of economic theory need not be tested. In a harbinger of the success of modern theories of artificial intelligence, Friedman's influential claim that a scientific methodology need only *predict* human actions proved that a divergence between the reality of human experience and the purported mathematical accounts of that experience was no vice.

In "Positive Economics," Friedman's most novel declaration was that the "assumptions" of economic hypotheses "cannot possibly" be tested by the actual working of the human mind, nor need they be "realistic."[93] While neoclassical economics had received significant criticism because actual businessmen did not report thinking as the models said they would—as narrow calculators who strove to maximize efficiency and profit—Friedman argued that there was a "basic confusion between descriptive accuracy and analytic relevance."[94] In essence, economic assumptions could be empirically wrong but theoretically correct. Institutionalists had produced hundreds of dense histories and ethnographies of business practices, but such rich accounts of actual economic behavior were unnecessary. If something like utility maximization predicted accurately what economic actors did, then it simply did not matter what actual businesspeople imagined themselves to be doing. Friedman gave a number of analogies to prove how a scientific theory could be "correct" even if its underlying ontology were false; perhaps the most illuminating was the example of a social science researcher attempting to study an "expert billiard player."[95] For Friedman, the best means to attempt to predict the behavior of the billiard player—i.e., what shots he would attempt—was to imagine the player "knew the complicated mathematical formulas that would give the optimum directions for travel, could estimate accurately by eye the angles . . . and could make lightning calculations from the formulas."[96] While Friedman admitted it was unlikely that actual billiard players performed such calculations, an economist writing *as if* the player did so was the best route to predicting each shot. Therefore, social scientists could proceed as if economic transactions were governed solely by individuals interested in maximizing returns, even if such individuals were motivated by a host of other reasons.

Though critiques of the "economic style" of utility maximization often focus on the problem as a matter of philosophy or public policy, others at the time cautioned that it might be a problem for the students themselves. After all, while a positivist approach based in mathematical formula may best predict the course of a game of billiards, this did not mean it was the best way to train expert billiard players. Indeed, as the British economist Lionel Robbins (1898–1984) worried two years after Friedman's essay was published, generations of students were being taught such high-minded theory without any real understanding of what they learned. In his Presidential Address to the Royal Economic Society in 1955, Robbins noted that economics was "a subject for grown-ups" and "no simple proposition in economics is likely to be true" without understanding "whole complex of assumptions."[97] Though Robbins did not doubt the gains of economic theory, he was concerned that "the unreality of its contents" was too much for many students to handle in British schools and universities. Suggesting that simple institutional economics, or even "civics," might be taught to most students instead, he asked why the mathematics that supported neoclassical economics was being taught at all: "What good does it do the [poor] student . . . to spend long days and laborious nights poring over the arcana of utility theory or the geometry of the analysis of the firm?" Repeating critiques of narrow education that stretch from Rousseau to modern champions of the liberal arts, Robbins declared, "A man is not necessarily a worse citizen if he is not at home in manipulating indifference curves, production opportunity slopes and the like." Instead, Robbins suggested that students not at elite schools should not be taught any theory at all and that teaching "mathematical economics" was "inadvisable."[98] In describing the struggles of students being dragged down by a subject they barely understood, and for which they could find no practical use, Robbins repeated the concerns of Eubank's fellow sociology students at the University of Chicago, even as he was addressing a particularly British problem. While Friedman may have been correct that the assumptions of theoretical economics need not have any connection to actual human behavior, Robbins's Presidential Address revealed a recurring theme in the history of reductive social science, one that has endured well into the twenty-first century: taught widely enough, such methodological approaches could create—rather than reveal—the exact kind of human material it had attempted to study.[99]

Ironically, although Robbins had called for a broad education for the average student, he had done as much as any economist to separate his field from other disciplines, declaring that economics was separate from the "classifications" of natural science. As historians have noted, Robbins's famous "definition" of economics—which held that economics could be applied to any field that dealt with resource scarcity—was prominent in textbooks throughout the second half of the twentieth century and helped contribute to the idea of economics as *the* serious social science.[100] While the implications for economics as a discipline are outside the scope of this study, what does matter is that the influential arguments of Friedman and Robbins in economics divorced social scientific theory from *any* connection to Veblen's "human material." In the words of one critic, by reducing all economic behavior to a simplistic intrinsic rationality and avoiding "any and all psychological sociological commitments," economics took "the path of operational meaninglessness."[101] Whereas past formulations of "economic man" and the "hedonistic calculus" could be found defective, the economist and future politician A.G. Papandreou (1919–1986) noted disapprovingly that "the construction in its present form is entirely beyond the reach of criticisms."[102] Mitchell had complained of the "thin and formal character" of previous neoclassical arguments, but for Friedman there need not be *any* character at all. Ironically, it was just this formless nature that allowed economics to be applied to all aspects of human life, even as its assumptions need not be based in the reality of those lives.

Conclusion

Theorists from Marx to Veblen to the present have argued that technology might change the nature of "human material," but Robbins did not need to invoke the machines of the Industrial Revolution or the Gilded Age to worry about "long days and laborious nights." Like Ogburn's plan for sociological practice, the routinized and mathematized program of economics education might have had an impact on the mind that rivaled the stunted education of children in British factories. As Fourcade has argued of the ability of social sciences to *create* the very world they attempt to describe, "economic technologies have been brought to bear on societies in increasingly meticulous ways, thereby manufacturing the social conditions that make economics calculation possible."[103] In this case, the "economic technology" was teaching neoclassical economics itself. Not only had social scientists erred in assuming a transhistorical constant that supported all economic and social action, but they failed to recognize that *they too* were products of a long and complex history of institutional culture and practice.

Here again, Veblen was prescient about how cultural and social institutions of science could produce, rather than describe, the objects of investigation. As he noted, the "canons of validity" for a scientist were "made for him by the cultural situation." Even more, what is often considered objective social science was for Veblen only "habits of thought imposed on him by the scheme of life current in the community in which he lives." While Veblen argued that the "scheme of life" was largely "machine made," he was well attuned to how the life of the social scientific researcher was transformed through such "habits of thought." Echoing the "dull" and "tired" work that Ogburn had called for, and the "long days and laborious nights" of the average economist at a second-rate school, Veblen noticed an odd feature of modern scientific life. Even as scientists seemed to be at the forefront of civilization, their "spiritual life" revealed no such grandeur. In Veblen's phrasing, "It may seem a curious paradox that the latest and most perfect flower of the western civilization is more nearly akin to the spiritual life of the serfs and villeins than it is to the grange or the abbey."[104] Like the Chicago sociologists who were expected to partake in the dull and tedious practice of verification, the social science of economics dragged down both the practitioners and the object of its study to the fallen state of mathematical theory.

The process of creating social scientific disciplines in America like sociology and economics therefore did not just introduce a new set of techniques to study human action and thought. Rather, like the sciences of man before them, such methodological choices had profound implications at several levels. First was simply the kind of "man" that could be studied. Ludwig Gumplowicz suggested that this abstract man was "sacrificed" at the "altar of sociology," in part because the actions of individuals seemed caught up in larger forces in the theoretical development of the field. Yet the methodological approach was recursive, influencing the researcher as much as the object of research, as sociologists like Earle Eubank were forced to abandon the kind of social work they had initially wanted to do.[105] So too in economics was the kind of "social gospel" work of Richard Ely banished, with ideas of morality, ethics, and purpose reframed as opposing the objectivity of the discipline. While past sciences of man—from those of Helvétius, Condorcet, and the Victorian reformers through the first American economists and sociologists—had been explicit about moral commitments and the connection of social science to practical reform, modern social science disciplines eliminated what had once been among the chief motivations to study humanity.

Perhaps unsurprisingly, the great energy and enthusiasm that had supported the early days of economics and sociology in America did not disappear completely. Rather, careers in social work and civil service allowed committed reformers opportunities to practice the kind of social science they had imagined outside of academia. Yet, as sociology and economics abandoned their previous moral commitments, it meant there were fewer academic social scientists deeply committed to studying how actual human beings lived and thought, whether through the practical social work of reformers or the deep empirical studies of institutionalists. Instead, in the first three decades of the twentieth century, sociologists had turned toward either dreary empirical work or the high theory of Comte and Spencer, neither of which had room for the kind of studies produced by Robert E. Park, Charles Henderson, Jane Addams, or the reformers who made up the first cohorts in sociology departments. For their part, economists by the time of Knight and Friedman had determined that investigation into the actual workings of the human mind was unnecessary for successful theoretical predictions. As will be seen in the next chapter, such a retreat left open the possibilities for new powerful theories of human thought and action. Two of the most successful were behaviorism and functionalism, which each experienced a tremendous surge of interest followed by a precipitous fall in the middle of the twentieth century. At the same time, however, there also emerged a more enduring approach to studying the human mind, one which seemed to capture the excitement of the Enlightenment sciences of man without having to bother with humanitarian commitments among its practitioners. It was called artificial intelligence.

8

Social Science *by* Other Means

Artificial Intelligence in Theory and Practice

> Even the social scientist, though . . . he studies his data
> objectively and statistically, has as yet very little power
> of prediction.
> —Frank Hankins, *American Sociological Review* (1939)

IN 1939, JUST A FEW YEARS AFTER EARLE EUBANK RETURNED FROM HIS
European odyssey, his good friend Frank Hankins (1877–1970), president
of the American Sociological Association, stood before his membership to
deliver another in a series of grave warnings to social scientists about the
failures of the discipline. Nearly two centuries after Fontenelle, Condillac,
Helvétius, d'Holbach, Smith, Condorcet, and others seemed to lay out the
real possibility of understanding a well-ordered world of human thought and
action through the deployment of the tools and methodologies of the natural
sciences, Hankins remained skeptical. In addition to believing that the "social
scientist . . . has very little power of prediction," Hankins also told the group
that just "one conclusion seems certain . . . that in the long run we reach a
wholly unexpected social condition."[1] Hankins, who was one of the leading
statistical sociologists in America and had written the first English-language
biography of Adolphe Quetelet, did not believe that the methodological approach of following the natural sciences was at fault. Rather, like William
Ogburn's speech to the ASA ten years before, he attacked what he called the
"emotional" strain that had distorted research, where social scientists inserted

their own political hopes for the realities of human existence. Criticizing the naive dreams of an earlier era, he challenged the presumption that had guided the sciences of man since Condorcet and the Enlightenment, that "increasing knowledge of social causation would lead to indefinite improvement in human affairs."[2] Like the economist Frank Knight, Hankins believed that dream had ended in the trenches of the First World War. As he claimed, it was "needless to say" that "prewar social thought now sounds strange and unreal. . . . [Its] outlook has darkened perceptibly."[3]

While the prospects for a progressive social science that could improve "human affairs" had certainly darkened, the same could not be said for a host of new fields that emerged around the same time, including the behavioral sciences, functionalism, modernization theory, and artificial intelligence. Those who tried to follow in the "emotional" spirit of the Enlightenment sciences of man found few supporters in economics and sociology departments—with a mathematized economics in particular seeming to distance itself from even the idea of a progressive social science—and these new approaches attracted significant resources from the federal government, private industry, and the nation's colleges and universities, helping to form what J. William Fulbright (1905–1995) would later call the "military-industrial-academic-complex."[4] Although significant work has been done on how these disciplines emerged in relation to the political realities of the Cold War and decolonization, less attention has been paid to what happened to the ideas that went away.[5] As this chapter will discuss, while the social sciences were stretched to their reductive and abstract limits in the methodologies of behaviorism and functionalism, in the middle of the twentieth century many of the animating ideas of the Enlightenment sciences of man were reborn in the work of artificial intelligence, only without the moral underpinnings of universalism, human rights, and equality.[6] By usurping the enthusiasm, hope, and telos of the classical sciences of man, and transferring the belief in certainty, prediction, evolution, growth, agency, and understanding from man to machine, early artificial intelligence researchers kept alive the fading dream of the classical sciences of man.[7]

The postwar era gave rise to a number of new ideas, disciplines, methodologies, and institutions dedicated to the social sciences and artificial intelligence, but this chapter focuses on three key figures: the Yale sociologist and psychologist Clark L. Hull, the Harvard sociologist Talcott Parsons, and the British mathematician Alan Turing. Although all three did groundbreaking work in the 1930s and 1940s, and all had extraordinary success and influence in trying to imagine a science that explained human action, they marked divergent paths for the social sciences. For Hull, social science research combined Ogburn's legion of verifying science workers with a return to the mechanical science of man of d'Holbach, as he believed that all human activity could be reduced to simple mechanistic stimulus and response behaviors derived from endless experimentation, data collection, and equation writing. Though at one point Hull was the most influential social scientist of the era, his program flamed out in spectacular fashion by the 1950s. For Parsons, a non-tenured economics professor who became head of Harvard's Department of Social Relations and was an advisor to dozens of government committees, the path was much slower. Rejecting the kind of laboratory work favored by behaviorists, Parsons developed perhaps the last paradigmatic social theory in the Western world, one that attempted to explain all human actions on earth *without* reducing them to biological, chemical, or physical actions. Although his ideas, which came to be known as structural functionalism, were foundational for Cold War "modernization theory," like Hull, he is rarely read today and was considered a pariah in the social sciences for decades. Unlike Hull and Parsons, of course, the fame of Turing continues to grow. While once ubiquitous books of social theory from the 1930s and 1940s gather dust in forgotten corners of libraries, Turing's work to create an "intelligent machinery" during these same decades endures in a steady stream of books, films, and academic conferences. Away from the confounding and unpredictable actions of humanity, the sciences of man could flourish.

"Never Think": Clark Hull, Behaviorism, and the Legacy of the Psychic Machine

In *Behavior of the Lower Organisms* (1906), the biologist H.S. Jennings (1868–1907) articulated what seemed to be yet another stunningly reductive model for the human mind. Echoing the organic science of his namesake—Herbert Spencer—Jennings claimed, "What we call intelligence in higher animals is a direct outgrowth of the same laws that give behavior its regulatory character in the Protozoa."[8] In work credited with giving rise to the use of "behavior" in psychology, Jennings imagined a singular organizing process for all organic material, beginning with the movements of unicellular organisms and extending all the way to the most complex elements of human thought and action. While Jennings was able to incorporate far more empirical data than Spencer, he shared the belief with the organic sciences of man that a universal process of selection and adaptation led to the development of all living action. Much as Spencer had seen intelligence in the coordinated activities of primitive life forms, Jennings defined intelligence as a coordinated system of "regulated behavior," noting that "intelligence is commonly held to consist essentially in the modification of behavior in accordance with experience." Though he admitted such work might seem "heretical," he reminded his readers that "there seems no reason to suppose that regulation in behavior [of intelligence] is of a fundamentally different character from regulation elsewhere."[9] In other words, there was no reason to believe that what went on in the mind was any different from what went on in a cell.

Jennings was not the first thinker to imagine the building blocks of intelligence in single-celled organisms like protozoa, but this did not necessarily have to lead to a simplistic reduction of human thought to biology. For example, Jennings was influenced by the organic ideas of the German naturalist and eugenicist Ernst Haeckel (1834–1919), who had imagined unicellular beings possessing a *Zellseele*, or "cell soul"; and he had also trained in the German laboratories of one of Haeckel's students, Max Verworn (1863–1921). For his part, Verworn claimed that "every elementary part of protoplasm has its own autonomous psyche" and Jennings's work in *Behavior* was connected to a much richer and more complex history of "physiological psychology."[10] As the historians Judy Johns Schloegel and Henning Schmidgen have shown, Jennings's work also followed the innovative and less reductionist work of the French *aliéniste* Alfred Binet (1857–1911), who believed in cellular psyches but also in a "sort of intelligence that could not be reduced" to a more simplistic level.[11] Binet, who would go on to create the first true intelligence test, had instead imagined intelligence as the complex process of responding to an environment. Microscopic organisms like protozoa offered only the simplest examples of learned behavior rather than the kind of rigid stimulus/response model suggested by later behaviorists.

As much as Jennings drew from earlier, non-reductive methodologies, his book was influential in later reductive social science for two reasons. The first was that he was an American biologist who could offer his social science colleagues a compelling metaphor for human thought and action—and scientific bona fides—in their quest to attract interest and resources from universities and foundations. The second, and most important consequence for the development of the social sciences, however, was that he used "behavior" in his title at a time when the term was basically unknown. "Behavior" was a strange word to use in 1906, especially for psychology. "Behavior," if used at all, referred to observable public actions; psychology, on the other hand, was a social science dedicated to studying *internal* and private acts, presumably "observable" to the researcher's own mind alone. For the psychologist of the nineteenth and early twentieth century, it was the perceiving subject that mattered rather than the resulting action. The tangible effects of what people actually *did* was the subject of other sciences of man like economics and political philosophy. Indeed, in his survey of the early years of psychology, the historian David Leary notes that the term "behavior" was not used in a 900-page edition of James Mark Baldwin's *Dictionary of Philosophy and Psychology* (1901/1902) and appeared only a few times in the 1905 edition.[12] Yet in the span of a few decades following *Behavior of the Lower Organisms*, "behavior*ism*" became the reigning idea for non-Freudian psychology—in part because of its adoption by Jennings's colleague at Johns Hopkins, John Watson (1878–1958). By the 1950s, the behavior*al* sciences would provide the social sciences with its most sustained surge of interest in American history.

The clearest shift in thinking about "behavior" occurred not with the biologist Jennings but with the psychologist Watson in his 1913 paper "Psychology as the Behaviorist Sees It." A classic in methodological ground clearing, this paper proposed that nearly the entire fifty-year history of his discipline had been in error, as psychology had been "warped" by studying metaphysical ideas like consciousness and psychic states. Instead, Watson argued that psychology should be a "science of mental behavior" and should "never use the terms consciousness, mental states, mind, content, introspectively veritable, imagery, and the like."[13] Like Jennings, Watson wanted to study the "observable fact that organisms, man and animal alike, do adjust themselves to their environment by means of hereditary and habit equipments."[14] But unlike Jennings, who thought the ideas of his Johns Hopkins colleague were "too absolute" and "strange and narrow," Watson believed people were fairly predictable beings. Rather than look toward introspection, he imagined that internal mental states could be revealed through observation and experiment.[15] As Watson wrote to Jennings in 1924, he could not see how "patterns" could be learned through experience alone: "Almost everything seems built in."[16] Drawing on a strategy that had been successful in the natural sciences, Watson's influential paper avoided more complex questions about what gave rise to the observed phenomena produced in controlled experiments, concentrating instead on counting what could be seen. Of the messy and unobservable complexity of the interior of the mind that had bedeviled psychologists, Watson claimed that the behaviorist needed only to "care as little about [a person's] 'conscious process' during the conduct of the experiment as we care about such process in rats."[17]

Though Watson helped popularize the term "behavior," it was the work of Clark L. Hull (1884–1952), a professor of sociology at Yale, that demonstrated the most sustained attempt to make behaviorism the foundation of a truly "value-free" science. B.F. Skinner (1904–1990) would later become the most famous name in the world of behaviorism, but historians have noted that Hull "dominated" the field in the decade following the end of the Second World War.[18] A former engineer, Hull imagined the mind operating as a system of deductive postulates and theorems that had the potential to explain all mental action in human beings and animals. In a 1937 article for the *Psychological Review* titled "Mind, Mechanism, and Adaptive Behavior," Hull even outlined a rough series of six postulates from which he derived dozens of theorems about how "reinforcement" conditions behavior.[19] Though psychology was still far from having the empirical data or mathematical eloquence of physics, Hull believed that his six postulates "appear to be phenomena of physical structures which most theoretical physicists believe will ultimately be derived, *i.e.*, deduced, by them from electrons, protons, deuterons, etc." Hull therefore thought it was just a matter of time before mental actions could be reduced to particle physics, which would result in "an unbroken chain extending from the primitive electron all the way up to complex purposive behavior."[20]

By descending from the cell to the level of the electron, Hull not only extended the spectrum of "life," but he also inverted the focus of physiological psychologists like Jennings and Haeckel. Rather than depict cells with complexity and a "soul," Hull declared all behavior was as predictable as physics. In a footnote, Hull even suggested that one might make an "experimental shortcut" through the tangle of deductive steps to derive all human action from the whirling of atomic particles. In a hypothetical that looks surprisingly like a proposal for artificial intelligence eighteen years prior to the Dartmouth Conference, Hull explained how a "psychic machine" could demonstrate that nonliving matter could perform "adaptive behavior":

> Suppose it were possible to construct from inorganic material . . . a mechanism which would display exactly the principles of behavior presented in the six postulates just examined. On the assumption that the logic of the above deductions is sound, it follows inevitably that such a "psychic" machine, if subjected to appropriate environmental influences, must manifest the complete adaptive phenomena presented by the theorems. And if, upon trial, this *a priori* expectation should be verified by the machine's behavior, it would be possible to say with assurance

and a clear conscience that such adaptive behavior may be "reached" by purely physical means.[21]

Although less eloquent and memorable than Turing's famous "test" for machine intelligence, Hull's thought experiment was no less provocative. If a "psychic machine" could demonstrate "adaptive behavior"—the cornerstone of all intelligent life—then life could presumably be created out of "inorganic material." Even more, by describing how the machine might learn by "trial" of "environmental influences," Hull's psychic machine anticipated later theories of machine learning that have become crucial to twenty-first-century artificial intelligence.[22] In fact, Hull's proposal would eventually be realized, at least in part, by Frank Rosenblatt's "perceptron," and his psychic machine emerged six years prior to Warren McCulloch and Walter Pitts's groundbreaking 1943 paper on "neural networks." Remarkably, Hull's postulates even look like Rosenblatt's earliest sketches, where a machine can be conditioned through "backpropagation" to learn how to behave in certain circumstances.[23]

Showing how little the geometric spirit had changed in almost three centuries, Hull initially considered calling his first major work *Psychology from the Standpoint of a Mechanist*, only to change it to *Principles of Behavior* (1943) after growing interest in the term "behavior."[24] Hull's most provocative move in his first book was to ask that his readers imagine a human being as "a robot" in order to avoid sentimental associations, claiming that by picturing man as pure machine, one would have a "prophylaxis against subjectivity."[25] Like Descartes, who in *L'Homme* three hundred years prior had asked his readers to think of a human being *as if* it were a machine, Hull proceeded to evaluate the mental process of human beings only by what one could perceive "objectively." Though many non-reductive social scientists had in intervening years tried to develop a more nuanced methodology, Hull reverted to the French sciences of man, arguing that human behavior should be studied in the same way Newton derived the laws of physics. Or, more precisely, he echoed d'Holbach's belief that a science of man *was* physics. In fact, so rooted were Hull's ideas in mathematics and mechanical philosophy that he referenced only six works in the opening chapter of *Principles of Behavior*: three by himself, one work each from Newton and Spinoza, and the *Principia Mathematica* by Alfred North Whitehead and Bertrand Russell.[26]

Similar to many social scientific works, *Principles of Behavior* began with an account of the scientific method itself, making it clear that when it came to deductive logic, "the difference between the physical and the behavioral sciences is not of kind but of degree."[27] While such a move had been common since the earliest suggestions of Fontenelle and Condillac to extend the geometric spirit, Hull followed with a far more novel claim. Rather than argue, as social scientists had for centuries, that the sciences of man were more challenging than natural sciences because human beings were more complicated than planetary movement, Hull returned to d'Holbach's position that human subjects had no more "inherent complexity" than the subjects of any natural science. The social sciences had only failed to reach the heights of the natural sciences because of prejudice and the subjective bias of human observers. Human minds really were no more complicated than the gears of machinery—they were just harder to see. Here again Hull echoed the radicalism of *L'Homme*, where Descartes had argued that human hydraulic springs were either too small or too hidden to be perceived. In fact, while Hull (like Descartes) had couched his earliest claims in the idea that a person should be treated *as if* he were a robot, it was likely unnecessary, as by the mid-twentieth century this was hardly heretical. Void of subjective thought, Hull's "man" resembled Helvétius's education machine, and he claimed that learning—the "highest and most significant phenomenon produced by the processes of organic evolution"—could be treated as "wholly automatic."[28]

While most of *Principles of Behavior* supported Hull's grand claim that human action does not lead to any "inherent complexity," readers may be surprised by a chapter late in the book titled, innocuously, "Behavioral Oscillation." Here, Hull introduced a significant caveat that had the potential to undercut nearly the entirety of the logic of mechanist behaviorism. Though his 1937 paper claimed that the predictable world of atomic particles could eventually be connected to human behavior, Hull was forced to admit that people (and even animals) did not act with *quite* the same regularity as inorganic matter, having instead what he called "variability, inconsistency, and specific unpredictability."[29] In fact, such traits were among the "chief . . . distinctions between organic and inorganic machines." Instead of admitting that this might have added a level of "inherent complexity" to the project, Hull attributed this difference between animals and machines to a mere "oscillatory force" that existed in all organic life.[30] After two major books and scores of articles on how psychology could be made a branch of physics, he conceded that "behavioral oscillation" indeed imposed a grave limitation on the social sciences. After noting the success of the central limit theorem in overcoming this obstacle, Hull was forced to admit what Laplace, Quetelet, and Galton had already noted about the sciences of man, that "a complete absence of blurring of the mean . . . is attained only when the number of measurements becomes infinite; this means that *absolutely exact* empirical laws are never attainable."[31] Echoing the plans of Quetelet and Galton, Hull reiterated that "even this approximation can be attained only at the cost of great care and vast labor in the massing of data."[32] Remarkably, in just one book, Hull managed to travel the path of the Enlightenment sciences of man from Descartes's mechanical certainty to Laplacian probability.

The point here is not to show the inconsistency of Hull's curious blend of deductive and probabilistic reasoning, as the ideas of *Principles of Behavior* have been discredited in psychology for three generations.[33] In 1954, for example, Sigmund Koch offered a devastating 170-page analysis of what he called the "quantificational" and "tortured" logic of Hull's work, and there is little to be gained from flogging the dead horse that was mid-twentieth-century behaviorism.[34] Rather, Hull's conflicted and at times incoherent methodical approach matters today because in spite of its pretentions to grandeur, and utter uselessness in the social sciences, it reads as a fairly rough approximation of the methodology that guides much work in modern artificial intelligence. Not only do Hull's postulates appear to imagine intelligence as a series of stimuli and responses, but he also recognized the crucial role of large data collection and statistical manipulation. In his outline for a perfect social science methodology, he even seems to be describing artificial intelligence, noting that the "culmination of the [general theory of individual and social behavior] . . . would finally appear [as] a work consisting chiefly of mathematics and mathematical logic."[35] Such a work, similar to the *Principia Mathematica* for human action, "would be formulated [as] a precise mathematical statement of the several postulates . . . which survive the intervening winnowing process." After establishing the basic mathematical rules, Hull suggested a "means of rigorous mathematical processes" through which researchers would derive "theorems paralleling all the empirical ramifications of the so-called social sciences."[36] In other words, the perfect understanding of human behavior would be accomplished through a set of mathematical equations trained and refined through experiment and large-scale data collection.

In referring to the "so-called social sciences," it was clear that Hull had truly stretched the sciences of man to their limit, with the methodologies of the natural sciences doing more than providing a handy tool to study people. For Hull, there really was no such thing as a separate field that studied human actions, only mathematical physics. Not only was his program well beyond what social scientists of the early twentieth century like Eubank or Ellwood might have imagined themselves to be doing, it was beyond what hardly *any* social scientist would have had in mind in the 1940s. For this reason, Hull concluded with a bracing and, to some, deflating vision of what work in the "so-called social sciences" might mean, a vision that outstripped even William Ogburn's proposal of science work as mental drudgery. While Hull had imagined this program for the behavioral scientist, the "arduous and exacting task" described below might make a better description of computer science or programming of the past three decades:

> Behavioral scientists must not only learn to read mathematics under-standingly—they must learn to *think* in terms of equations and the higher mathematics. The so-called social sciences will no longer be a division of *belles-lettres*. . . . Progress in this new era will consist in the laborious writing, one by one, of hundreds of equations; in the experimental determinations, one by one, of hundreds of the empirical constraints contained in the equations; in the devising of practically usable units in which to measure the quantities expressed in the equations; in the objective definition of hundreds of symbols appearing in the equations; [and] in the rigorous deduction, one by one, of thousands of theorems and corollaries from the primary definitions and equations.[37]

As seen in the previous chapter, Hull's vision was not the only iteration of scientism that seemed to equate research with endless and repetitive labor, or one that asked researchers to only "think in terms of equations." And, as the counterrevolutionaries had complained 150 years earlier, removing *belles-lettres* from social science would only lead to "dull and industrious laborers" and the "laborious writing" of hundreds of equations.[38] Between the two, Ogburn and Hull threatened to turn the social sciences into an intellectually and emotionally barren process of endless equations, algorithms, and data collection.

Given his many robot analogies, belief in a "psychic machine," and affinity for equation writing, it may be unsurprising that Hull's ideas found more lasting success in artificial intelligence and computer science than psychology. For example, perhaps the most cited paper in "neural network" artificial intelligence literature, and one of the foundational papers for the electronic computer—"A Logical Calculus of the Ideas Immanent in Nervous Activity" (1943)—was explicitly behaviorist. Walter Pitts (1923–1969) provided the dense logical framework and notation for the paper, though the core of "Logical Calculus" began with a simple question that had been lodged in the mind of Warren McCulloch (1898–1969) for years: What if neurons in the brain operated like "either/or" propositions similar to those in symbolic logic? For example, what if neurons were either "excited" completely or not all, similar to the idea that a proposition was either true or false? Since both neurons and propositions were "all-or-none," it could be imagined that for "each reaction of any neuron there is a corresponding assertion of a simple proposition."[39] While later neurophysiology has disproved the idea that neurons are "all-or-none," the idea allowed for McCulloch, with the considerable assistance of Pitts's mathematical skill, to compose a set of postulates by which one might be able to show how a grouping of neural activity—or net—could change over time.[40] Crucial to the theory was the behaviorist idea of "learning," where, as McCulloch and Pitts put it, "activities concurrent at some stimulus which would have previously been inadequate [are] now adequate."[41] Meaning: neural on/off switches could be trained through external stimuli, and learning would consist of the strengthening of certain connections. Critical to later connectionist AI, such a process was described by McCulloch and Pitts as, tellingly, the "behavior of nets."[42]

McCulloch and Pitts claimed in "A Logical Calculus" that their work derived from Alan Turing's papers—noting that it offered a "psychological justification" for Turing's definition of computing—but McCulloch had already seen in the social sciences how to connect logic and biology well before he encountered Turing.[43] McCulloch had spent his early career as a psychologist looking for ways to escape the dominance of Freudian psychoanalysis and had been eager to "scientize psychiatry." As a recent biography by Tara Abraham explains, McCulloch's early work was also driven by a belief that "quantitative techniques should be applied to biological problems."[44] At the same time, McCulloch was also working at Yale's Laboratory of Neurophysiology under the direction of Johannes Dusser de Barenne (1885–1940).[45] In an indication of the kind of methodological influences McCulloch drew upon from 1934 to 1940, it might suffice to quote one of Dusser de Barenne's maxims: "Never think, if you can experiment."[46] At the same time at Yale, McCulloch had attended a series of lectures in 1936 given by Hull that outlined many of the specific postulates that found their way into *Principles of Behavior.*[47] Hull's work on postulates and experimentation would have been a revelation, as it offered a way to both scientize psychiatry *and* quantify biology. As the zoologist and popular science writer Matthew Cobb noted, perhaps with understatement, attending Hull's talk "encourage[d] McCulloch to think more about applying logic to biology."[48] While "A Logical Calculus" only cited three sources—two classics of mathematical logic by Rudolph Carnap and David Hilbert as well as *Principia Mathematica*—the "psychological" model for "nets" in fact looked much like Hull's robot.[49] Just a few years before Hull's mechanical methodologies would be roundly rejected in psychiatry, they had found a permanent home in one of the foundational papers for artificial intelligence.

The Embrace of Socio-Astrology: Or, How Talcott Parsons Learned to Stop Worrying and Love Quantification

Of course, not all social scientists of the twentieth century were willing to accept the idea of a value-free observational social science that studied people as if they were robots. For example, just ten pages into his epic synthesis of European sociological ideas, *The Structure of Social Action* (1937), the sociologist Talcott Parsons (1902–1979) felt the need to quote Alan Marshall on the fallacy of neutral observation. Marshall had claimed, "The most reckless and treacherous of all theorists is he who professes to let facts and figures speak for themselves."[50] Like Hull, Parsons in *Structure* was, in his own way, trying to respond to the kinds of critiques of social science made by William Ogburn, Frank Knight, and Frank Hankins. While Hull had tried to eliminate reform and moral thinking from the social sciences through re-mechanizing the mind, Parsons's groundbreaking book offered a completely different methodological approach, specifically warning of the dangers of a return to mathematical and quantitative social science. Countering Ogburn, Knight, Hankins, and Hull's belief that morality and ethical considerations had led social scientists astray, Parsons instead pointed to the failures of a long history of atheoretical and reductive methodology that had "dominated Western European social thought."[51] In his development of the idea of what came to be called "functionalism," Parsons saw behaviorists as just the latest iteration of a form of anti-intellectualism that held to the "treacherous" belief that facts could indeed "speak for themselves."

When *The Structure of Social Action* was published in 1937, Parsons began with the declaration "Spencer is dead," believing that the kind of sociology associated with nineteenth-century organic ideas was finished. Although this is one of Parsons's most quoted lines, he had actually borrowed it from the historian Crane Brinton (1898–1968). Brinton was a colleague at Harvard who was among the most severe critics of the kind of sociology Parsons was trying to dethrone. In the same year Parsons's book arrived, Brinton attacked the first tome of Pitirim Sorokin's four-volume work *Social and Cultural Dynamics*, declaring that the book could only be called "socio-astrology" because of its Spencerian attempts to make long-term predictions out of the hazy experiments of history.[52] Chief among Brinton's concerns was the deployment of statistics throughout the book, as Sorokin attempted to quantify the level of destructiveness of various wars and revolutions. While Brinton admitted that the question of the relative levels of destructiveness was "well worth trying to answer," he mocked the level of precision that Sorokin accorded historical events. For example, Brinton scoffed at the idea that in France, the February Revolution of 1848 rated a 20.32 while an attempt by Louis Napoleon to overthrow the French government in 1836 rated a 2.46. In Brinton's words, "There is no more sense in saying that the February Revolution was ten times more serious than Louis Napoleon's attempted coup than there would be in saying that the Pulitzer prizes are ten times more important than the World's Series."[53] Earlier, Brinton had claimed that Sorokin's statistics were used "perhaps a little deceptively," and it was Sorokin's combination of precision and prediction (rather than accuracy) that was part of the problem with the approach.[54] As a model of how *not* to do sociology, Sorokin's work was almost a perfect foil for Parsons's planned revolution in the field.

Although Sorokin would be named chair of Harvard's sociology department in 1940, Brinton and Parsons's anti-quantitative and anti-reductive vision had several allies at Harvard in the 1930s.[55] For example, both had been occasional members of a group that became known as the "Pareto Circle," a loose set of seminars and lectures that centered around the work of the Italian thinker Vilfredo Pareto (1848–1923).[56] The goal of these meetings, led by the biologist and chemist L.J. Henderson (1878–1942), was to develop a methodology for the social sciences that challenged the vision of transparent experimental data found in sociology like Sorokin's, and Henderson's status as a natural scientist gave the interdisciplinary group the kind of authoritative cover Parsons was looking for in his own social science. In these meetings, Parsons and Brinton relied mostly on Henderson's view of Pareto, who had proposed that a scientific study of society could not be reduced to mechanical procedures of behavior. Although Brinton recalled Pareto being dismissed as "the Karl Marx of the bourgeoisie," Pareto (and Henderson) offered a strong challenge to reductive models because they focused on the "interaction" and interdependent nature of human beings in society.[57] In Pareto's view, reductionism was impossible, as two events could only occur in "mutual dependence rather than of cause and effect."[58]

Parsons initially did not embrace these ideas, but after the seminar Pareto ended up being one of four crucial theorists—along with Marshall, Durkheim, and Weber—that Parsons relied on in *The Structure of Social Action*, the book that would eventually bring a synthetic anti-reductive sociology to America. Crucial to Parsons's work was the insight that experimental and statistical data can only be interpreted within a theoretical framework. Furthermore, he dared to suggest that such a framework needed to be produced *philosophically* in tandem with scientific research. In the opening pages of *Social Action*, for example, Parsons lamented, like Brinton, that there was a "deep-rooted view that the progress of scientific knowledge consists essentially in the cumulative piling up of 'discoveries' and 'fact.'" For Parsons, this meant that "knowledge is held to be an entirely quantitative affair" and that social science risked being reduced to the simplistic act of data accumulation.[59] Quoting Henderson on Pareto, Parsons insisted that "all empirical observation" can only be seen in terms of a conceptual scheme, and "facts cannot be described except within a schema."[60] At the same time Frank Hankins was calling for value-free sociology—and behaviorists were producing reports of "objective" experimental results—Parsons criticized as incoherent ideas like "pure sense data" and "raw experience." Although he included the necessary caution that the "distinction of science from all the philosophical disciplines is vital," *Social Action* was at its core a new philosophy of science. As he claimed, in full Paretian spirit, "every system of scientific theory needs philosophical assumptions."[61]

In the history of sociology, it is hard to overstate the initial influence of *Social Action*, which not only introduced Weber and Durkheim to most American sociologists but also helped put them in the canon of sociological thought where they, but not Parsons, still reside. Parsons's book also became a standard textbook for decades, creating what Charles Camic termed a "charter" for sociology, and was foundational to what has come to be called (mostly by its critics) the Standard Social Scientific Model. As Camic noted, much of Parsons's success was due not to new philosophical insight into the nature of society, or even the introduction of a productive research methodology, but rather to the fact that it provided a defense against "the assault by behaviorist psychologists on the social-scientific domain."[62] Indeed, Parsons's own counterassault on behaviorism was withering, and his chief concerns echoed the earliest critics of mechanical philosophy. He wrote, in a faint echo of d'Alembert's concern about the perils of *décomposition* and *généralisation*, "A machine can actually be taken apart. An organism cannot be taken apart in the same sense, at least without permanently destroying the organism."[63]

Summing up the profoundly anti-reductive nature of what would be called "functionalism," Parsons claimed, "Those features of organic systems which are emergent at any given level of the complexity of systems cannot, by definition, exist concretely apart from . . . the more elementary units of the system." In the work of European theorists like Pareto, Weber, and Durkheim, Parsons stressed that such features "cannot be reduced to a single quantitative scale of variation and still fit the conceptual scheme employed."[64]

"Emergent" ideas have only recently been discussed in the literature of artificial intelligence, where even simple computer programs cannot be "reverse-engineered" to their constituent parts, but Parsons's vision was not far from d'Alembert's decomposition, Bonald's social intelligence, Veblen's organic institutionalism, or many other critical approaches of reductive methodologies in the social sciences.[65] For Parsons, this meant that the behavioralist program was doomed not because it worked within a faulty conceptual scheme but because it denied even the existence of such a scheme. Anticipating the revolutionary philosophy of science of a later Harvard colleague, Thomas Kuhn (1922–1996), Parsons declared that even "facts" were similarly constructed, arguing that "a fact is understood to be an empirically verifiable statement about phenomena in terms of a conceptual scheme."[66] Or, more bluntly, "the facts do not tell their own story."[67] Lacking such a scheme meant that behaviorism in particular was "not a theory of action,"[68] and Parsons agreed with Hull—for very different reasons—that behaviorism was not even a *social* science, as it "tends to reduce psychological to biological considerations."[69] Like social physics, eugenics, and social Darwinism—and even the sciences of man of d'Holbach and Helvétius—behaviorism as practiced simply did not require much thinking, as the reduction of everything to concrete biological acts left nothing of the social or even human left to investigate. In summing up the failures of most of social science from the previous one hundred years, Parsons concluded that "*all of the versions of positivist social thought constitute untenable positions.*"[70]

Renewed interest in conformity to social laws after the Second World War necessitated a new printing of *Social Action* in 1949. As has been demonstrated in a number of studies, the Cold War represented a remarkable opportunity for social scientists to contribute to the defeat of international communism and the suppression of internal dissent, as political and military leaders became increasingly fearful of irrational, or at least unpredictable, mass action.[71] It was an opportunity Parsons was prepared for, and it would help redirect the nature of his anti-reductive theory. For example, he argued in a 1946 letter requesting funding from the National Science Foundation (NSF), "The urgency of the practical needs for rational control of social processes is so great and so obvious as scarcely to need discussing."[72] Parsons further claimed in "Science Legislation and the Social Sciences" that the natural sciences had largely done their job, but "the great problems of our time are not those of the control of nature but of the stability and adequacy of the social order."[73] For him, it was particularly important that such questions be addressed at the level of government, because there was "a widespread idea that the common man is his own social scientist; that any ordinary intelligent person is qualified to understand the operation of the social processes."[74] Noting that this was "very far from the truth," Parsons directly pleaded for technocratic knowledge at the heart of the modern democratic nation, one that took the opposite approach of social scientists like Frank Knight and Frank Hankins: "To a very substantial and rapidly increasing degree, the actual functioning of our social order is dependent on a social technology which is in fact applied social science. Technically trained persons are playing a larger and larger part, for instance, in the administrative process, in the adjustment of industrial relations, in the field of communications, in the control functions of the economy as through the central banking system, in the operation of foreign trade, and various such fields."[75] Although Uta Gerhardt noted that his letter "found no friendly audience" at the time, the report on the successes of sociology would become the first draft of *The Social System* (1951), perhaps the most influential social science text of the Cold War era.[76]

Only a decade prior, Parsons had been skeptical about directing social science research toward practical application, data collection, and statistics, but the lessons of the war and the possibility of massive funding for social sciences likely helped the evolution in his thought. Behaviorists had shown that in the social as well as the natural sciences data collection attracted far more public and private interest than theorizing, and Parsons could not fail to notice Hull's Institute of Human Relations at Yale or Skinner's array of "teaching machines" and "learning boxes" at Harvard. As Joel Isaac's study of social thought at Harvard has shown, Parsons was particularly adept at shifting his research interests to meet institutional needs, managing to move from a untenured economics lecturer to eventual chair of the sociology program, working his way through a number of "interstitial" groups at Harvard.[77] Lawrence Nichols also noted that Parsons maintained "shifting but always expanding intellectual and professional networks."[78] While his first book had been based on a theoretical synthesis of a handful of authors, requiring little more than time, pencil, and paper, in his plea to the NSF, Parsons spoke proudly of social scientists who "have emerged from the library and taken up the direct empirical study of the phenomena about them." This meant they had a "need for facilities which have the same function as laboratory equipment in the natural sciences." Even more, because social behavior was so much more complex than the natural world, Parsons argued that it has "become necessary to have facilities even more elaborate and expensive . . . from those used in other fields."[79] Perhaps most surprisingly, Parsons asked for more "statistical information," noting that it was now "feasible to collect statistical data for the specific purpose of testing scientific hypotheses as such."[80] Last, Parsons repeated a line that had been claimed in favor of large data collection since the time of Quetelet: "The essential case for support of the social sciences thus does not rest on their previous record or their present state; it rests on their potentialities."[81]

Such arguments were perfectly tailored to the NSF, which had bolstered the credentials of social science by including them with the natural sciences, but only if they adhered to a particular methodology. There likely would have been no funding for Parsons's anti-reductive synthesis of European theory, but his new project was better tailored to what Mark Solovey called the "particular epistemological ideals such as value neutrality, objectivity, and generalizability" and "large-scale social science databases" favored by the NSF.[82] The "potentialities" for Parsons's ideas, therefore, *were* enormous, and Parsons's two major books from 1951—*The Social System* and the edited volume *Towards a General Theory of Action*—were watershed moments for the creation of a new grand science of man: modernization theory.[83] As Edward Shils (1910–1995), Parsons's coeditor on *General Theory*, wrote, "It is the aim of general theory to become genuinely universal and transhistorical" in order to make it "equally applicable to all societies of the past and present."[84] While many anti-reductive visions like Veblen's were nonnormative and treated social actions as embedded historical and cultural creations, functionalism provided a more monolithic and totalizing idea. Combined with behaviorism, modernization theory suggested malleable and ahistorical individuals who could be shaped into new modern peoples. In the context of the Cold War, this meant all nations and peoples could learn to adapt to the normative equilibrium of Western liberal capitalism, thus providing the central social scientific paradigm for the messier politics of assassinations, militia funding, and propaganda that spread throughout the world from the 1950s through the late 1970s.

Remarkably, the theorist who had delivered a broadside against "anti-theoretical positivism" at his 1950 Presidential Address to the ASA was *attacked* as a positivist.[85] Initiated as a theoretical rejection of Spencerian ideas, functionalism became, in books like Alvin Gouldner's *The Coming Crisis of Western Sociology* (1970), an apology for Western dominance and imperialism. Although it has been claimed that "Gouldner's incessant wailing" was the greatest "disservice" in the history of sociology, the tag stuck to Parsons, as the initial anti-reductive and anti-positivistic thrust of *The Structure of Social Action* was ignored in many discussions of his later work.[86] In an example of how far this critique has endured, Parsons is even described in a recent history as a "founding figure of behavioral science."[87] Parsons's first book had helped destroy one paradigm, but his second, along with the coedited *General Theory*, was the instauration of a new guiding principle for the social sciences.

In the three decades after Gouldner's book, Parsons virtually disappeared from many sociology syllabi. Although the fall of functionalism and modernization theory was not as swift as the collapse of Hull's behaviorism, at least at the disciplinary level, Parsons's grand hope for a *Principia* of the social sciences was unsuccessful. Though Gouldner and others in the cultural turn had actually taken up a version of Parsons's initial critiques of scientism, no new synthesis has since replaced his ideas in sociology, and certainly not in the social sciences as a whole. As Isaac claimed of the social sciences of the past fifty years, "the balkanized intellectual landscape" of modern social science "resulted from the stalled interpretive revolution against the behavioral sciences."[88] While this may overstate the case slightly, it is hard to argue that social science in the twenty-first century has anywhere near the level of enthusiasm or interest that it did in Cold War America, and sociology has never returned to the prominent position in American academia that it experienced in the first decades of the twentieth century. As seen in the previous chapter, economics, or at least the "economic style," largely succeeded in embedding itself into American institutions by *rejecting* other social sciences. In part, the "deconstruction" of grand narratives was a necessary corrective, as the recognition that social scientists were substituting the modern West as a transhistorical and transcultural standard of social action was an important challenge to a centuries-long conflation of science and normative values.[89] Yet, as functionalism and modernization theory joined the list of failed attempts to construct a *Principia* of human action and thought—along with social physics, classical political economy, eugenics, and Hull's radical behaviorism—with it went perhaps the final attempt to offer a non-reductive and all-encompassing social science. The attempts to articulate a theory of human thought and action that could fulfill the ambition and progressive hopes that had attended the sciences of man since the days of the Enlightenment seemed at an end.

Paper Machines and Office Work: Alan Turing's Social Science

While American psychologists and sociologists spent the first half of the twentieth century trying to imagine a form of science that could account for human action, Alan Turing (1912–1954) took a different approach to observing his fellow humans. Turing was a great skeptic of the social sciences, claiming that "the works and customs of mankind" were not amenable to "scientific induction." Instead, beginning in 1936 with his seminal paper in mathematics, "On Computable Numbers," Turing helped create the idea for mechanical computers and intelligent machines. While he did not build any artificial intelligence machines, he has rightfully been called the "founding father" of AI for his many thought experiments about how machines *might* go about thinking.[90] In part because of two papers that laid out the possibility of thinking machines—"Intelligent Machinery" and "Computing Machinery and Intelligence"—Turing has influenced how generations of computer scientists, mathematicians, and philosophers framed the problem of how a machine might be able to think.[91] In addition to his idea of a thinking machine and his definitive role in breaking the German Enigma code during the Second World War, his subsequent persecution and tragic death have made Turing one of the rare mathematicians to influence others across generations through both personal biography and genius. As one collection of studies recently claimed, "Turing's universal machine transported us into a world where many young people have never known life without the Internet."[92]

In spite of his broad influence in a number of fields and scores of books and journal articles, no one seems to have imagined Turing as a social scientist. Yet all three of his foundational papers on thinking machines relied on observation and reflection on human action. Although he claimed that an inductive account of human behavior was unlikely, and had no training in social scientific practice, none of his work would have been possible without generalizations and theorizations about how human intelligence worked. Indeed, Turing's initial formulation of thinking machines was highly ambiguous and contradictory, a muddle that borrowed from the social sciences even as it was helping to displace them.[93] Turing clearly anticipated a number of major developments in electronic computing, but few studies in computer science history have recognized that Turing's novel insights were themselves embedded in a much longer history of ideas and implicit assumptions about human behavior. While the past two chapters have shown social scientists engaged in fierce debates about methodology and praxis in trying to discover the laws that ruled human action, experience, and intelligence, Turing never needed to engage in such debates nor defend his methodology. As the founder of a new field, computer science, he was able to create a discipline where such questions never needed to be asked.

Given his skepticism over the social sciences, it would be hard to call Turing himself a behaviorist, but it is worth investigating just what kind of inductive work he did in observing and reflecting on human action. It might be imagined that Turing worked only in the abstract realm of mathematics and symbolic logic, but his first major mathematical paper—"On Computable Numbers"—also deployed inductive logic drawn from observing human actions. The paper, a highly technical and complex proof demonstrating that a complete axiomatic system of mathematics was impossible, began with a simple intuition or reflection. Turing needed to define a "computable number," so he reflected on how humans think when they "compute" numbers.[94] At the time of Turing's writing, "computers" were still people, and Turing had to essentially try to get inside the head of human calculators. Perhaps more nuanced than Watson's and Hull's complete dismissal of the conscious process, Turing's inductions had their own limitations. In fact, Turing referred not to the "actions" of a hypothetical computer but to a person's "behavior" in response to external events. In his monograph on Turing's paper, Chris Bernhardt explains that for Turing, "the behavior of the [human] computer at any moment is determined by the symbols which he is observing and the 'state of mind' at that moment."[95] And, as would be clear, the "state of mind" was determined by previous symbols and observation, making computing a mostly passive human experience.

While this was clear enough, aside from the introduction to "Computable Numbers," Turing never defended (or even mentioned) what methodology he used to determine how a human being thinks through the process of computation. Instead, like Descartes in *L'Homme* and Hull in *Principles of Behavior*, Turing began by imagining a person *as if* it were a machine. In asking readers to consider how a machine might make calculations, "Computable Numbers" began by claiming, "We may compare a man in the process of computing a real number to a machine which is only capable of a finite number of conditions . . . which will be called *m*-conditions." From there, Turing explained how the "machine" might process a "magnetic tape," segmented into "squares" and filled with "symbols." As the machine was fed tape, Turing imagined the process by which it could either "remember" certain symbols, shift the tape, or "mark" blank squares with new symbols. Such a machine would also have the ability to alter its "*m*-configurations," meaning that the entire "configuration" of a machine at any point was the combination of its own *m*-condition with whatever the tape happened to say. As Turing put it, "the configuration determines the possible behaviour of the machine."[96] Though Turing was not explicit, the obvious parallel was that the "behavior" and *m*-condition of a human computer was similarly determined by the process of observed symbols.

In "Intelligent Machinery" (1948), however, Turing's approach to observing human thought and action became more complex than simple behaviorism. In this paper, in which he summarized a discussion of the "possible ways which machinery might be made to show intelligent behavior," Turing drew on a long strain of thinking in the social sciences about learning. As he noted of one possible route through which a machine might learn, seemingly a return to Jevons's hedonistic calculus, "The training of the human child depends largely on a system of rewards and punishment, and this suggests that it ought to be possible to carry through the organizing [of a machine] with only two interfering inputs, one for 'pleasure' or 'reward' and the other for 'pain' or 'punishment.'"[97] Yet Turing qualified this statement, saying children also needed "initiative" to learn. Employing a term often used in the Pareto Circle at Harvard, Turing claimed that such initiative leaves a "residue" that comes about during human interactions. Suggesting this was more important than simply adjusting pain and pleasure receptors, Turing at this point claimed he wanted "to discover the nature of this residue as it occurs in man, and to copy it in machines."[98] To find the "residue"—or what others might call the "emergence"—of social action in the mechanistic world, Turing proposed two solutions, both of which would dominate artificial intelligence thinking for decades. Either computer scientists would need to build a fully disciplined machine and see how it "learns," or they would have to create "unorganized machines" akin to the "infant human cortex."[99] Turing almost certainly had no idea of "residue" in its Paretian sense, or even of Parsons's work, but his inadvertent alternation between behaviorism and functionalism managed to traverse the entire spectrum of social scientific thought of the age in a few paragraphs.

As nuanced (or confused) as his inductive approach to human learning was, by the time of Turing's third major paper on thinking machines, "Computing Machinery and Intelligence" (1950), the behaviorist language became much stronger. Here Turing introduced his famous "imitation game," or what has come to be known as the "Turing Test." In his thought experiment, Turing proposed the question: "Can machines think?" Rather than try to answer such an "absurd" question, he imagined a "game" whereby a human examiner was given the opportunity to question both a machine and a person. If the examiner cannot determine which respondent was the machine, the machine (presumably) can think.[100] While few AI researchers today view the Turing Test as a productive path toward artificial intelligence, the test inspired legions of sci-fi stories and shaped research in the field for generations. Therefore, it is notable that while obsessive focus has been dedicated to how the machine might fool the examiner, hardly any discussion or reflection occurs over how or why the *human* examiner and respondent might think in these situations, and Turing himself says almost nothing about the people involved.

However, there are some clues to how Turing imagined a human competitor might behave. For example, in describing how computers and humans might act, Turing uses the term "behaviour" at least twenty-three times, most often in referring to how a person or machine might respond to an "input" during the test. Later in the paper, Turing acknowledged that "intelligent behaviour" might be a "departure" from pure calculation, but it was a "rather slight one."[101] In contrast to this use of behavioral language, Turing deployed the more Parsonian term "action" only three times: twice in reference to a machine and only once when referencing a person. Whereas histories of artificial intelligence and computer science have rarely mentioned Turing's explicit and implicit social scientific assumptions, psychological texts have done much better in discussing the Turing Test's origins in the methodologies of the social sciences. One history of psychology claimed that Turing "proposed a perfectly behaviorist test for 'thought,'" while another noted that "Turing's test is a paradigm of behaviorism."[102]

Although Turing needed to inductively reason over how human beings thought, none of his three foundational papers give any insight into how and why he came to believe that people acted in a machine-like and behavioralist manner. Yet his comments on human computers in "Intelligent Machinery" can give some indication of where he got this view of the human mind. Here Turing introduced his influential idea of a "universal machine" that could be programmed to simulate the work of any other machine. While seemingly an ambitious technology, the reality was that the wonderous "universal machine" could barely exceed the tedious tasks of a clerical worker. As Turing wrote in 1948, "It is possible to produce the effect of a computing machine by writing down a set of rules and procedures and asking a man to carry them out. Such a combination of a man with written instruction will be called a 'Paper Machine.' A man provided with paper, pencil, and rubber, and subject to strict discipline, is in effect a *universal machine*."[103] "Universal machines" were not a thought experiment, therefore, but a metaphor, and it is worth asking where Turing got his ideas for how people "think."

It is possible that Turing first encountered his idea of a "paper machine" while working for the British Government Code and Cypher School at Bletchley Park. It was here that Turing and others did much of the important codebreaking work on the German Enigma machine from 1939 through 1945. While this is usually (and justifiably) told as a story of heroic adventure because of its importance for the British war effort, the actual structure and organization of Bletchley Park looked more like an enormous series of "paper machines" than the cinematic portrayals of cryptologists at work. In a much-needed account, Chris Smith shows how Bletchley Park "mechanised and bureaucratised its processes and increasingly came to resemble a factory based on production-line principles."[104] The production line occurred in the collection, evaluation, and dissemination of data and had been based on a long history of administrative thinking about data in England and Europe. Had he visited, it would have been Clark Hull's dream for social science. As historians of science have demonstrated, "paper machines" and human "computers" had been used at various statistical and scientific institutes in Europe for decades, dating back to the Greenwich Observatory under the direction of George Biddell Airy (1801–1892), where individuals with little training or skill were given the mundane task of filling in forms. As human computers, they were essentially asked to follow simple algorithmic processes to return arithmetical answers for which they themselves had no context to understand.[105] Even those individuals who "erred" in their predictability were standardized through equations.[106] Such work endured as a feature of computer science labor and, as will be seen in the next chapter, has become a crucial element of AI labor as well.[107]

It seems likely that in his major papers anticipating artificial intelligence, Turing made inductively reasoned conclusions not from observing the way in which *all* human beings thought or "computed" but rather from observing how a very peculiar and particular kind of person behaved: the British computer or bureaucrat in a large administrative endeavor. From the perspective of a social scientist, this would have seemed to be a fairly limited inductive base on which to summarize human action and, from the perspective of artificial intelligence, a fairly limited goal to attain. While Turing cannot be said to speak for all later thinking about artificial intelligence and computer science, his influence on the field was extraordinary. At the aptly named "Mechanization of Thought" conference in 1959, the American computer scientist and one of the founding fathers of artificial intelligence Marvin Minsky (1927–2016) explained just how simple it would be to replicate this kind of limited bureaucratic thinking. Referencing Turing's work from "Computable Numbers," Minsky said, "The behavior of any machine is always explicable in terms of its past states, external contingencies, and the causal or problematic relations between them."[108] Turing's initial hedge about the relative role of the "residue" of "initiative" had gotten lost. As Jamie Cohen-Cole has pointed out, such "paper machines" lasted even into the work of Herbert Simon (1916–2001), the psychologist-cum-computer scientist who envisioned a "computer constructed of human components."[109] Just as "universal machines" were not visionary ideas but mechanical clerks, artificial intelligence too seems to have begun as a metaphor for mundane science work.

While Turing rarely reflected on the kind of people his machines might be trying to replicate, he did predict some of the social consequences that would result from the creation of computing machines. Anticipating concerns about human labor in the age of machines, Turing worried that a universal learning computer would largely leave individuals in the role of rote "servants" of the machines. In "Intelligent Machinery," he speculated that though building intelligent machines might begin with creative engineering projects, engineers in the future would "be replaced by the office work of programming the universal machine to do these jobs." He reiterated that such programming would be "pure paper work."[110] Given that the earliest dreamers of computing machines from Leibniz through Babbage had been motivated by *eliminating* rote work, such machines seemed only to push the drudgery onto a new class of workers. While biographies of Turing often envision this reeducation of mathematicians and statisticians as another astonishing prediction of a "new field of industry and employment," it is hard to imagine Turing was excited by the prospects involved in this work.[111]

Like much else, Turing also predicted the coming competition between man and machine, claiming in a 1947 lecture at the London Mathematical Society that "it is expected therefore that the large-scale hand-computing will die out." Human workers would still be needed, "but whenever a single calculation may be expected to take a human computer days of work, it will presumably be done by an electronic computer instead."[112] Work would therefore be shifted to "masters" and "servants."[113] Though long hailed by later computer scientists as a visionary for the creative work of engineers, Turing's work seems closer to imagining the "future of work" as mundane maintenance, identical to later stories in the *I, Robot* series from Isaac Asimov (1920–1992), where human labor largely exists in a contented stupor in order to keep the machines running.[114] In fact, the biggest future challenge Turing imagined in the world of intelligent machines was fairly quotidian. He noted that in the future world of machines we might need "a number of librarian types to keep us in order."[115]

Conclusion

The period between 1930 and 1970 promised to be an extraordinary time for the growth of the social sciences in America, a time when they could finally emerge out of the depths of metaphysics, Enlightenment morality, European theory, unnecessary abstractions, and Christian reform movements to produce a scientific account driven by new tools of research.[116] When Frank Hankins asked his social science colleagues to drop their progressive leanings and to focus on the creation of a purely "scientific" science of man, social scientists in psychology and sociology responded in at least two separate ways: Clark Hull's behaviorist approach and Talcott Parsons's functionalist theory of action. While these schools had different objects of study, methodologies, philosophical underpinnings, and overall goals, they both offered the opportunity for a vast expansion of the research program in the sciences of man beyond philosophical speculation and casual induction. Hull's vision of behaviorism offered the possibility of a massive project of equation writers and observers, with thousands of theorems "laboriously" deduced in order to translate human action into particle physics. For Parsons, sociology in the 1930s had meant reflecting on a handful of European writers, but by 1951, it involved requests for funding and laboratory space that exceeded what the nation had spent on natural science research. Had the social sciences received what both Hull and Parsons had asked for, this would be one of the most powerful fields in the world today, with social scientists playing the role of careful technocrats overseeing a steadily progressive and improving world.

Of course, with some exceptions, the triumph of the social sciences never occurred. While initially attracting research from a number of independent foundations, the grand project of the "behavioral sciences" collapsed.[117] In psychology and sociology, neither Hull's legions of mathematicians nor Parsons's massive institutions to study a "theory of action" materialized. Hull's program had a precipitous fall after achieving dominance in the late 1940s and early 1950s, in part because the program could never emerge from its endless methodological statements and repetitive and inconclusive animal experiments. Although B.F. Skinner (1904–1990) maintained a public profile for behaviorism for decades, his final works recycled the same methodological questions and theorizing from half a century before, and thousands of experiments on rats and pigeons perhaps unsurprisingly led to few important scientific theories of people. In *Beyond Freedom and Dignity* (1971), for example, Skinner could still claim, almost 350 years after Fontenelle first invoked the geometric spirit, that "the methods of science have scarcely been applied to human behavior."[118]

Whereas Parsons had more enduring success, his ideas too began to fade, as the modernization theory built from his work was challenged by a generation of social thinkers who claimed to unmask his ideas as apologies for a particularly Western, male, and Protestant value system that privileged the status quo. By the 1980s, in fact, the idea of a technocratic elite, trained in vast social scientific laboratories sponsored by state governments and working out the intricacies of human behavior through algorithms and experiments, had almost completely given way to the positivist belief of economic science in individual markets and the flow of capital.[119] The great dream of the Enlightenment sciences of man—understanding humanity through scientific means to bring about progress—seemed over.

Except that it was not, at least not entirely. While the era from 1930 to 1970 failed to produce an *Origin* or *Principia* for the social sciences, it did produce a nonhuman computer, along with the idea that such a computer could be made to think and "behave" like a person. During this time, much of the enthusiasm—and certainly the resources—that had attended the social sciences were transferred into making computers as dreams of progress moved from people to machines. The thousands of mathematical thinkers and equation writers did not emerge in psychology laboratories, as Hull wanted, but in computer science classes. The resources that Parsons dreamed of went instead to STEM training, computer labs, and DARPA (the US Defense Advanced Research Projects Agency). So too was the process of studying computers imagined as an easier process than studying people. Whereas Hull had imagined the removal of complexity and interpretation as the key to psychology, Minsky recognized that it could be done much more easily with computers.

He claimed, "We could avoid the problem of interpretation—of understanding—if we could specify, along with the statement of the rules, the details of the mechanism that is to interpret them. That would leave no ambiguity."[120] Taken out of context, Minsky's description of computer science would also make for a perfect statement of "scientism" in sociology and psychology in the 1930s and 1940s. "Interpretation" and "understanding" could be avoided not through eliminating ambiguity in the object under investigation (people) but through "specifying" the "mechanism" that interpreted them (the observers). If social and computer scientists could be mechanized, the objects of their observations might naturally follow as well.

While one might have thought the failure of the social sciences to understand intelligent acts would have subsequently made artificial intelligence impossible, the past seventy years have demonstrated that it was far easier to make progress with thinking machines then it was to figure out how and why people think. Rather than artificial intelligence as a complex set of research tools to build a machine based on a proven theory of human action, computer science took the much easier route of elevating the most mundane of human behaviors to the realm of intelligence. In doing so, they managed to build off of a methodological approach they had inherited from three centuries of reductive and quantitative work that brought intelligence down to the level of the computer. As will be seen in the next chapter, for some, the development of computer processing power and the internet even meant that Hull's and Parsons's great programs of experimentation were superfluous. Online, people could conduct all the necessary social science experiments on themselves, and at rates far greater than could have ever emerged with the top-down programs Hull and Parsons imagined. In his surface observations of human computers, Turing had helped invent a machine that could perform more experiments on "behavior" in a few minutes than Hull's laboratories could have performed in a lifetime. While an extraordinary transformation in AI work resulted from the ability of mass data accumulation, some things about studying human behavior and thought remained the same. As it had been for centuries, the supposedly atheoretical approach to data collection and accumulation was accompanied by two enduring ideas: the bias of supposedly objective numbers and the diminution of intelligence required for working with such numbers.

9

Second-Rate Mathematicians

Statisticians, Data Analysis, and the Persistence of Bias

> To unlock the analysis of a body of data, to find a good
> way or ways to approach it, may require a key, whose
> finding is a creative act.
>
> —John Tukey, *Exploratory Data Analysis* (1977)

TO RETURN TO FONTENELLE'S CLAIM FROM THE FIRST CHAPTER OF THIS book, the first few decades of the Cold War era might have marked the peak attempt of a centuries-long process to "transport . . . the geometric spirit" to "other forms of knowledge" like "morals" and "politics."[1] Or perhaps it was the best example of Laplace's dictum to "apply to the political and moral sciences the method founded on observation and on the calculus [of probabilities] . . . which has served us so well in the natural sciences."[2] Never had such financial resources, intellectual energy, and public attention been lavished on the sciences of man, as governments, universities, and private institutions built labs, ran experiments, and trained thousands of researchers in the best way to observe, collect data, and produce statistically sound conclusions on human beings. It was an era that the philosopher of social science Charles Taylor (1931–) labeled "empiricist," where "brute data" and "mathematical inference" would allow social scientists to build "knowledge . . . which could be anchored in a certainty beyond subjective intuition."[3] In a confirmation of Parsons's early warnings about the anti-intellectualism inherent in this kind of science work, Taylor's "Interpretation and the Sciences of Man" (1971) defined brute data as "data whose validity cannot be questioned by offering another interpretation or reading, data whose credibility cannot be founded or undermined by further reasoning." The criticisms of Taylor and others, to say nothing of the reality of American foreign policy, helped end this era of the social sciences, challenging the unity of brute data and mathematical inference as the guiding methodology for a science of man. As Taylor pointed out in 1971, such a vision was "largely a thing of the past."[4]

Had that been the final word, then the nearly 400-year-long attempt to ground predictable and certain knowledge of human action in the reductive methodologies of the natural sciences—the course charted in this book—would have come to an end in 1971. Yet as Taylor also noted, almost offhandedly, the strict "empiricist" view "no doubt . . . goes marching on . . . as a theory of how the human mind and human knowledge actually function." In fact, Taylor knew of at least one place where the "epistemology" of brute data and mathematical inference could still find a home. "In a sense," he wrote, the "contemporary period has seen a better, more rigorous statement of what this epistemology is in the form of computer-influenced theories of intelligence." Taylor pointed out that such theories "try to model intelligence as consisting of operations on machine-recognizable input which could themselves be matched by programs which could be run on machines." Even when the "brute data" approach had failed in studying human actions, "the machine criterion provides us with our assurance against an appeal to intuition or interpretation."[5] Although Taylor did not return to the idea of "computer-influenced theories of intelligence" for the remainder of his seminal paper, he was correct in seeing that something like artificial intelligence provided the best, and perhaps only, means to salvage the vast pool of failed theories of human action and behavior that had been attempted over the previous centuries. It is for this reason that this book has one more chapter.

The funneling of brute data and mathematical verification into modern AI research, as well as its persistence in some corners of the social sciences, was not inevitable. Ten years before Taylor critiqued what he called the "obsession" with the "empiricist" tradition, the mathematician and statistician John Tukey offered an alternate path for quantitative analysis, one that recalled the "creative" approaches of d'Alembert and Wesley Mitchell, where most thinking is done *prior to* data collection. As he wrote in "The Future of Data Analysis" (1962), mathematics and statistics should be used for "judgement" rather than "stamps of validity," and it was "better" to search for "an approximate answer to the *right* question, which is often vague, than an *exact* answer to the wrong questions."[6] Like d'Alembert's critique of mathematical reduction in the Enlightenment, Tukey's concerns were especially valid because he was an institutional insider: chair of the Princeton Department of Statistics, longtime researcher at Bell Telephone Laboratories (Bell Labs), and a member of count-less committees formed by the federal government and private foundations, including the Ford Foundation and Stanford Center for Advanced Studies in the Behavioral Sciences. As revealed in his notes, letters, and publications, Tukey was skeptical of most quantitative products, AI included, and believed that, rather than rely on brute data, statisticians needed to take an early role in the process of shaping data collection, analysis, and presentation.

Tukey hoped for statisticians to be "detectives" of data analysis, but his vi-sion was not often realized. Rather than the "creative act" that would provide the "key" to data analysis, much of statistics, especially in the social sciences in the twenty-first century, have become almost exclusively "confirmatory." While the recent field of computational social sciences (CSS) has attempted to offer an update on quantitative social science, the overwhelming "deluge" of data has left little room for Tukey's approach. Much like the expansion of connectionist AI neural networks in the second decade of the twenty-first century, the interest of CSS in recent years has been fueled not by new con-ceptual ideas or theory, but by the growth of data accumulated through the internet and various tracking devices in phones, cars, "wearables," and other connected machines. Like AI systems, CSS researchers have also often offered up conclusions drawn from purportedly "raw data," in spite of the recognition that the vast majority of this data is collected by private companies whose methods and means of acquisition are largely proprietary. Unsurprisingly, both AI and CSS have had to deal with significant bias despite relying on "objective" processes. While attempts to address bias in the world of CSS have largely focused on technical fixes, the history of social science suggests that these biases are far more deeply rooted and that the lessons Tukey of-fered have gone unheeded. Statisticians, whom Tukey believed could serve as the leading scientists of data, have become in the age of computers, AI, and CSS exactly what Alan Turing had predicted: clerical workers consigned to marginal labor and "librarian types" tending to the purported intelligence in the machine.

John Tukey and the Creative Act of Data Analysis

As one of the most important mathematical statisticians of his era, John Wild-er Tukey (1915–2000) seemed to be ideally placed to carry on the dreams of thinking machines and vast data collection that had emerged in the 1950s and 1960s.[7] Tukey overlapped with Turing's brief stint at Princeton, worked with Claude Shannon at Bell Labs, and corresponded regularly with Marvin

Minsky, Norbert Wiener, Warren McCulloch, and Herbert Simon. At Princeton, he became a professor of mathematics in 1949, created the Department of Statistics in 1965, and remained at the university until his retirement in 1985. At the same time, he worked at the famed Murray Hill campus of Bell Labs during its peak influence in the world of computers, helping to start the Department of Statistics and Data Analysis and later becoming a consultant for NBC News, Xerox, the Educational Testing Service, and dozens of other corporations. Outside of his work at Princeton and Bell Labs, Tukey also managed to contribute to a number of high-profile government committees—from an investigation of Alfred Kinsey's statistical methodology to arms control and the census bureau—and participated in a number of research groups created by private foundations, including the Ford and Rockefeller Foundations. In the 1950s, he was given the "highest clearance" at the Atomic Energy Commission, and in 1973 he was awarded the National Medal of Science in recognition of a lifetime of service in statistics and mathematics.[8] As his official National Academy of Science "Biographical Memoir" notes, Tukey "merged the scientific, governmental, technological, and industrial worlds more seamlessly than perhaps anyone else in the 20th century."[9]

Given the direction of AI research in the twenty-first century, which involved the "seamless" merger of academia, industry, and government, Tukey might have been a foundational figure for modern machines built on data. His contributions to creating the Cooley-Tukey fast Fourier transform, an algorithm that sped up data analysis, was essential to modern data manipulation, and his technical contributions to statistical analysis were a fundamental part of how modern institutions and machines transformed "Big Data" into manageable pieces for evaluation.[10] As examples of his influence on the computer era, Tukey coined the term "bit," was the first person to use "software" in a published paper, and essentially invented the concept of "data analysis." And yet he has almost no presence in histories of machine thinking or Cold War social science, with few mentions in the academic literature outside of his many technical contributions.[11] Tukey has likely been overlooked in the literature for at least three reasons: his general antipathy to machine learning, his expansive vision of what statistics could do, and his lifelong concern about the dangers of naive quantification. As a symbol of Tukey's struggle into the headwinds of reductive brute data and the "wisdom of the crowds," it is notable that one of his crowning successes—the establishment of an independent Department of Statistics at Princeton—was short-lived. Though one of the most successful programs in the country, the department's independent status ended in 1985 with Tukey's retirement, when it was returned against his wishes to the supervision of the Department of Mathematics.

Although Tukey had many interests, artificial intelligence was not among them. A mathematician interested in data collection and evaluation, he shared none of the AI founders' interest in modeling the brain, behaviorist ideas, or the introduction of biological metaphors. When, late in his career, he joined some of his Bell Labs colleagues for a presentation on thinking machines in California, he wrote at the top of his transparencies, "AI, Yes! Artificial Intelligence . . . No?" Though that line did not make it into the formal presentation, the paper was quite skeptical about the prospects of a thinking machine, calling it a "bundle of attitudes and hope," and it made no mention of the grand goal of AI to model human thought. Instead, the report, coauthored with two associates, focused on what they believed to be the limited role of a computer. Tukey and his team thought perhaps a machine could serve as a "programmer's apprentice," which could "assist in coordinating a large programming team" or "do simply coding tasks." Even this narrow range of work could only be accomplished, the group argued, if there was a "mixed human-computer diagnostic system" and a "working group for AI *not* composed of managers." In a sign of just how much he rejected bureaucratic structures, Tukey had written in his notes that "managers" was just a "polite term for housemothers, for both programmers and programs."[12] In his folder from the trip, Tukey even kept a note from former Bell Labs colleague M. D. McIlroy, who attacked the current AI as a "dreamland" and "pie in the sky" vision, when it should be a simple "collection of rule-based programming techniques."[13] Coming at the end of his career, it reflected Tukey's lifelong vision that any technical system required active and engaged "human programmers."

Though his comments occurred during the height of an "AI winter" in 1984, they reflected a much longer history of Tukey's skepticism about work that relied on mathematics or the natural sciences to make claims about human actions.[14] In spite of being a correspondent and friend of many of the founders of AI, cybernetics, and the behavioral sciences, he was ambivalent about these fields throughout most of his career. As early as 1947, just as Tukey was coming up for tenure at Princeton, he expressed reservations about an early draft of Norbert Wiener's "Time, Communication, and the Nervous System" sent by Warren McCulloch, "at Wiener's request." Tukey responded, Wiener's "description of the social sciences must be enlarged," and "it reduced too much of human activity to the biological or individual." In a surprising note for a statistician and mathematician to send to a writer interested in communication, and one similar to the anti-reductionism of critics from d'Alembert to the early Talcott Parsons, Tukey reminded Mc-Culloch (and Wiener) that "surely the study of the behavior of social groups of all sizes as interlinked, reacting units must be included."[15] In 1954, he had also written a letter of recommendation for Marvin Minsky for a teaching fellowship at Princeton, noting that the future AI pioneer "may be expected to make real and continuous contributions to the area where the Theory of Automata touches on Neurophysiology," even arguing that his "total contribution" would exceed that of John Nash (1928–2015), the famous mathematician who had been Tukey's PhD student seven years prior. While Minsky thanked Tukey for helping him during "the weak position I have at the present," the two spoke infrequently, if at all, after that.[16] Even though his work overlapped with the dawning of the computer age, Tukey kept a large file from 1955 entitled "Machines," which included dozens of pictures of analog computers and notes on "slow computing."[17] Without showing outright hostility to McCulloch, Wiener, Minsky, or computer science, this correspondence contrasts with Tukey's ebullient and warm conversations with statisticians, and he rarely returned to these subjects (or correspondents) throughout his career.

In addition to showing only limited interest in cybernetics and Minsky's later work, Tukey also showed tepid support for another field that had recently emerged: the newly named "behavioral sciences." In 1959, for example, Tukey began a fractured stint with the "Behavioral Sciences Division" of the National Academy of Sciences–National Research Council (NAS-NRC). For someone who was skeptical of reducing "too much of human activity to the biological or individual," it proved a difficult fit. In the minutes from the July 8 meeting, held at the Rockefeller Institution in New York, the differences between the behavioral sciences and the classical science of man were made clear: the behavioral sciences "explicitly include . . . human biology (especially genetics, neurology, and physiology)," while also "exclud[ing] . . . areas that are dominantly historical or philosophical." In particular, sociology was singled out as a discipline that, decades after Eubank's struggles and the polemics of Ogburn and Hankins, "still embraces much that completely neglects the biological dimension in human behavior and is . . . reformist in tone." For Tukey, who was just two years away from publishing "The Future of Data Analysis," a guide to a priori theorizing about data sets, a note in the minutes from the executive secretary Glenn Finch might have seemed particularly jarring. "Logical analysis," Finch wrote, "always takes off from and returns to observation and experiment."[18] This would be the opposite approach from what Tukey would eventually recommend in his paper and textbook on data analysis. While Harry Shapiro, the committee chair, later noted that "the behavioral sciences do not necessarily need to ape the physical sciences," the direction of the committee seemed set.[19]

While Tukey did not comment in the minutes from the July 8 session, a year later he seemed finished with the Behavioral Sciences Division. Writing to fellow members of the committee from a bench in Penn Station ("where I missed the train to Princeton"), Tukey laid out in clear language what he considered the limitations of the group: "I believe that the responsibility of the committee is to determine what actions in the field of the behavioral sciences should be taken for the good of the world, the United States, [and] the behavioral sciences . . . and I had thought that the members of the committee had similar beliefs. I now begin to wonder. . . . If the committee were to take certain actions, each of us might find it appropriate to resign."[20] Tukey did not mention his specific objections, but he noted that it was a "well established result" of "behavioral science . . . to try to alter . . . the basic presuppositions of that culture or organization . . . to obtain the desired end." Tukey, who thought the initial stage of any scientific endeavor should begin with open-ended questions and multiple approaches, also complained that there was too much concern with "symmetry" and "completeness" in the group's aims, indicating that the Behavioral Sciences Division had preemptively defined its scope.[21]

Tukey did not resign, and it is possible that he influenced the first draft of the group's 1960 report from June 9, which noted several problems with the narrow approach adopted the previous year. The first problem listed was a familiar theme in the history of the sciences of man and the social sciences. The minutes noted dryly, "The introduction of scientific knowledge about people is always more difficult than knowledge about things or organisms."[22] Rather than a conceptual problem, however, the report complained that such difficulty made funding scarce. As historians have noted, and Talcott Parsons experienced, resources would generally only flow to social sciences that relied on "value-free" empiricism and methodological imperatives borrowed from the natural sciences. Tukey's concern about a broader definition of the field may have made the project less palatable for future funding.[23] Before even laying out the planned work of the future, the report warned of a "failure" of the project due to the lack of resources. "Under such circumstances," the group reported, "it would be impelled to suggest that it be disbanded or maintained merely as a stand-by committee."[24] Two years later, Tukey seemed uninterested and wrote to Shapiro after a delay of many months (the letter had "worked its way deep into the stack") that he was "unclear as to the need for a meeting of the NAS-NRC committee. Has anyone produced a good need?"[25]

By 1964 American foreign policy in Southeast Asia had produced such a "need." In writing to the disbanded committee, Finch explained a new set of practical goals for the re-chartered behavioral sciences group, including "measuring indicators of internal war possibility" and determining the "effectiveness of various courses of governmental action upon . . . the indigenous culture to which the action is directed."[26] In other words, behavioral scientists were being asked to consider how to prevent internal dissent, wage a neocolonial war, and occupy a foreign land, exactly the kind of "altering" of "culture" Tukey had worried about on that bench at Penn Station. Although he did not know it at the time, Tukey was being asked to join the notorious Project Camelot, a failed Department of Defense campaign to develop psychological counterinsurgency techniques and computer models to "predict" revolution in foreign countries like Vietnam. Not only were its antidemocratic ideals revealed in Congress in 1965 as antithetical to university work, but Project Camelot also exposed the limits of the objectivity of social science, where militaristic rather than reformist values held sway.[27] If the military aims were not clear enough to members of the group, Finch also told the committee that they would now be part of the Human Factors and Operation Research Division "at the request of the Army Research Office." Though Finch sent his letter in September 1964 and asked for a "reply within a few days," Tukey did not even respond to such an extraordinary request until seven months later. Whatever his reasons were, he wrote Finch on April 16, 1965, to say only that he could not join such a project, and Finch should have assumed that when the letter "fell into a wordless silence, you realized I was much too busy."[28] Though Tukey would later rejoin the group after Project Camelot was over, his files and letters show little interest in the Behavioral Sciences Division after 1965.

Tukey's fading interest in the behavioral sciences coincided with the emergence of one of the most novel approaches to quantitative data from the twentieth century: data analysis. Though it is a relatively recent term, Tukey first applied the idea in a 1961 paper that called for statisticians to play a more robust role in science. One of Tukey's most cited papers, "The Future of Data Analysis" (1962), combined a technical guide of statistical techniques with a manifesto to encourage statisticians to do more than simply report regression analyses or run "goodness-of-fit" tests for social and natural scientists. While it is speculation to suggest that the reductive work Tukey witnessed in the Behavioral Sciences Division *inspired* data analysis, his article provided a full-throated defense for any statistician who imagined themselves as more than confirmatory actors for the data collection of endless behavioral experiments. While it was full of mathematical details about how to handle particularly tricky data sets, as well as how to graphically represent what he called "spotty data," at the core of the paper was the belief that "large parts of data analysis are incisive, laying bare indications which we could not perceive by simple and direct examination of the raw data."[29] Though Tukey put the matter a bit more delicately, this was an echo of Alfred Marshall's claim that facts do not "speak for themselves." Tukey counted among "classical" data analysis techniques and operations like Gaussian distribution—which Quetelet had used—as well as Galton's and Pearson's chi-squared and regression analysis, and noted that most of these techniques had reached "a respectable antiquity."[30] But as statistics had grown in the years since, the ability for data analysis—including techniques for gathering, interpretating, and presenting data—had not kept up.

While few would argue that statistics needed better ways to gather, interpret, and present data, what made Tukey's work "heretical"—or, in the affectionate words of Chris Wiggins, the Chief Data Science for the *New York Times*, "kooky"—was his belief that this process should be "regarded as a science" rather than a helpful skill.[31] Tukey quickly qualified that it was a science based on "a ubiquitous problem" rather than a "concrete subject," but he nevertheless believed that it passed all possible tests for what constituted a science. Tukey argued that data analysis was even more of a science than mathematics, for example, because it included a "test of experience as the ultimate standard of validity," while in mathematics "validity is an agreed-upon sort of logical consistency and provability." This was an appeal to the occasionally pejorative idea that a "pure science," like Euclidian geometry, was one so perfect that it cannot be confirmed in the real world. Therefore, in his words, "data analysis . . . must take on the characteristics of a science rather than those of mathematics." Tukey's decision to have large areas of statistics break free from association with mathematics—a move he formalized by moving the Princeton Department of Statistics out of the Department of Mathematics in 1965—had significant consequences for its methodology, of which he listed three:

1) Data analysis must seek for scope and usefulness rather than security.

2) Data analysis must be willing to err moderately often in order that inadequate evidence shall more often *suggest* the right answer.

3) Data analysis must use mathematical argument and mathematical results as bases for judgement rather than as bases for proof or stamps of validity.[32]

Taken together, this was a considerable challenge to much of the quantitative work being done in twentieth-century social science. In specifying "usefulness," the preference for the "right" answer, and the marginalization of mathematic "validity," Tukey stressed precisely the kind of methodological tools that social scientists like Ogburn, Hankins, Knight, and Hull wanted to abandon. Put simply, Tukey was the rare quantitative voice of the Cold War era who believed in the classical vision of the science of man, where practical utility and moral values mattered. Tukey did not identify the social sciences specifically as areas where data analysis should search for "usefulness," "err moderately," and use "judgement" over "validity," but this was a far cry from Glenn Finch's note from the Behavioral Science group that "logical analysis . . . always takes off from and returns to observation and experiment." On the contrary, Tukey imagined data analysis as preceding (and therefore shaping) observation and experiment, similar to Parsons's "conceptual scheme" that shaped all subsequent actions. In stressing "usefulness" over "security," he also resurrected an idea that had been central to the science of man from the Enlightenment through Earle Eubank's time at the University of Chicago. In light of the scientism of behaviorism, the suggestion that the data analyst should "err" in order to "suggest the right answer" seemed almost profane.

For Tukey, however, the false presentation of statistics as a tool of certainty had done significant damage to the field, turning statisticians into nonthinking confirmatory automatons rather than careful scientists. As the British statistician R.M. Loynes lamented in 1969, the field was in danger of being reduced to mathematics, or what Loynes called a "too facile writing down, in pseudo-exact form, of relations which merely express our own ignorance."[33] Because Tukey did not write as a social scientist, hoping to apply the methods of the natural sciences to the study of human thought and action, he was able to recognize that science work need not be unreflective regression analysis of statistical data garnered from experiment. He claimed in "The Future of Data Analysis," "A scientist's actions are *guided*, not determined, by what has been derived from theory or established by experiments. . . . They know that they will sometimes be wrong; they try not to err too often but they accept some insecurity as the price of wide scope."[34] As an example of the kind of misleading ethos that had pervaded statisticians' thinking, Tukey singled out a 1955 article from the *Journal of the Royal Statistical Society* that decried the "appalling position . . . in which one can get any answer one wants if only one goes around to a large enough number of statisticians." Tukey saw nothing "appalling" in ambiguity and believed the defensive response of the article "nearly typifies a picture of statistics as a monolithic, authoritarian structure designed to produce the 'official results.'"[35] In Tukey's view, the fact that different statisticians gave different answers was not a condemnation of the field but a reminder that statisticians were still thinking.

To show that Tukey wanted to strengthen rather than banish quantitative analysis, his popular textbook *Exploratory Data Analysis* (1977) condensed the sometimes difficult language of "The Future of Data Analysis" to a kind of cat-echism for future students.[36] Its preface began (boldface and capitalization in-cluded) with a note that it was based on a "simple principle": It "**is important to understand what you CAN DO before you learn to measure how WELL you seem to have DONE it.**"[37] In the introduction Tukey assumed that all students of statistics were familiar with what he called "confirmatory techniques"—chi-squared fitness test, P-value significance tests, least squares, and regression analysis—but stressed that statistics had once meant much more. In referencing a world before Quetelet, Gauss, Galton, and Pearson, Tukey noted, "Once upon a time, statistics only explored," but such a process had been diminished to "mere descriptive statistics." While not denying the importance of confirmatory techniques, Tukey believed they had created a prison for statisticians, as "exact confirmation" led to a lack of "flexibility."[38]

Contrasted with the "quasi-judicial" role of confirmatory statistics, Tukey offered his students the chance to act in another role, as "detectives" who probed the first sets of data to look for clues to potential meanings, inter-pretations, and courses for later analysis. In a 1969 paper he asked, simply, whether "analyzing data" was "sanctification or detective work."[39] Conceding that the numbers include numerous "accidental and misleading" indications, Tukey believed no good detective would preemptively dismiss any clue. In other words, Tukey had an affinity for outliers, randomness, mistakes, and accidents in the data that disappeared in the normal confirmatory processes, and *Exploratory Data Analysis* was meant as a way for statisticians to regain their role as interpretive, analytical, and creative—in other words, *thinking*—be-ings.[40] At a celebration of Tukey's life and work in 2014, the mathematician Peter Huber remarked that Tukey had taught him to value "judgment" rather than "proof," and that "data analysis should take on the characteristics of a science rather than those of mathematics."[41] In rueful tones that contrasted with the overall celebration, Huber quoted the British statistician George E.P. Box (1919–2013). Addressing a room largely sympathetic to Tukey, Huber reminded his audience of Box's lament that "we've settled to be second-rate mathematicians when we could aspire to be first-rate scientists."[42]

After the Flood: Bias and Meaning in Artificial Intelligence and the Computational Social Sciences

To judge by twenty-first-century data collection, manipulation, and presentation, it appears statistics as a profession has accepted its lot. As the statistician Loynes said in his lament for the work of his colleagues, "The system was a black box, and so long as we could measure the relationship between stimulus and response . . . it was unnecessary to probe inside the black box to see what was going on."[43] Rather than the proliferation of trained experts in data analysis and a professional class of scientist-statisticians, the decades following *Exploratory Data Analysis* were awash in visions of revelatory Big Data and the almost theological belief among America technology companies that raw data collection alone was sufficient (and profitable).[44] Almost by accident, it was these same companies that largely took over the project of artificial intelligence, with the convenient belief that raw data would triumph where state expertise and regulations had failed.[45] As many scholars of the Cold War have shown, current thinking about data emerged in the social sciences of the mid-twentieth century, when social scientists operated as if the previous century had not existed. Joel Isaac, for example, demonstrated that since the Cold War, there was a push in all fields to "accumulate raw, 'value-neutral' empirical data" in a wide range of social sciences, and Hunter Heyck found that the "systematic" era from 1955 to 1970 was underwritten by Herbert Simon's assumption that "there was a 'universal man' about whom one could construct a universal social science."[46] The technocracy that Hull, Parsons, and other Cold War social scientists had imagined was also replaced by a more simplistic ethos borrowed from economics, usually tagged under the imprecise heading of "neoliberalism," which believed that numbers—like information and capital—should be free. James Surowiecki pointed out in the defining book (or at least title) of the movement, *The Wisdom of Crowds* (2005), that when it came to quantitative data, random "group intelligence" seemed to beat expert opinion almost every time.[47] As Tukey might have noted, this was not detective work. It was sanctification.

Though the perils of naive data collection might have been apparent since Quetelet, one academic discipline has attempted to revive the geometric spirit once more to create a science that can explain "real world" behavior through the use of social data "traces." When the authors of a 2009 report in *Science* announced the emergence of the field—called the computational social sciences (CSS)—they claimed it would be able to create "enormous progress for the public good" through combining academic research with data compiled by American technology companies.[48] Even a cursory review of the recent history of CSS reveals that data analysis as Tukey understood it was not part of the process and that the careful attention to theorizing, interpretation, and presentation has been overwhelmed by data. For example, in the breathless enthusiasm of *Social Physics* (2014), one of the founders of CSS, Alex Pentland, declared a revolution in science from the lessons of data collection. Although Adolphe Quetelet and Auguste Comte had vied over the title *physique sociale* in the first decades of the nineteenth century, the subtitle of the book when it was first published claimed it offered "Lessons from a New Science."[49] As Pentland, former head of the MIT Media Lab, claimed, "Social physics is a *new science*" that uses the "digital bread crumbs we all leave behind us."[50] In precisely the kind of anti-intellectual move Tukey warned about, Pentland explained how "data tell us the story of everyday life by recording what each of us has chosen to do."[51] Not only did Pentland avoid any methodological issue of data analysis, but he did not seem to recognize that there was once a science that attempted the very same thing—with the very same title—170 years prior.

Similarly, the authors of *Reality Mining* (2014) claimed that data analysis was dedicated to "chang[ing] people's lives for the better" by showing users how to access and acquire "mineable data" before "applying" the data.[52] Focusing purely on instrumental techniques to find and acquire data, and listing the many ways by which data is acquired, the book included no discussion of the kind of detective work that precedes such mining operations, nor any acknowledgment of the history of data evaluation. In another recent book attempting to apply "Big Data" to psychiatry, the author laments that he "has no quantitative measures to determine whether and how my treatment works," wondering whether it was possible to get "real-time" data on the "neurons and receptors" of his patients.[53] Yet as a review of the book noted, "the quest to find quantifiable measures . . . seems more descriptive than scientific," as there is little theoretical work to understand what numbers might be meaningful.[54] More historically informed approaches to the field of CSS, such as

Frank Schweitzer's review "Sociophysics" in *Physics Today*, recognized that the history of previous attempts at a "science of man" have been fraught with problems, from Comtean positivism through mid-twentieth-century hopes for sociobiology and eugenics, but such historically informed pieces have been in the minority.[55] Though written a decade after the original science article that introduced CSS, Schweitzer conceded that in 2019 "foundations are still being laid" for the field, a refrain quite common to the social scientist of the first half of the twentieth century.[56] As an indication of current ambivalence toward CSS, David Lazer, the primary author of the *Science* article that announced the set of disciplines, has spent the last few years warning *against* the enthusiasm for Big Data, noting that once-touted breakthroughs have been misleading.[57] In 2020, the original group who founded CSS returned with a list of "obstacles." Though they offered a number of solutions, nothing like Tukey's "creative act" was included.[58]

As Lazer's work indicates, enthusiasm for CSS work has cooled, mirroring the fading hype of many "data-driven" approaches discussed in this book. Rather than offering a panacea, dissatisfaction with progress in CSS has even led to soul-searching on the part of corporate leaders who were expected to provide the data for the new "social physics." A 2019 report in the *New Yorker* attests that the connection of technology companies and large data collection is at the heart of a waning enthusiasm in Silicon Valley for a socially beneficent internet and AI, as dispirited tech founders head to resorts like Esalen to find peace with the many destructive and harmful effects of their products.[59] Like social scientists before them, AI visionaries and data collectors have seen early enthusiasm fade, even as the profitability of generative AI grows. Examples of regretful innovators include Chris Hughes, an early tech founder who now calls for monopolistic social media companies to be broken up, and the engineer-turned-ethicist Tristan Harris's Center for Humane Technology, which has tried to replace screen time with "Time Well Spent." In a *mea culpa* posed as *volte face*, Nir Eyal, the 2009 author of a bestseller that urged data collection tools to prompt addiction in users, now promotes a book meant to free people of their addiction.[60] On the academic side of CSS, physicist Pawel Sobkowicz has made a "plea" to computer scientists and physicists to engage with the real-world effects of their models and simulations, a role Tukey had taken for granted. As Sobkowicz rightly noted, there is already a history of the "deliberate use of . . . social simulations to promote ethically questionable goals," and the kind of slight "nudges" of social scientists can be brought to bear on much more adverse forms of manipulation.[61] Most recently, Geoffrey

Hinton, the "Godfather of AI" who helped shape modern generative AI, spent much of 2023 warning of the dangers of his invention.[62] In perhaps the most baffling comment—but one that betrayed the positivist and amoral stance necessitated by modern data science—Hinton claimed that he had previously followed the mantra of Robert Oppenheimer, that "when you see something that is technically sweet, you go ahead and do it," regardless of the consequences.[63] Hinton may be applauded for his newly found commitment to the human costs of technology, but one wonders why neither the example of Oppenheimer—depicted on thousands of movie screens that very same summer—nor the *atomic bomb itself* did not previously call into question his commitment to value-free science.

Perhaps the most obvious example of the failure of "raw data" and generative AI to produce clear scientific knowledge of human behavior can be found in the extraordinary level of bias in the machines. As documented in a host of bestsellers and academic studies, modern AI systems, trained on millions of bits of shared data, have reproduced (and introduced) bias, potentially hardwiring human prejudices into the supposedly objective world of statistics, mathematics, and computer science.[64] To take just a few of the now hundreds of examples, Amazon facial recognition software, trained on more images of whites than of minorities, consistently misidentifies the latter. Predictive software routinely returns images of men for searches of "doctor" and of women for "nurse." A 2013 paper found that searches for names coded as "Black" had a far more likely chance of including an advertisement for locating a criminal record, and even when white-coded names that *did* have criminal records were entered, no such ad appeared.[65] A 2016 White House report on Big Data that expressed optimism that "data systems will contribute to removing inappropriate human bias where it has previously existed" also warned that the rise of machine learning posed "great risks" that such "innovations could perpetuate discrimination and unequal access to opportunity as the use of data expands."[66] Safiya Umoja Noble traces all this and more in *Algorithms of Oppression* (2018), beginning with her disturbing revelation that a simple search for the phrase "black girls" produced a stream of pornographic sites.[67] The list goes on, as Amazon and others push forward with biased algorithms.[68]

In a sign of concern over the naive approach described in *Social Physics* and other books that linked corporate algorithms to academic research, in 2017 a variety of European companies and think tanks launched a three-year series of symposia on challenges in the computational social sciences. In 2018, the group met in Cologne, Germany, to address "Bias and Discrimination," bringing together engineers, computer scientists, quantitative social scientists, and at least one historian. Though the speakers varied in their concerns, the consensus was that bias had gone from a relatively ignored feature of AI to an overwhelming concern, as one poster reported a "massive increase" in papers on biased algorithms between 2014 and 2017.[69] In particular, Christo Wilson's talk on bias at monster.com and indeed.com showed that white men were significantly overrated on the sites and were much more likely to appear in the algorithms.[70] Other talks found that women were well underrepresented in Wikipedia and that darker-skinned women were correctly identified by Google Vision only 68 percent of the time.[71] Elizaveta Sivak found that men posted 45 percent more tweets about their sons than daughters and that women liked these posts more often.[72] Several presentations focused on COMPAS, a supposedly neutral AI system used by the legal system to predict recidivism rates in parole cases, but which instead seems to act mostly as a proxy for race. As Sarah Hajian put it, it was a textbook example of "how to make a racist AI without really trying." Summing up much of the weekend, Hajian declared that "Big Data is unfair" and bias "a blot on the profession."[73]

When it came to solutions, it was clear that the conference attendees had abandoned the early, naive vision of CSS, where the wisdom of the crowd and "group intelligence" would root out any bias. As was noted by one CSS researcher over lunch at the Cologne conference, the "entire point" of AI search recommendations and predictive software "is to be discriminatory" in the nonprejudicial sense of the word. And indeed, one of the first significant papers in the world of CSS—one that controversially used proprietary and nonvoluntary data from a social media site—was based upon successful discrimination, correctly guessing the gender, race, and sexual preference of a high percentage of users based on their social media habits.[74] Yet nearly all the "solutions" involved ad hoc technical fixes rather than interpretive and a priori "creative acts." In a paper entitled "Where Does Data Bias Come From?," Emre Kiciman of Microsoft Research noted that in some cases bias needed to be included. For example, he noted that it would be important to log the race or gender of individuals in medical data so that one particular group was not overrepresented. While previous attempts to root out data have been focused on trying to attain "neutrality" in the data collection process, Kiciman argued that such efforts could never be "rigorous." Instead, he recommended a "technical solution," one that bypassed the presumably important question of *why* data had become biased. In the same way that machine learning AI does not really "know" how to play the games it dominates, Kiciman suggested that in statistics actually knowing the causal structure was irrelevant to reducing bias. Instead, he suggested that combating bias in algorithmic design should look only at outcomes, an approach diametrically opposed to data analysis but in keeping with the methodologies of past positivist social science.[75]

Where many CSS advocates adopted a positivist approach, researchers at the Alan Turing Institute have suggested that bias could be controlled "by explicitly modeling the causal structure of the world."[76] While "modeling the causal structure of the world" might seem to be a difficult project in itself, it is even more challenging in studying AI given that the vast amount of data necessary to build models are contained within the "black box" of proprietary corporate servers and algorithms. In late 2023, as it became clear that AI was becoming *less* transparent as it became more profitable, a group of researchers at Stanford launched what they called the "Foundation Model Transparency Index." One of the participants noted of the major AI developers, "there's no information about what these models are, how they're built and where they're used."[77] The major corporations who hold this data have little incentive to release such information, and even when people can see the specific information gathered about them as they use the internet—as in the case of many European countries—they do not have access to the metadata, or "shadow text," created from trace data on the internet. Crucially, this means that all of the technical plans drawn up by computer scientists to remove bias require even more extensive data collection and evaluation about the tastes and traits of different races, genders, and religious groups. Recognizing this problem of proprietary algorithms, a number of computer scientists have sought the aid of more traditional social scientists to try to crack the code—or reverse engineer—the hidden mechanism. At the Data Science Institute at Columbia University, for example, a team of researchers created fake email accounts and populated them with a wide variety of messages, essentially testing AI systems to see which advertisements and replies they would receive based on different messages. While the experiment was successful on a technical level, the group has also called for help from non-quantitative social scientists in order to frame the right kind of messages to send to these accounts.[78] To literalize Ian Hacking's notable phrase, social scientists were now needed to "make up people" to study algorithms—algorithms that had themselves been based on artifacts of human communication.

Another example where humans have been trained to act like machines in order to fool algorithms occurred in recent work at Northeastern University to "audit" search results. Rather than create fake users, the team assembled a group of real people to evaluate the results of a number of searches for political information. Such work overlaps with the same group's finding that web search results can be highly influential in elections, and in fact play a much stronger role in influencing voting patterns than social media or "fake news."[79] Their project to "score" the results of algorithms was built on the work of the communications researcher Christian Sandvig and others, who in 2014 laid out a number of possible ways to audit search engine results.[80] Though one might expect that computer scientists could design a computer program to conduct a review of how search queries are returned, the authors note that it is currently only legal to use *actual people* in these audits. As Sandvig's research group put it, "while it might seem absurd that it would be illegal to use a computer program to act as a user, yet legal to hire a person to act like a computer program, indeed this seems to be the case."[81]

The Columbia and Northeastern groups have offered ways to crack the underlying code of the shadow text, but such solutions may create even more problems. Though as academics their methodologies are open, they require substantial efforts on the parts of real people to submit themselves to quantitative interrogation. Relying on the "ghost work" of thousands of largely low-paid and anonymous workers, these "taskers" ironically must engage in repetitive labor to crack the mysteries of AI algorithms trained on repetitive tasks.[82] At its most basic level, such research involves extraordinarily long and automated human labor, even for the researchers themselves. Those in academia who seek to uncover bias in the field often report in their methodologies of having to sit and watch thousands of online videos in order to "score" their content. In just one example, a researcher at the CSS conference in Cologne endured over one thousand videos of hateful speech and graphic violence to "uncover" the recommendation algorithm, an experience not unlike the low-paid content filters who sit through a daily eight hours of pornography, violence, and offensive speech to "clean" out the world largest social media sites.[83] To understand the "shadow text" requires, in other words, long work in the shadows for thousands of actual people.[84]

In addition to the basic irony (or absurdity) of using thousands of hours of poorly paid and mundane human labor to crack the code of AI algorithms built on unpaid and mundane human labor, there has also been surprisingly little discussion or agreement on the ultimate aims of this work. One thing researchers have agreed on is that the ultimate goal was to reduce bias and discrimination in the real world and to make sure that recommending algorithms employ what is known as "ecological or external" validity, meaning that outcomes in the digital world match up to fairness in the real world. One major problem with this model is that of course the real world itself is far from fair, a fact brought home in an extraordinary paper from 2016 that found bias to be latent in language and culture itself. As the authors put it, they "document machine prejudice that derives so fundamentally from human culture that it is not possible to eliminate it."[85] While some might argue (or joke) that this means AI bias counts as a *success* in achieving human-level status, one wonders if the massive data collected by social media sites and technology companies that fuels such bias truly reflects some transhistorical "human culture."

Even more profound a challenge may be that ecological validity tests are based on a rather old-fashioned assumption, which still maintains that our actual, real, or offline worlds are in some sense represented in the online world. It may, however, no longer be tenable to imagine that social media and internet users fashion some form of "avatar" online and that this being is fundamentally different from their biological self. Indeed, it is becoming increasingly clear that online selves and online artifacts seem to be more important—or at least more valuable from an economic perspective—than the "real" people whose actions created these beings. In the world of social media, some have argued that "social and machine learning . . . are reinforcing and shaping one another in practice."[86] While earlier rules-based and symbolic AI had tried (like Turing) to imagine what the human mind was doing while it was computing, "naive" machine learning has proceeded successfully with no assumptions or inductive conclusions whatsoever about how and why human beings operate. In this account, people, not machines, now exhibit data hunger and the "searching disposition," leading to a "new sense of selfhood."[87]

If this is the case, then much of the handwringing over AI bias and machine learning issues may be moot. While the computer scientists seeking remedies to AI's ills are at the cutting edge of technical development, their assumption of "ecological validity" and the distinct ontologies of real and virtual worlds seem an almost quaint relic of the earliest days of the online world. Ironically, by employing so many to spend so much time tracking these algorithms, many contemporary solutions to the problem of bias only increase the speed at which obliteration of "ecological validity" occurs. As Shoshana Zuboff claimed in *Surveillance Capitalism*, "It is no longer enough to automate information flows *about us*. . . . The goal now is to *automate us*."[88] Yet this might not go far enough, as it still assumes an "us" that could be automated. In a world where automated assistants make purchases and schedule meetings, the "algorithmic self" or "data double" buying the item is not the external self but an intelligent artifact created through the accumulation of online data sharing.[89] Pushed to the extreme, it might even be argued that the algorithmic self is no more real (or false) then the 1950s consumerist "self," an imagined being created by postwar consumer culture and advertising, and that AI transcendence of the "human level" would only be swapping out one imagined self for another.[90] Though computer science researchers may be celebrated for their attempts to promote real-world fairness in the machines, a review of attempts to combat bias suggests that ad hoc and technical fixes have not succeeded in removing bias from the data.

Given the general lack of historical understanding in modern statistical research, it should not be a surprise that the abuses of data and bias from the past have endured in other forms. A 2022 review of a book in the field of "sociogenetics" noted, "Biological essentialism, aimed at demonstrating an innate hierarchy of intelligence, is going strong after more than two centuries of empirical failure." As the reviewers state of the methodological approach in *The Genetic Lottery* (2021), such studies utilize a "passive voice" to imply that "no one actively does all [the] estimating, measuring, counting weighting [and] correlating."[91] In other cases, "new" biometric data has inevitably meant the return of statistical racism, as supposedly value-free conclusions are made based on the data alone. The repackaging of eugenics has proved remarkably enduring across the world, and statistically based claims of racial difference persist, making the controversies over Charles Murray's *The Bell Curve* seem quaint.[92] In what may be a harbinger of the dangers of new data collection of physical and mental features, a popular work by a once-accredited journalist now essentially lays out a case for bringing statistical racism back, as quantitative archaeologists still naively argue that they cannot deny what the data tells them.[93] In one unintentional irony, the author of *The Genetic Lottery* argued in the opinion pages of the *New York Times* that "progressives" should "embrace" statistical research into "links between genes and educational success," failing to mention that one of the major criticisms of early progressives of the twentieth century was their embrace of just such research.[94] While the long use and abuse of statistics and science in support of discrimination has been well documented, most popularly by Stephen Jay Gould (1941–2002), such lessons can be learned only by those who read history.[95] Though historians, humanities scholars, and non-reductive social thinkers have recognized that the problems of naive data collection and bias have been at the heart of the sciences of man for centuries, their lessons seem not to have been learned by many quantitative social scientists, computer scientists, and AI researchers of today.

Conclusion

As seen in this survey of recent uses and abuses of AI and CSS, much statistical work remains confirmatory, with little role for statisticians (or anyone) to shape the earliest questions and processes of online data collection, and it seems fair to claim that Tukey's idea of a data detective did not materialize. While the methodologies of the social sciences did not necessarily triumph, positivist approaches toward data collection and examination remain in many major technology companies and research institutes. The "value-free" experimenters, verifiers, and equation writers that William Ogburn, Frank Hankins, and Clark Hull wanted arrived, although they are no longer housed in social science departments. In fact, in the decades since *Exploratory Data Analysis*, data has instead become "sanctified" as AI machines have learned primarily through the brute application of thousands and sometimes millions of pieces of data gathered for no ostensible reason other than to increase the profitability of major technology corporations. In "The Future of Data Analysis," Tukey had hoped statisticians would be taught to *think* about data in more expansive ways, but given the profusion of STEM classes and recent concerns and confusions over the "P-value" crisis, the problem of statistics as mere "confirmatory techniques" continues. As a result, the creative act of statistics is reduced to the kind of blunt confirmatory correlation claims popularized in books like *Freakonomics* and recounted in podcasts like *Hidden Brain*. Although several new forms of data evaluation are emerging, rejecting the homogenous averages of the past, absent worker revolts in Silicon Valley or government-mandated unpacking of proprietary technologies, there is still little statisticians can do to shape the kind of data collection that currently powers most AI systems.[96]

As a result, the field of AI now seems close to where the social sciences in America were in the 1960s, fresh from decades of tremendous enthusiasm and resources, but without a consistent methodology or purpose beyond exposing more and more neural networks to more and more data. In a recent "brief" history of AI, the computer scientist Michael Wooldridge admitted that the idea to build "machines that are self-aware, conscious, and autonomous in the same way that people" are may seem an "intuitive and obvious" goal, but that such a goal is now in dispute: "The truth is we don't remotely understand what it is we want to create or the mechanisms that create it in people."[97]

Reflecting the failure of brute-data neural networks is the fact that researchers have begun deploying a new term—artificial general intelligence (AGI)—to refer to what was originally meant by artificial intelligence. As Stuart Russell noted of the proliferation of a *new* idea, "Artificial General Intelligence . . . is really just a reminder of our real goals in AI. . . . In that sense, AGI is actually what we've always called artificial intelligence."[98] As seen by the enthusiasm over generative AI in 2023, the potential financial benefits of chatbots may render the actual debates over various AI models moot. Yet more than seven decades after Turing imagined the idea of a thinking machine, AI would seem then to be no closer to understanding what the machine is supposed to imitate, still mired in the endless definitional questions that plagued the social sciences for centuries.

As has occurred so often in the past four hundred years of social scientific research, initial enthusiasm about AI has been reduced to more prosaic concerns, and like many of the behavioral sciences, AI uses have found their greatest success in advertising. Gary Marcus, the founder of Uber's AI Labs, noted in a recent interview, "AI research and implementation right now is mostly about ad placement."[99] In her 2019 book *Artificial Intelligence*, Google's former head of ethics, Melanie Mitchell, pointed out that even the most ambitious programs for AI—such as IBM's Watson and, more recently, the late Douglas Lenat's project to build common sense in machines—usually devolve into a handful of helpful but overhyped "applications." As she pointed out, "foundational AI research efforts with vast scope and ambitious" devolved into a "set of commercial products with elevated marketing claims."[100] Mitchell has even gone so far as to critique modern AI as being unable—and, given its current trajectory, unlikely—to break through the "barrier of meaning" without changing its approach.[101] Sounding more like an interpretive social scientist of the 1970s concerned with behaviorism than a figure with a long and distinguished career in building AI machines, Mitchell worried that the field had "hit a wall" because it was unable to build machines that can move beyond brute-data collection toward abstraction and meaning. In perhaps a sign of the field continuing to struggle with bias and meaning, both Mitchell and another leading Google ethicist critical of the company's approach toward bias—Timnit Gebru—were fired in the years after Mitchell's book was published.[102]

While the future of AI research is far from settled, the astonishing rush of capital into generative AI may settle the matter for the foreseeable future, creating a freeze in new methodological discussions of what AI is supposed to be doing. Like the social sciences, AI could lose any sense of unity beyond the name, where the field becomes splintered into groups who point to the success of practical AI versus those who are increasingly looking for meaning, true AGI, and reliable insights into human intelligence and consciousness. In many senses, AI has already reached this point, with a recent debate between Gary Marcus and one of the leading developers of neural networks, Yoshua Bengio, recycling almost the entire history of mechanical philosophy, social physics, probability theory, evolution, and behaviorist theories of the mind.[103] Given the centrality of brute-data neural network AI to the business model of the world's richest corporations, practical neural nets will endure for some time, as the "wall" these nets have hit is only conceptual, not financial. Yet for those interested in AGI, one possible solution might be to return to a vision like Tukey's, where data analysts in AI—engineers, computer scientists, and mathematicians—are given broader training than just mathematical techniques and therefore are able to imagine creative solutions beyond technical repairs. Such was the suggestion of d'Alembert, Diderot, Bonald, Maistre, Veblen, Wesley Mitchell, Parsons, and scores of social thinkers who worked against reducing human intelligence to a single process.

If Melanie Mitchell is right—and it is meaning that is missing—there are several possibilities for how AI might better approach the human level. One course of action would be for the social sciences to continue the trajectory traced in this book and reduce human actions even further into the world of natural sciences so that they are more easily modeled. Indeed, many of the works mentioned above do just this, and such an approach might find its most popular audience in the world of neurobiology, where a 2023 work declares that "there is no meaning" in human life and that all thoughts and actions are due entirely to "cumulative biological and environmental luck."[104] If so defined, the machines' task would be all that much easier. Yet there is another solution, one that might elevate the complexity of machine learning systems rather than one that once again simplifies the human level. Perhaps AI workers would benefit from rigorous training in fields that have long dealt with such questions of meaning, including history, literature, philosophy, fine arts, and the humanities. The jettisoning of insights from the liberal arts and humanities may bring about initial success and increased resources, but as the fates of many sciences of man from the Enlightenment to the Cold War seem to demonstrate, there is a cautionary tale about relying too much on the dreams of data collection and mathematical eloquence alone.

CONCLUSION

The Intelligent Artifact

It is comforting . . . to think that man is only a recent
invention, a figure not yet two centuries old, a new wrin-
kle in our knowledge, and that he will disappear again as
soon as that knowledge has discovered a new form.
—Michel Foucault, *The Order of Things* (1966)

Some days, helping men solve problems they had created
for themselves, I felt like a piece of software myself, a bot:
instead of being an artificial intelligence, I was an intelligent artifice, an empathetic text snippet or a warm voice,
giving instructions, listening comfortably.
—Anna Wiener, *Uncanny Valley* (2020)

BEFORE OFFERING A FEW CONCLUSIONS ABOUT THE RELATIONSHIP BETWEEN
the history of the social sciences and the state of current AI research, it might
be a good idea to briefly restate the main argument and narrative arc of the
book. The book has traced the descent of the idea of artificial intelligence
through attempts over the past four hundred years to reduce the complex
thought and behavior of human beings to a level where they might be studied
by a variety of tools and methodologies borrowed from the natural sciences.
As seen in the stories of key AI papers by Alan Turing, Warren McCulloch,
and Walter Pitts, the earliest approaches to modeling the human mind were
rooted in popular behavioralist approaches of the mid-twentieth century,
which themselves were based on scientific methodologies borrowed from hundreds of years of work in the sciences of man. Just as importantly, however, the
sciences of man helped shape the perception of the "human level" since the
age of Descartes and the geometric spirit, as social sciences researchers—and

even the subjects of such research—were expected to conform to the narrow categories suggested by reductive methodologies. While such methodologies often began in tandem with Enlightenment ideas of human rights, liberty, equality, and evolutionary progress, moral considerations were often eroded in fidelity to "objective" methodologies, which abandoned many of the humanitarian values professed by Condorcet and the "social scientists" of the Scottish Enlightenment. Of the many examples of reductive approaches, the ones covered in this book include Descartes's mechanical philosophy, the geometric spirit of Fontenelle and Condillac, Adolphe Quetelet's social physics, William Stanley Jevons's "hedonistic calculus," the "laws" of classical political economy, Francis Galton's eugenics, William Ogburn's verification fetish, Milton Friedman's positivist economics, and Clark Hull's behaviorism. Whereas traditional stories of AI involve smarter and smarter machines "rising" to the "human level," the account here suggests rather that AI might result from the descent of the human level to another kind of artificial intelligence: simple people.

For the critics of these approaches, simple people were not mere abstractions. In fact, concerns about the adoption of mathematical, quantitative, and empirical tools from the natural sciences went beyond worries that the sciences of man failed to capture the rich diversity of human action and behavior. Rather, critics worried that such approaches to studying humanity, in the form of education, political intervention, acculturation, and diffusion, might *create* the very kind of people social scientists purported to be studying; or, as Diderot claimed, the "metaphysics" of the geometric spirit meant that "bodies are stripped of their individual qualities." From the very beginning of the geometric spirit—which Fontenelle imagined could only be practiced through "right conduct"—French authors from Gabriel Daniel and Pascal through Rousseau, d'Alembert, and Diderot worried about what effect such a simplified approach might have on the mind of a "geometer" or practitioner of the sciences of man. The reactionary "prophets of the past" who followed argued that exclusive education in the sciences, particularly mathematics and physics, limited the full scope of intelligence for a generation of Frenchmen. In the nineteenth century, the future Belgian prime minister Pierre de Decker countered Quetelet's determinist critics by claiming social physics only determined the lives of those who participate in state and science, but in the political economy of Harriet Martineau, Charles Knight, and Jevons (and the eugenics of Galton), full public participation was *necessary* for social scientific theories to succeed. And in just the past century, the critiques of Thorstein

Veblen, Earle Eubank, Charles Ellwood, Wesley Mitchell, H.S. Jennings, Talcott Parsons, and Charles Taylor revealed that the methodologies and disciplinary practices of sociology, economics, and psychology often shaped, rather than revealed, the nature of human thought and action, with social science researchers frequently being the first victims, forced to give up personal and moral motivations as they became professional social scientists. To trace a history of the development of reductive social scientific methodologies and its critics is therefore to see that the path to creating something like "artificial intelligence" required more than the technical achievements of mathematics and machine building. It also required the construction of an entirely new methodological approach to studying humanity.

As the roboticist Illah Reza Nourbakhsh stated at the outset of this book, echoing Diderot's worry that the geometric spirit would eliminate "individual qualities," "We may accomplish the AI dream by stripping humans of" their "singular identity," and no group seemed closer than those researchers caught up in the work of quantitative social science and engineering. Indeed, Turing's lack of explicit interest in theorizing about human thoughts and behavior directly led to the birth of AI, and modern-day researchers in AI and CSS can be seen as yet another methodological extension of the geometric spirit into the study of man, one that traces from Quetelet and Galton through Hull and Ogburn, who all believed that more data and quantitative tools alone would reveal the laws of human thought and behavior. And, as philosophers of science have shown, such simple motivations and lack of reflection are often part of the scientific method borrowed from the natural sciences. For example, the philosopher of science Michael Strevens recently explained that such "simple minds" are in fact central to modern science, noting that the best path to scientific success would be to "equip students of science with only empirical modes of reasoning and only empirical knowledge" and "to give them the capacity for empirical thought alone."[1] For those who might argue that the creative types of d'Alembert's "analyst" and Tukey's "detective" would help those invested with the geometric spirit, Strevens has a stark reply: scientists' "obliviousness is the greatest guarantee that they will follow without deviation the empirical path laid out."[2] If there is anything to add to this thesis, it is that perhaps such rigid thinking is even more common in the social sciences modeled on natural science methodologies than in the natural sciences themselves, as fears of funding and insecurity over legitimacy drive such social science researchers to embrace extreme examples of the empirical path.[3]

As the concerns and responses to bias in AI machines demonstrate, such "simple" thinking endures in many corners of the business world and academia, where the creative a priori work of statisticians in constructing neural nets is vastly outweighed by the massive "data exhaust" accumulated through social media and technology companies. Statisticians, a group that Tukey imagined serving as scientific detectives and crucial to research design prior to data collection, instead have often become the ad hoc "sanctifiers" of large data sets, with the careful creation and arrangement of limited numbers from the 1960s giving way to black box and ad hoc interpretations drawn from the overwhelming number of actions and behaviors that have been tracked indiscriminately on the internet. While much of the discussion has rightfully concerned how technology companies have violated individual privacy, at a more basic level, this work has also imposed a far more reductive and repetitive task upon those who deal with data. In the era of Big Data, d'Alembert and Tukey's hoped-for "data analysts" have instead become "data scientists," a recent coinage for a position that lacks nearly all of the preemptive investigative and creative work suggested in Tukey's *Exploratory Data Analysis*. As one data scientist remarked about his work for an AI company in a collection of anonymous interviews with tech workers, the "people who are the most confident about self-identifying as data scientists are almost unilaterally frauds."[4] Although *Harvard Business Review* called data scientist "The Sexiest Job of the 21st Century," accounts of working through data at AI companies often sounds far more like Turing's "librarian types" than Tukey's detectives, and the headline may say more about twenty-first-century work than it does about data science. While Tukey had imagined data analysis at every level of hypothesis and interpretation, that same data scientist noted that nearly all the work comes after the fact, claiming, "There's information in the correlation structure of the variables that can be revealed, but only through really huge amounts of data."[5]

The limitations of intellectual life in tech work were confirmed in Anna Wiener's 2020 memoir of working in Silicon Valley, where the feeling of becoming "an intelligent artifice" or "bot" was felt by those even at the heart of the industry. She noted how the grand excitement of working for technology companies soon gave way to a feeling of rote existence and meaningless drudgery among her colleagues, with "everyone . . . stuck in a feedback loop with themselves."[6] Given that Wiener's description of Silicon Valley covers the time when major technology companies were investing and transitioning to AI research in the 2010s, it is worth noting that the practices she observed also share origins in the methodological approach to studying humanity using the tools of the natural sciences. For example, Fontenelle's right conduct stressed stoic sacrifice in service of the geometric spirit, and Quetelet's vision of science workers included the idea of interchangeable and trained *savants*, concepts borrowed from the standardization of human computers and star watchers at European and American observatories. Like modern tech workers, who are expected to find purpose outside of work and yet maintain cheer and optimism in the office, the workers that Martineau and Knight lectured on political economy were also expected to find meaning outside of their routine, as the division of labor with machines had long ago been extended from physical to mental labor. While many social scientists of the twentieth century tried to retain a broader sense of culture, history, and practice in their work, dictums like the "fetish of verification" and "never think" became cornerstones of reductive social science practice. Indeed, over twenty years before Wiener's description of ennui, when Silicon Valley still seemed synonymous with creativity, the computer programmer Ellen Ullman had already declared the "dumbing down of programming."[7] And of course this is just the work of programmers and technical workers, to say nothing of the legions of unpaid or underpaid workers discussed in the previous chapter who labor in the shadows in order for AI systems to function.

Such work certainly seems far from the easy and very *canny* feeling of the artisan at home in his workshop that Adam Smith described. So too does it seem far from the vision Karl Marx and Friedrich Engels presented in *The German Ideology* (1862), where they famously imagined the worker able "to hunt in the morning, fish in the afternoon, rear cattle in the evening, [and] criticize after dinner . . . without ever becoming hunter, fisherman, herdsman or critic."[8] As seen in the stories of Martineau and Knight, however, those like "Old Armstrong" who tried to live their lives in this way were assailed for their lack of industriousness, as were the British citizens who partook of the "amusements" that Jevons found to be "irrational." In the same way, reformist-minded sociologists and economists found themselves marginalized in their fields, as their attempts to directly alleviate poverty and intercede in the lives of those they studied was deemed outside the bounds of acceptable practice. As Veblen and the institutionalists also learned, an economist who imagined a theory of human action that incorporated an understanding of culture, history, and actual economic practice was all but forgotten by the middle of the twentieth century. And Tukey's idea of the "creative act" as the cornerstone of statistical research is far from the practice of most quantitative research today, as statisticians join the parade of workers over the centuries whose scope of work has been curtailed.

What might be most remarkable about such labor practices in the social sciences (as opposed to the natural sciences) is that they have not come anywhere close to producing a "correct," or scientifically verified, vision of what it means to be an intelligent human being. There is in fact no more agreement over the human level today than there was in the seventeenth century, when Descartes first took up the task of how the tools of the natural sciences might apply to human beings. There remains no *Principia* of human thought and action, and the "glimpse of a weak light" has, if anything, dimmed in the years since d'Alembert criticized the geometric spirit. Instead, most of the attempts at a science of man described in this book have been *failed* efforts to offer a comprehensive understanding of how and why human beings think and act. From "statue men," the "invisible hand," and the "man machine" through "average men," the "psychical machine," and *homo economicus*, every attempt to offer a lawlike and consistent theory of human action and thought based on the tools of the natural sciences has been wrong, often spectacularly so. Perhaps this is why many of those who were most familiar with practices borrowed from the natural sciences—such as Pascal, d'Alembert, La Mettrie, Laplace, Darwin, and Tukey—were the *least* interested in applying those practices to a science of man. As Turing claimed, scientific induction has indeed proved a poor guide to understanding the "works and customs" of humanity.

To go by Martin Ford's recent collection of interviews with two dozen "architects of intelligence," leading AI researchers continue to be beset by the problem of not knowing exactly what AI is supposed to recreate. For example, Yoshua Bengio, one of the key researchers in neural networks who helped revolutionize image recognition in the 1990s, noted that "we don't yet have a full picture of how the brain works" and suggested neural nets could in fact serve as a "testable hypothesis for brain science."[9] Even such a supposedly "simple" project such as understanding the brain seems far away, and one comprehensive history says "we may be approaching the end of the computational metaphor" for human intelligence.[10] Stuart Russell, author of the leading AI textbook, surprisingly argues that this lack of understanding is a matter of resources, pointing out that trillions of dollars have been spent on creating a smartphone but "almost nothing" is spent "on understanding how people can live interesting and fulfilling lives."[11] Long after Turing's initial skepticism, Russell claimed in the same interview that "if we want to build artificial intelligence, we'd better figure out what it means to be intelligent."[12]

Although it might seem obvious, the ability to play chess, win at *Jeopardy!*, drive cars, schedule appointments, transcribe speech, and write college essays are not examples of transhistorical and transcultural intelligence; they are instead achievements deeply bound to specific and historically created notions of human purpose. While AI success in object recognition and translation reflects a far more inclusive vision of what it means to be human, it is hard to see how "general intelligence" will not come to embed the biases of the specific and recently imagined intelligence delineated by four hundred years of Western social science, as the earliest reports of bias in AI seem to foretell. To his credit, Ford recognized the inherent problem of reducing the human level to a very particular and contingent form of intelligence and behavior. For example, in his interview with Nick Bostrom, author of *Superintelligence* (2014), Ford pointed out, "It's always a concern about how the machine is not going to do what we want, where 'we' applies to collective humanity as though there's some sort of universal set of human desires or values. Yet, if you look at the world today, that's really not the case. The world has different cultures with different value sets."[13]

Ford hinted at a potentially devastating idea for AI—that there was no fixed form of human behavior, thought, and intelligence for AI to replicate or surpass—but Bostrom's answer reveals the kind of limitations AI has inherited from its origins in the reductive social sciences. Rather than engage in the question of values, Bostrom echoed the explanations of CSS researchers for the persistence of bias in AI: "You need to solve the technical problem to get the opportunity to squabble over . . . whose values should guide the use of this technology."[14] Although as director of the Future of Humanity Institute at Oxford University Bostrom has been positioned as one of the more astute thinkers about potential misuses of AI, historians of science have long made clear that the separation of "technical problems" from "values" is impossible.[15] As with any technology, but especially with a technology being built *to mimic human thought and behavior*, values would seem to be a crucial part to address at the beginning.

The story of the emergence of AI told in this book may not match the excitement of the "quest" narrative common to many histories, yet it may at least help lessen concerns about the potentially disastrous consequences of something like a "superintelligence." It is not, of course, that machines cannot be harmful, only that there are plenty of machines already in existence far more damaging than a superintelligence. In a world with nuclear weapons, nearly any machine technology (or person) connected to their storage, use, and communication networks carries potential dangers far exceeding AI. As historians and philosophers have shown, however, technophobia results mostly from a sense of technological determinism, so imagining AI as a product of culture and history rather than untethered technological ingenuity might help abate such fears.

Seen another way, the idea of AI as a dual product of technological progress *and* latent values from failed social sciences has the potential to destabilize current understandings of AI discussions, from science fiction to philosophy to politics to the predictions of "futurists." Both enthusiasts and alarmists, technophiles *and* technophobes, are debating not some great union of man and machine but the union of two intelligent artifacts: the "man" that emerged in the social sciences since the Enlightenment and the pale reflection of that being seen in AI created by major American technology companies in the past twenty years. Neither, of course, is wholly the creation of humanity or nature. Instead, as seen in the many attempts to create a science of man, both are traces of ideas developed in a very specific culture and history in the past four hundred years. While the historian and philosopher Michel Foucault (1926–1984) believed that "man" of the Enlightenment was "only a recent invention . . . and . . . will disappear again as soon as that knowledge has discovered a new form," he may have been discomforted to know how substantially this "man" has been embedded into the machines of the future.[16]

As this book has argued, the idea of an AI quest is therefore misplaced because the supposed human level it is after is an artifact of history, subject to the cultural, social, and intellectual worlds in which it was produced. It is therefore able to be reborn in any number of new cultural, social, and intellectual contexts should people reject the belief that their thoughts and behaviors are amenable to understanding through the tools of the natural sciences. Indeed, the legacy of four centuries of failed sciences of man would provide good reason to do just that. This then might be why the AI "dream" can only be accomplished through stripping people of their individual identities. Simple people, not smart machines, are required for the successful artificial intelligences of the future. By then, perhaps, it might even be possible to study them scientifically.

NOTES

Introduction. Turing's Paradox and the Failure of the Sciences of Man

1. Nourbakhsh, *Robot Futures*, 105.

2. While no single term can encompass the centuries-long process of applying the tools of the natural sciences to the study of humanity, for the sake of clarity the book will mostly rely on "sciences of man" for approaches prior to the twentieth century and "social sciences" for those after. Terms such as "human sciences" or "behavioral sciences" will be used when necessary. In France, such efforts were usually called *sciences morales* or *sciences humaines* and in Germany *Geistewissenschaft*. For a discussion of the historical use of these terms (and others), see the invaluable Porter and Ross, "Writing the History of Social Science." For an excellent single-author history, see R. Smith, *Norton History of the Human Sciences*.

3. Turing, "Computing Machinery," 443.

4. Ford, *Architects of Intelligence*, 6.

5. Kapor and Kurzweil, "Wager." See also Proudfoot, "Turing Test"; French, "Dusting Off."

6. Newcomb, "Meet the Computer." See also Larson, *Myth*, 193.

7. Poniewozik, "TV's War with the Robots."

8. For a recent iteration, see Carr, "Is Google Making Us Stupid?"

9. For the benefits of a longer approach, see Guldi and Armitage, *History Manifesto*. For the limits of this approach in the history of science, see the "Viewpoint" articles in *Isis* 107, no. 2 (June 2016), especially Heilbron, "Are Historians Fit to Rule?"; and Kevles, "What's Manifest." For more discussion of larger narratives around the scientific revolution, see also "Viewpoint: Clocks to Computers."

10. For the most popular accounts, see Nilsson, *Quest*; and Dormehl, *Thinking Machines*. More scholarly "quest" accounts include Rose, *American Quest*; Ekbia, *Artificial Dreams*; and Roland and Shiman, *Strategic Computing*. For the ultimate insider narrative that reveals the "secrets" of modern learning machines, see Domingos, *Master Algorithm*. All of these books include "quest" in either their titles or subtitles.

11. Turing, "On Computable Numbers." For a helpful guide to working through the mathematics of the paper, see Bernhardt, "Turing's Vision."

12. Turing, "Computing Machinery"; Turing, "Intelligent Machinery."

13. McCarthy et al., "Proposal." The electronic version was scanned by McCarthy from the original mimeographed proposal.

14. Lighthill, "Artificial Intelligence." For a contemporary challenge to this report, see McCarthy, "Artificial Intelligence."

15. McCulloch and Pitts, "Logical Calculus."

16. Rosenblatt, "Perceptron"; Minsky and Papert, *Perceptrons*.

17. Mitchell, *Artificial Intelligence*, 81–96.

18. For the current state of the field, see Rouse, *Computing Possible Futures*, 131–50.

19. The most popular and enduring technical guide is the oft-printed Russell and Norvig, *Artificial Intelligence*. On personal memoirs, see McCorduck, *Machines Who Think*; Turkle, *Second*

Self. For philosophical reflections, see Hayles, *Posthuman*; Haraway, *Simians, Cyborgs, and Women*. The polemics are numerous, but recent notable popular examples include Carr, *Glass Cage*; Lanier, *You Are Not a Gadget*; Townsend, *Dark Side*; Foer, *World without Mind*. The liveliest account of the personalities involved in creating artificial intelligence is Markoff, *Machines*.

20. Nilsson, *Quest*, 525.

21. Galison, *Image and Logic*.

22. Daston, "Enlightenment Calculations," 186.

23. Dick, "Artificial Intelligence."

24. Horvitz, "Opening Statement," 7.

25. Somers, "One-Trick Pony"; Lohr, "Researchers"; Fjelland, "General Artificial Intelligence." Also see "Gary Marcus," in Ford, *Architects of Intelligence*, 305–32.

26. Boden, *AI*, 121.

27. Lu, "Generative AI"; Griffith, "They Fled San Francisco."

28. Dick, "Artificial Intelligence."

29. The Unabomber's warnings about technology can be found in Kaczynski, *Technological Slavery*. See also Barrat, *Our Final Invention*; Bostrom, *Superintelligence*.

30. Joy, "Why the Future Doesn't Need Us." In 2018 Elon Musk and Stephen Hawking joined AI experts in expressing the dangers of robot killers. See "Open Letter." For the glorious transformation, see Kurzweil, *Singularity*; Chu, *Human Purpose*. Although speculative works like these are often dismissed in AI literature, they are frequently offered as counterpoints to the dystopian vision. See Basulto, "Artificial Intelligence."

31. For classic accounts of the perils of technological determinism, see Marx, *Machine in the Garden*; Edgerton, *Shock of the Old*; McCloskey, *Bourgeois Dignity*.

32. Quoted in Maas, *William Stanley Jevons*, 141.

33. Crawford, *Atlas of AI*, 8–9.

34. Grier, *When Computers Were Human*, 28.

35. Grier, *When Computers Were Human*, 98.

36. The literature on scientific praxis is enormous, and citations can be found throughout the book. On "industrial science," see Shapin, *Scientific Life*, 95. For a survey of nineteenth-century work practices, see the discussion in Donnelly, "Boredom of Science," 481–86.

37. Lohr, "Janitor Work."

38. Block and Riesewieck, *Cleaners*.

39. Irani, "Justice." See also Brynjolfsson and McAfee, *Second Machine Age*.

40. On low-paid internet laborers, see Gray and Suri, *Ghost Work*. On the larger economic stakes, see Zuboff, *Surveillance Capitalism*.

41. Geissler, *Seasonal Associate*. See also Bruder, *Nomadland*.

42. Ullman, *Life in Code*; Wiener, *Uncanny Valley*.

43. Kuhn, *Structure*, 24.

44. For recent works revealing bias in the machines, see discussion in chapter 9.

45. Burrell and Fourcade, "Society of Algorithms," 219.

46. Adas, *Machines*, 4–7, 59–67.

47. Radin, "'Digital Natives,'" 46; Donnelly, "We Have Always Been Biased."

48. For example, even projects like Fei-Fei Li's "AI4ALL" group attempt to assimilate diverse groups into an existing paradigm of AI research.

49. Gilman, *Mandarins*; Heyck, *Age of System*; Latham, *Modernization*.

50. O'Neil, "Ivory Tower."

51. Boden, *Mind as Machine*; Jones, *Reckoning with Matter*; Riskin, *Restless Clock*.

52. Kang, *Sublime Dreams*.

53. Mayor, *Gods and Robots*, 3.

54. Voskuhl, *Androids*; LaGrandeur, *Intelligent Networks*, 164.

55. Krüger, Daston, and Heidelberger, *Probabilistic Revolution*.

56. Daston, *Classical Probability*; Hacking, *Taming of Chance*; Porter, *Rise of Statistical Thinking*; Stigler, *History of Statistics*.

57. Poovey, *Modern Fact*; Daston and Galison, *Objectivity*.

58. De Chadarevian and Porter, "Scrutinizing the Data World," 549; Gladwell, *Tipping Point*; Surowiecki, *Wisdom of Crowds*; Anderson, *Long Tail*.

59. Jacket copy on Stone, *Counting*.

60. Bouk, *How Our Days*, xvi; Harcourt, *Exposed*.

61. De Chadarevian and Porter, "Histories of Data and the Database."

62. Aronova, von Oertzen, and Sepkoski, "Historicizing Big Data."

63. Pasquale, "Algorithmic Self"; Bouk, "History and Political Economy"; Bouk, "National Data Center."

64. Hacking, "Making Up People"; Müller-Wille, "Making and Unmaking Populations." For the most forceful and influential accounts of how data collection can shape societies, see Scott, *Seeing Like a State*.

65. This has been driven in part by the emergence of groups like the Data and Society Institute at Columbia University and the British Special Group on Artificial Intelligence located at Cambridge.

66. Dick, "Models and Machines"; Heyck, *Herbert A. Simon*.

67. Lemov, *Database of Dreams*; Lepore, *If, Then*.

68. Lupton, *Quantified Self*; Neff and Nafus, *Self-Tracking*. A notable exception is the discussion of early modern pedometers in Wernimont, *Numbered Lives*, 91–128. As strong as the historical work is in the *Osiris* volume "Data Histories," apart from Stevens, "Feeling for the Algorithm," the collection does not engage with artificial intelligence.

69. Mitchell, *Artificial Intelligence*, 262.

70. Taylor, "Interpretation."

71. Friend, "Superior Intelligence," 44.

72. Although the historian and philosopher Michel Foucault announced genealogy as the province of "relentless erudition," the project here is not quite as ambitious as that and makes no pretense of following the role of the "historian" that Foucault describes. Rather, it simply seeks to challenge the teleological account of smarter machines naturally progressing to the level of humanity. Foucault, "Nietzsche," 77, 89–91. The idea of a "genealogical" history as opposed to a "clear" and "tidy" account dates back to Nietzsche, *Genealogy of Morals*, 17. See also Williams, *Truth*, 12–19.

73. Blakely, *We Built Reality*, 48–49.

74. In specifically looking for stories that seem to resonate in the world of twenty-first-century AI, the story does of course risk the charge of presentism. However, no attempt is made to impose moral and cultural values of today onto the actors of the past or to assume these stories could not have been otherwise. For a recent account of the problems of presentism, see Sweet, "Is History History?"

75. Hamman, quoted in Berlin and Hardy, *Roots of Romanticism*, 50.

76. Von Helmholtz, "Relation of Natural Science," 8.

77. Yonay, *Soul of Economics*, 42–48; Ross, "Changing Contours," 211–14.

78. Barkan, *Retreat of Scientific Racism*; Gould, *Mismeasure of Man*.

79. Cowles, *Scientific Method*, 8.

80. Strevens, *Knowledge Machine*, 258.

81. Larson, *Myth*, 23.

82. Larson, *Myth*, 83.

83. Hacking, "Looping Effects." See also Noë, *Entanglement*, 22.

84. Desch, *Cult of the Irrelevant*.

85. Scott, "When the Movies Pictured AI."

1. Intelligence Lost

1. Barth, "Defector," 28.

2. Laplace, *Essai philosophique*, 33. Unless otherwise noted, all translations from the French are my own. For a recent summary of the influence of Laplace's Demon, see Canales, *Bedeviled*, 29–40.

3. Oldenburg to Spinoza, September 27, 1661. Quoted in Sytsma, *Richard Baxter*, 48–49.

4. Laplace, *Essai philosophique*, 33. Unless otherwise noted, all translations from the French are my own. For a recent summary of the influence of Laplace's Demon, see Canales, *Bedeviled*, 29–40.

5. For quotation and discussion, see Koyré, "Significance," 21.

6. For the most sustained argument that puts Spinoza, rather than Descartes, at the heart of the story, see the works of Israel, especially *Radical Enlightenment*. For an argument for the emergence of Enlightenment out of the freethinkers of the British Isles, see Jacob, *Radical Enlightenment*.

7. Meeker, *Voluptuous Philosophy*, 2.

8. For the best recent guide to this era, see Riskin, *Restless Clock*, ch. 2. The classic study of the era is Westfall, *Construction of Modern Science*. See also Garber and Roux, *Mechanization of Natural Philosophy*; Shapin, *Scientific Revolution*, 181–82; Bertoloni Meli, *Mechanism*.

9. Avramov, "Apprenticeship in Scientific Communication," 198.

10. Sarasohn, *Gassendi's Ethics*, 190.

11. McDonough, "Freedom and Contingency."

12. Henry, "Primary and Secondary Causation"; Gleick, *Isaac Newton*, 108–11.

13. Quoted in Gottlieb, *Dream of Reason*, 205.

14. Quoted in Henry, "Primary and Secondary Causation," 549.

15. Wilson, *Ideas and Mechanism*, ch. 10.

16. Quoted and translated in Browne, "Descartes's Dreams," 259.

17. Clarke, *Descartes*, 58–65.

18. Gaukroger, introduction to *Descartes' Meditations*, 1.

19. Descartes, *Œuvres philosophique*, 63 (hereafter cited as *OPD*).

20. Spallanzani, "First Philosophy."

21. Riskin, *Restless Clock*, 46–49.

22. Bennett, "Mechanics' Philosophy," 24.

23. Garber, "Descartes," 191.

24. Van Berkel, *Isaac Beeckman*, 106; Buzon, "Beeckman."

25. Descartes to Beeckman, March 26, 1618, *OPD*, 40.

26. Garber, *Descartes*, 30–62.

27. Doyle, "Descartes' *Regulae*."

28. For an argument against the importance of the *Rules*, see Ariew, "Mathematization of Nature."

29. Quoted in Doyle, "Descartes' *Regulae*," 3.

30. Kraus, "Universal Mathematics," 160.

31. For discussion of the Rule IV controversy, see Sasaki, *Descartes's Mathematical Thought*, 333–58; Rabouin, "Mathesis Universalis"; Weber, *La constitution du texte*; Gaukroger, *Descartes' System*, 8–10.

32. Des Chene, *Spirits and Clocks*, 2.

33. Garber, "Descartes," 190.

34. Phillips, "Pascal's Reading," especially 31–33. In the same volume, see Hammond, "Pascal's *Pensées*," 246. See also Donnelly, *Adolphe Quetelet*, 25.

35. Sytsma, *Richard Baxter*, 14.

36. Sytsma, *Richard Baxter*, 24.

37. Quoted in Sytsma, *Richard Baxter*, 69.

38. Quoted and translated in Roux, "Empire Divided," 87.

39. Quoted and translated in Roux, "Empire Divided," 87.

40. Descartes, *Meditations*, 17 (italics added).

41. Roux, "Empire Divided," 93.

42. Lennon, *Battle of the Gods*, 33.

43. On Fontenelle's stubborn defense, see Rioux-Beauline, "What Is Cartesianism?"

44. Fontenelle, *Digression*, 308.

45. Fontenelle, *Digression*, 309.

46. Fontenelle, *Digression*, 310.

47. Fontenelle, *Digression*, 310.

48. Fontenelle, *Digression*, 310.

49. Seguin, "Histoire," 323.

50. Paul, *Science and Immortality*; Hahn, *Anatomy*; Adkins, "When Ideas Matter."

51. Paul, *Science and Immortality*, 99.

52. Fontenelle, *Digression*, 313.

53. Fontenelle, *Digression*, 313.

54. Fontenelle, *Digression*, 313.

55. Koyré, "Newton and Descartes"; Aiton, *Vortex Theory*; Gaukroger, *Descartes*, 382–83.

56. Albury, "Order of Ideas," 205.

57. Condillac, *Traité*, 13.

58. Knight, *Geometric Spirit*.

59. Condillac, *Traité*, 17.

60. Condillac, *Traité*, 24.

61. Condillac, *Traité*, 217.

62. McMullin, "Newton's *Principia*."

63. Condillac, *Traité*, 13.

64. Condillac, *Traité*, 13–14.

65. Condillac, *Traité*, 24, 55–56.

66. Knight, *Geometric Spirit*, 52.

67. Quoted in Condillac, *Traité*, 67.

68. Condillac, *Traité*, 79.

69. Condillac, *Traité*, 79.

70. Cunning, "Malebranche," 477.

71. O'Neal, *Authority of Experience*, 61–83.

72. Hayes, *Reading the French Enlightenment*, 126–28; Hine, *Condillac's Traité des Systèmes*, 58–71.

73. Hulliung, *Autocritique of Enlightenment*.

74. Black, *Rousseau's Critique of Science*, 10.

75. Rousseau and Launay, *Œuvres complètes*, 2:57–58 (hereafter cited as *OCR*).

76. Vartanian, *Diderot and Descartes*, 7.

77. Diderot, *Œuvres philosophiques*, 285 (hereafter cited as *DD*).

78. Quoted in *DD*, 1193fn45.

79. *DD*, 320; Vartanian, "Diderot and Maupertuis," 49.

80. Vartanian, *Diderot and Descartes*, 271.

81. Cassirer, *Philosophy of the Enlightenment*, 4.

82. Quoted in Cassirer, *Philosophy of the Enlightenment*, 4.

83. For more on d'Alembert's ambivalence, see Riskin, *Science in the Age of Sensibility*, 122.

84. D'Alembert, *Mélanges*, 704 (hereafter cited as *DML*).

85. Schaffer, "Scientific Discoveries."

86. This is not to say that Descartes himself would have imagined a thinking machine. See Alanen, *Descartes's Concept of Mind*, 94–99.

87. Osler, *Divine Will*, 179.

88. Quoted in Sailor, "Cudworth and Descartes," 135.

2. At the Bleeding Edge

1. Robespierre, *Œuvres*, 310.

2. Robespierre, *Œuvres*, 309.

3. Quoted in Carson, *Measure of Merit*, 53.

4. Popkin, *New World Begins*, 379–417.

5. For a discussion of Robespierre's motivations, see Ozouf, *Festivals*, 107–15. See also the epilogue to Israel, *Radical Enlightenment*, 714–19.

6. Quoted in Popkin, *New World Begins*, 351.

7. Israel, *Enlightenment That Failed*, 84.

8. For two classic accounts, see Daston, *Classical Probability*; and Hacking, *Taming of Chance*.

9. Gillispie, *Science and Polity*.

10. On the gendered role of such a transformation, see Daston, "Enlightenment Calculations."

11. Abbas, *Thinking Machines*, 34. Limber, "Language in Child and Chimp?," 198; Popkin, "Leonora Cohen Rosenfield."

12. Voskuhl, *Androids*, 218–19.

13. McCorduck, *Machines Who Think*, 43.

14. Wellman, *La mettrie*, 172.

15. Vartanian, *Science and Humanism*, 57.

16. La Mettrie, *L'Homme machine*, 153 (hereafter cited as *HM*). In his 1960 analysis Vartanian reprinted the final 1751 edition as an appendix. I have retained La Mettrie's unique capitalization in order to capture his prose style.

17. *HM*, 159. For context, see *HM*, 211fn42.

18. *Larousse dictionnaire de française*, s.v. "portefaix," accessed March 30, 2021, https://www.larousse.fr/dictionnaires/francais/portefaix/62742.

19. Fossati, "Maximum Influence," 49.

20. Helvétius, *Œuvres complètes*, 2:40 (hereafter cited as *OCH*).

21. Cumming, *Helvétius*, 200–217.

22. Smith, *Helvétius*, 103–14.

23. O'Neal, *Authority of Experience*, 83.

24. Smith, *Helvétius*, 13.

25. Quoted in Halévy, *Radicalisme philosophique*, 1:289–90.

26. Cumming, *Helvétius*, 129.

27. Wade, *Structure and Form*, 291.

28. Israel, *Radical Enlightenment*; Cumming, *Helvétius*, 85.

29. Wade, *Structure and Form*, 287–90.

30. Schøsler, "Rousseau et Diderot."

31. Riskin, "Defecating Duck."

32. Vartanian, "La Mettrie and Diderot Revisited," 177.

33. Carson, *Measure of Merit*, 24.

34. Kors, *D'Holbach's Coterie*; Lilti, *World of the Salons*.

35. Curran, *Diderot*, 234–36.

36. Kors, *D'Holbach's Coterie*.

37. D'Holbach, *Œuvres philosophique*, 168 (hereafter cited as *OPH*).

38. Kors, *D'Holbach's Coterie*, 201.

39. Daston, *Classical Probability*, 10.

40. *OPH*, 196; Israel, *Enlightenment That Failed*, 213.

41. Dupré, *Enlightenment*, 265.

42. Quoted in Cushing, "Baron d'Holbach," 51.

43. Strugnell, *Diderot's Politics*, 131.

44. Cushing, "Baron d'Holbach," 49.

45. Daston, "D'Alembert's Critique."

46. Curran, *Diderot*, 244.

47. Hahn, *Laplace*, 32–34.

48. Gillispie, Fox, and Grattan-Guinness, *Laplace*, 124–26.

49. Gillispie, *Science and Polity*, 445–58.

50. Hahn, *Laplace*, 120.

51. Laplace, *Essai*, 31.

52. Laplace, *Essai*, 32.

53. Laplace, *Essai*, 33.

54. Laplace, *Essai*, 117.

55. Laplace, *Essai*, 40.

56. Laplace, *Essai*, 40.

57. Laplace, *Essai*, 43.

58. Laplace, *Essai*, 167–68.

59. Quoted in Mykhailova, "D'Holbach's Legacy," 305.

60. Lanfrey, *Napoleon the First*, 335. See also Hayward, *Fragmented France*, 79.

61. Grier, *When Computers Were Human*, 46–54.

3. Warnings of a New Barbarism

1. Newton to Hooke, February 5, 1675. Quoted in Brewster, *Memoirs*, 142. For context, see Gleick, *Isaac Newton*, 79–98.

2. Barbey, *Les prophètes*, 129.

3. Barbey, *Les prophètes*, 129.

4. Barbey, *Les prophètes*, 135.

5. Pius IX, "Syllabus of Errors."

6. See further citations for critical discussions of Bonald and Maistre.

7. Johannessen, *Prophètes*, 12.

8. For a full discussion of the term, see Woudenberg, Peels, and de Ridder, "Putting Science on the Philosophical Agenda."

9. Gildea, *Children of the Revolution*, 6.

10. Grendler, "Universities Warring," 100; Anderson, *European Universities*, 23.

11. Popkin, *New World*, 428.

12. Woloch, *New Regime*, 197–207.

13. Popkin, *New World*, 429.

14. Grattan-Guinness, *"Ecole polytechnique,"* 238; Pannabecker, "Technocracy," 620.

15. Crosland, "Science Empire," 30.

16. Porter, "Objectivity and Authority," 256. See also Porter, *Trust in Numbers*.

17. For an overview of the debate, see Pannabecker, "Technocracy."

18. Schubring, "Napoleonic Structural Reforms," 436.

19. Gengembre, *La contre-révolution*, 10.

20. Smith, *Irrationality*, 3. For another popular look at modern strains of the counter-Enlightenment, see Mishra, *Age of Anger*.

21. McMahon, *Enemies*, 48. On Berlin and the term "counter-Enlightenment," see Wokler, *Rousseau*, 244–59.

22. Sternhell, *Les anti-lumieres*, 30.

23. Seidman, *Liberalism*, 48.

24. Sternhell, *Les anti-lumieres*, 14.

25. Maistre, *Examen*, 1:318.

26. Klinck, *Louis de Bonald*, 37–38.

27. Toda, introduction, 6.

28. Quoted in Toda, introduction, 37.

29. Bénéton, foreword, xxv.

30. Berlin, *Crooked Timber*, 105.

31. Bonald, *Lettres à Joseph de Maistre*, 100.

32. On Bonald's limitations, see Nisbet, "De Bonald."

33. Muller, "Conservatism."

34. Toda, introduction, 9.

35. Bonald, "Préface," 173.

36. Klinck, *Louis de Bonald*, 14.

37. Klinck, *Louis de Bonald*, 56.

38. Barbey, *Les prophetes*, 58.

39. Quoted in Klinck, *Louis de Bonald*, 61.

40. Gengembre, *La contre-revolution*, 155.

41. Gengembre, *La contre-revolution*, 177.

42. Bonald, "Sur la guerre," in *Œuvres complètes*, 3:1072–73 (hereafter cited as *OCB*).

43. Anderson, *European Universities*, 4.

44. *OCB*, 3:1141. For the change in approach, see Gascoigne, "Study of Nature."

45. Gengembre, *La contre-revolution*, 70.

46. Gough, *Terror*, 87.

47. Quoted in Barnard, *Education*, 131.

48. Harrison, *Romantic Catholics*, 8.

49. "Varieties," 3.

50. "Varieties," 4.

51. "Varieties," 3.

52. "Varieties," 3.

53. "Varieties," 4.

54. Siegfried, "Politicization," 26fn31.

55. Byrnes, *Catholic and French*, 81–82; Burrows, *French Exile Journalism*; Cabanis, *La presse*; Klinck, *Louis de Bonald*, 86.

56. Klinck, *Louis de Bonald*, 7.

57. "Prospectus." Also see Compagnon, *Les antimodernes*, 156.

58. Garrard, *Counter-Enlightenments*, 72.

59. "Quelques réflexions."

60. For biographical information on Génisset, see Rousset, *Dictionnaire géographique*, 384. It is possible the review was in fact written by Chateaubriand, who had not yet left the *Mercure*. See Segala, "Jean-Baptiste Biot's Collaboration," 116.

61. While Génisset's entire speech does not seem to have survived, it was extensively quoted in a review from *Le Conservateur*. See "Discours." On another response to Génniset, see Zékian, "Siècle des lettres."

62. "Quelques réflexions," 391.

63. "Quelques réflexions," 393.

64. "Quelques réflexions," 393.

65. "Quelques réflexions," 393.

66. "Quelques réflexions," 396–97.

67. "Quelques réflexions," 394–95.

68. "Quelques réflexions," 397.

69. Guairard, review.

70. Guairard, review, 404.

71. Guairard, review, 405.

72. Guairard, review, 408.

73. Guairard, review, 409.

74. Segala, "Jean-Baptiste Biot's Collaboration," 116. Chateaubriand's full critique is found particularly in *Genie du Christianisme*. Heilbron, *Social Theory*, 153. Also see Byrnes, "Chateaubriand and Destutt de Tracy."

75. Guairard, review, 409.

76. "Quelques réflexions," 396.

77. Pearson, *Unacknowledged Legislators*, 140.

78. Garrard, *Counter-Enlightenments*, 49.

79. Berlin, "Joseph de Maistre."

80. Holmes, *Anatomy of Antiliberalism*, 14, 23.

81. On Maistre's darkest side, see Bradley, "Maistre's Theory."

82. Lebrun, "Joseph de Maistre."

83. Darcel, "Roads of Exile," 24.

84. Lebrun, "Joseph de Maistre," 214.

85. Armenteros, *French Idea*, 9.

86. Armenteros, *French Idea*, 37, 44.

87. For the contrary notion that "Maistrian thought has never made a school," cf. Pranchère, "Maistrian Thought," 292. There have been admirable efforts to explore Maistre's international reputation. See, for example, Armenteros and Lebrun, *Joseph de Maistre*. On the close affinity of Barbey to Maistre's ideas, see Kevin Michael Erwin, "*Le mystique de la tradition*: Barbey Worships at the Altar of Joseph de Maistre," in Armenteros and Lebrun, *Joseph de Maistre*.

88. Garrard, *Counter-Enlightenments*, 52.

89. Maistre, Œuvres *complètes*, 8:300 (hereafter cited as *OCM*).

90. Maistre, *Examen*, 1:4.

91. Maistre, *Examen*, 1:9.

92. Maistre, *Examen*, 1:76.

93. Maistre, *Examen*, 1:101.

94. Maistre, *Examen*, 1:102.

95. Maistre, *Examen*, 1:160.

96. Maistre, *Examen*, 1:289.

97. Maistre, *Examen*, 1:9 (italics in the original).

98. Maistre, *Examen*, 1:21.

99. Maistre, *Examen*, 1:37.

100. Maistre, *Examen*, 1:318.

101. Ippolito, "Avant-Propos," 13.

102. Bonald, *Lettres à Joseph de Maistre*, 85.

103. Gengembre, *La contre-revolution*, 151.

104. Horkheimer and Adorno, *Dialectic*, 20.

105. Horkheimer and Adorno, *Dialectic*, xiv.

106. Pranchère, "Maistrian Thought," 323.

107. Anonymous, quoted in King, *Les doctrine littéraires*, 38.

108. "Quelques réflexions," 395.

4. Progress to the Mean

1. Adolphe Quetelet, *Correspondance mathématique et physique* (1832). See Stigler, *History of Statistics*, ch. 5; Hacking, *Taming of Chance*, 95–104.

2. For a valiant attempt to make sense of Quetelet's ideas and free will, see Lottin, *Quetelet*.

3. Wittrock, Heilbron, and Magnusson, "Rise of the Social Sciences," 3.

4. Hacking, *Taming of Chance*, 96–97; Daston, *Classical Probability*, 371.

5. Gillispie, "Probability and Politics," 2.

6. Condorcet himself preferred the singular "social mathematic," but the plural version is more commonly used. Condorcet, *Tableau*, in Œuvres *de Condorcet*, 1:541 (hereafter cited as *OC*).

7. Sarton, "Preface," 10.

8. Maas, *Jevons*, 175.

9. Mirowski, *More Heat than Light*.

10. Condorcet, *Outlines*, 316.

11. Condorcet, *Essai*, i.

12. On the difference between Condorcet and Laplace in conceiving of a social mathematics, see Gillispie, *Science and Polity*, 40–49.

13. Williams, *Condorcet*, 17–18.

14. Huet, *Mourning Glory*, 20.

15. Williams, *Condorcet*, 18.

16. Popkin, *New World*, 320–22.

17. Marty and Amirault, *Nicolas de Condorcet*, 2; Jurt, "Condorcet et les colonies," 21.

18. Baker, *Condorcet*, 371–73.

19. The most detailed account can be found in Crépel and Rieucau, "Few Tables," 245–47.

20. Condorcet, *Outlines*, 360.

21. Translated and quoted in Williams, *Condorcet*, 97.

22. Williams, *Condorcet*, 113–16.

23. Williams, *Condorcet*, 32.

24. Condorcet, *Outlines*, 65.

25. Donnelly, *Adolphe Quetelet*, 28.

26. Quetelet, *Sciences mathematique*, 7.

27. Desrosières, *Politics of Large Numbers*, 79.

28. Stigler, *History of Statistics*, 161–220. Stigler also provides a more freewheeling account of Quetelet's *l'homme moyen* in *Statistics on the Table*, 51–65. For a recent history of data that begins with Quetelet, see Wiggins and Jones, *How Data Happened*, ix.

29. Mailly, *Essai sur la vie*, 67.

30. Donnelly, "Redeeming Belgian Science."

31. Quetelet, *Notice sur Alexis Bouvard*.

32. Collard, "Adolphe Quetelet et l'astronomie."

33. Quetelet, "Arago," 14.

34. *Correspondance mathématique et physique* 2, no. 3: 177 (hereafter cited as *CMP*).

35. Gutin, "In BMI We Trust."

36. Quetelet, "Mémoire," 496n.

37. Stigler, *History of Statistics*, ch. 6. As Stigler points out, Quetelet actually got the numbers wrong.

38. Aubin, "Principles of Mechanics," 207. Quotes refer to Aubin's English translation.

39. Aubin, "Principles of Mechanics," 220.

40. Aubin, "Principles of Mechanics," 220.

41. Quoted in Beirne, "Adolphe Quetelet," 1140.

42. Aubin, "Principles of Mechanics," 205.

43. Aubin, "Principles of Mechanics," 204.

44. Aubin, "Principles of Mechanics," 218.

45. Aubin, "Principles of Mechanics," 219.

46. Aubin, "Principles of Mechanics," 219.

47. Lottin, *Quetelet*.

48. Hacking, *Taming of Chance*, 127.

49. Quetelet, "Sur la statistique," 4. For one of the most sustained critiques of Quetelet's ideas, see Michotte, *Études*, 434–43.

50. Van Meenen, "De l'influence."

51. De Decker, "De l'influence," 77.

52. For the full context, see Donnelly, *Adolphe Quetelet*, 154–57.

53. Crary, *Techniques of the Observer*, 6.

54. Diaz-Bone and Didier, "Perspectives."

55. Mirowski, *More Heat than Light*, 35.

56. Donnelly, *Adolphe Quetelet*, 203fn2.

57. Mirowski, *More Heat than Light*, 217.

58. Schabas, *World Ruled by Number*, 18; Mosselmans, "Adolphe Quetelet."

59. Turner, *Search*, introduction.

60. Mosselmans, *William Stanley Jevons*, 6–12.

61. Maas, *Jevons*.

62. Maas, *Jevons*, 62–72; Snyder, *Reforming Philosophy*; Turner, *Search*, 29–38; Goldman, "Origins"; Yeo, *Defining Science*, 178–85.

63. Quoted in Maas, *Jevons*, 70.

64. Cannon, *Science in Culture*, 77.

65. Maas, *Jevons*, 74.

66. Jevons, "Mechanical Performance," 497.

67. Jevons, "Mechanical Performance," 498.

68. Peart, "W. S. Jevons's Methodology," 436.

69. Maas, *Jevons*, 141.

70. Jevons, "Mechanical Performance," 500.

71. Schaffer, "Babbage's Intelligence."

72. Maas, *Jevons*, 284.

73. Maas, *Jevons*, 170, 217. On Jevons's interest in Jennings, see Maas, *Jevons*, 207; and White, "Richard Jennings."

74. Jevons, *Principles of Science*, 5.

75. Jevons, *Principles of Science*, 11.

76. Jevons, *Principles of Science*, 720, 730–31.

77. Jevons, *Principles of Science*, 734.

78. Jevons, *Principles of Science*, 759–60.

79. Jevons, *Principles of Science*, 761.

80. Jevons, *Principles of Science*, 737.

81. Jevons, *Principles of Science*, 735.

82. Peart, "On Making and Remaking Ourselves," 222.

83. Jevons, "Amusements of the People," 2–3.

84. Jevons, "Amusements of the People," 1.

85. Roncaglia, *Age of Fragmentation*, 27.

86. Stallo, *Concepts and Theories*, 69.

87. Jevons, *Principles of Science*, 769.

88. Stuckenberg, "Relation of Science to Religion," 344.

5. Tuning the Mind

1. Peacock, *Crotchet Castle*, 135.

2. O'Brien, *J. R. McCulloch*.

3. Peacock, *Crotchet Castle*, 138.

4. Peacock, *Crotchet Castle*, 135.

5. Redman, *Rise of Political Economy*, 63–82.

6. Peacock, *Crotchet Castle*, 156.

7. Ferguson, *History of Civil Society*, 4.

8. Hollander, "Retrospectives"; King, *David Ricardo*.

9. Quoted in Schabas, *Natural Origins*, 14.

10. Berg, *Machinery Question*; Jacob, *Scientific Culture*; Schabas, *Natural Origins*; Horn, *Path Not Taken*; and Mokyr, *Enlightened Economy*.

11. Thompson, *Making of the English Working Class*; Frey, *Technology Trap*.

12. An exception is Berg, *Age of Manufactures*.

13. Hobsbawm, *Age of Revolution*; Archer, *Social Unrest*.

14. Norman, *Adam Smith*, x.

15. Quoted in McVickar, *Outlines*, 41. McVickar's work was a commentary on McCulloch's *Supplement to the Encyclopedia Britannica*. See Davenport, *Unrighteous Mammon*, 231fn11.

16. Hetherington, "Isaac Newton's Influence," 497–501; Schabas, "Debts to Nature." Elmslie, "Retrospectives," 210–12.

17. C. Smith, "Essays."

18. Foley, *Social Physics*, 12.

19. C. Smith, "Essays," 91.

20. Schabas, "Debts to Nature," 264.

21. Thomson, "Philosophy of Science"; Montes, "Newton's Real Influence"; Diemer and Guillemin, "L'économie politique."

22. The best account of these essays is Kim, "View of Science."

23. Smith, *Essays*, 327–28.

24. Smith, *Essays*, 348.

25. Smith, *Essays*, 334.

26. Smith, *Essays*, 336.

27. Smith, *Essays*, 348.

28. Smith, *Essays*, 344–45.

29. Smith, *Essays*, 352.

30. Smith, *Essays*, 353.

31. Smith, *Essays*, 359.

32. Van Lunteren, "Clocks to Computers," 768.

33. Aspromourgos, "Machine."

34. Smith, *Wealth of Nations*, 81 (hereafter cited as *WN*).

35. Sutherland, "Explanatory Notes," in *WN*, 471–72.

36. Montes, "Das Adam Smith Problem."

37. West, "Two Views."

38. Martineau, *Thirty Years' Peace*, 1:42.

39. Darval, *Popular Disturbances*. Quoted in Hobsbawm, "Machine Breakers," 68. Frey references this same statistic in *Technology Trap*, 9.

40. Frey, *Technology Trap*, 130. For a good discussion of the literature, see Horn, "Machine-Breaking," 146–49.

41. Martineau, *Thirty Years' Peace*, 1:40.

42. See Freedgood, *Victorian Writing*. The most sustained work on this era is Oražem, *Political Economy*.

43. Sanders and Weiner, *Harriet Martineau*.

44. Martineau, *Thirty Years' Peace*, 1:41.

45. Martineau, *Autobiography*, 1:135–37.

46. Peterson, "French Revolution," 417.

47. Peterson, "French Revolution," 424.

48. McNally, *Against the Market*, especially ch. 4.

49. Martineau, *Rioters*.

50. Quoted in Kestner, *Protest and Reform*, 32.

51. Martineau, *Rioters*, 4.

52. Martineau, *Rioters*, 20.

53. Berg, *Machinery Question*, 293.

54. Martineau, *Rioters*, 27–28.

55. Martineau, *Rioters*, 36.

56. Martineau, *Rioters*, 34.

57. Martineau, *Rioters*, 44.

58. Hobsbawm and Rudé, *Captain Swing*.

59. Freedgood, "Banishing Panic."

60. Martineau, *Autobiography*, 1:109.

61. Freedgood, "Banishing Panic," 34; Dzelzainis, "Radicalism and Reform," 430.

62. Martineau, *Illustrations*, 1:xiii.

63. Martineau, *Illustrations*, 1:xix.

64. Martineau, *Illustrations*, 1:15.

65. Martineau, *Illustrations*, 1:16.

66. Martineau, *Illustrations*, 1:37.

67. Martineau, *Illustrations*, 1:41.

68. Martineau, *Illustrations*, 1:90.

69. Martineau, *Illustrations*, 1:91.

70. Martineau, *Illustrations*, 1:91.

71. Martineau, *Illustrations*, 1:94.

72. Martineau, *Illustrations*, 1:119.

73. Martineau, *Illustrations*, 1:133.

74. Martineau, *Illustrations*, 1:132.

75. Martineau, *Illustrations*, 1:138.

76. On Martineau and Knight, see Easley, *Literary Celebrity*, 97–106.

77. Gray, *Charles Knight*, 17.

78. Brougham was assumed to be the author of *The Working Man's Companion* series well into the twentieth century. See Gilbert, "Work of Lord Brougham." See also Gray, *Charles Knight*, 59; Kennedy, "Lord Brougham," 70.

79. Gray, *Charles Knight*, 125.

80. Gray, *Charles Knight*, 132–38.

81. [Knight], *Results*, 6–7.

82. [Knight], *Results*, 26.

83. Frey, *Technology Trap*, 129.

84. [Knight], *Results*, 179.

85. Knight, *Rights of Industry*, 167.

86. [Knight], *Results*, 193.

87. Gagnier, "Social Atoms," 339–40.

88. Glickstein, *Concepts of Free Labor*, 60.

89. Glickstein, *Concepts of Free Labor*, 67.

90. McCulloch, *Principles*, 132.

91. Knight and Martineau, *Mind amongst the Spindles*, vii (hereafter cited as *MAS*).

92. Emerson, *Method of Nature*, 5.

93. Emerson, *Method of Nature*, 5.

94. Frawley, "Behind the Scenes," 143–44.

95. Gagnier, "Social Atoms," 339.

96. Dodd, *Narrative*, 31.

97. This is not to confuse their intentions with those of the girls and women of Lowell. See Cook, *Working Women*, ch. 2.

98. Peacock, *Crotchet Castle*, 243.

99. Peacock, *Crotchet Castle*, 250.

6. The Descent of Man (and Intelligence)

1. Spencer, "Nebular Hypothesis," 1:286.

2. Spencer, *Principles of Psychology*, 1:567.

3. Godfrey-Smith argued that "Spencer was heading . . . towards a view of life that we now might call . . . cybernetic." "Spencer and Dewey," 317.

4. Adas, *Machines*; Peart and Levy, *"Vanity of the Philosopher"*; McAleer and MacKenzie, *Exhibiting the Empire*.

5. Quoted in Kevles, *In the Name of Eugenics*, 33.

6. Richards, *Darwin*; Bowler, *Evolution*; Ruse, "Evolution before Darwin."

7. Jacob and Stewart, *Practical Matter*; Mokyr, *Enlightened Economy*; Cantor, "Science, Providence, and Progress."

8. Tischler, *Introduction to Sociology*, 11.

9. Hofstadter, *Social Darwinism*.

10. Richards, *Darwin*, 11.

11. Beck, "Social Darwinism," 196–97.

12. Haines, "Spencer, Darwin."

13. Francis, *Herbert Spencer*; Offer, *Herbert Spencer*.

14. Elliott, "Erasmus Darwin."

15. Smith, "Evolution," 58.

16. Ruse, "Evolution before Darwin," 39.

17. Spencer, *Social Statics*, v.

18. Spencer, *Social Statics*, 11.

19. Spencer, *Social Statics*, 12.

20. Spencer, *Social Statics*, 18.

21. Spencer, *Social Statics*, 42.

22. Paul, "Selection"; Claeys, "Origins of Social Darwinism."

23. Spencer, *Social Statics*, 61.

24. Spencer, *Social Statics*, 62.

25. Spencer, *Social Statics*, 61.

26. Spencer, *Social Statics*, 59.

27. Elliott, "Erasmus Darwin," 28.

28. Spencer, *Theory of Population*, 5.

29. Spencer, *Theory of Population*, 35.

30. Spencer, *Theory of Population*, 7.

31. Spencer, *Theory of Population*, 11.

32. Quoted in Spencer, *Theory of Population*, 15.

33. Spencer, *Theory of Population*, 16.

34. Spencer, *Theory of Population*, 32–33.

35. Spencer, *Theory of Population*, 32.

36. Spencer, *Theory of Population*, 33.

37. Claeys, "Origins of Social Darwinism," 227.

38. Elliot, "Erasmus Darwin."

39. Francis, *Herbert Spencer*, 193; Smith, "Evolution," 62.

40. Francis, *Herbert Spencer*, 225.

41. Bowler, "Herbert Spencer," 204; Adams, "Anarchist Sociology," 62.

42. Spencer, *Autobiography*.

43. Marx to Engels, June 18, 1862, Marx/Engels Archive, https://marxists.architexturez
.net/archive/marx/works/1862/letters/62_06_18.htm.

44. For the best survey of the literature of this exchange, see van Wyhe, "Alfred Russel
Wallace."

45. Wallace, "Origin of Human Races," clxi.

46. Wallace, "Origin of Human Races," clxii.

47. Wallace, "Origin of Human Races," clxii.

48. Wallace, "Origin of Human Races," clxiii.

49. Wallace, "Origin of Human Races," clxvii.

50. Wallace, "Origin of Human Races," clxviii.

51. Raby, *Alfred Russel Wallace*; Shermer, *Darwin's Shadow*; Flannery, *Nature's Prophet*.

52. Wallace, "Origin of Human Races," clxv.

53. Wallace, "Origin of Human Races," clxix.

54. Wallace, "Origin of Human Races," clxix.

55. Wallace, "Origin of Human Races," clxix–clxx.

56. Darwin to Spencer, November 25, 1858. Quoted in Haines, "Spencer, Darwin," 420.

57. Richards, *Darwin*, 189.

58. Darwin, *Descent of Man*, 1:3 (2). The text follows the 1877 second edition published by
John Murray (London). Page numbers reference the 1989 edition, with 1877 pagination in
parentheticals.

59. Darwin, *Descent*, 1:4 (2).

60. Darwin, *Descent*, 1:52 (48).

61. Darwin, *Descent*, 1:66 (62).

62. Darwin, *Descent*, 1:69 (65).

63. Darwin, *Descent*, 1:10 (6).

64. Darwin, *Descent*, 1:58 (54).

65. Darwin, *Descent*, 1:107 (103).

66. Darwin, *Descent*, 2:618 (619).

67. Darwin to Hooker, December 10, 1866. Quoted in Haines, "Spencer, Darwin," 419.

68. Darwin, *Descent*, 1:139 (144).

69. Weikart, *Darwin to Hitler*.

70. Himmelfarb, *Darwinian Revolution*.

71. Gould, "So Cleverly Kind."

72. Darwin, *Descent*, 1:139 (144).

73. Quoted in Smith, "Evolution," 85.

74. Milo, *Good Enough*, 65.

75. Brookes, *Extreme Measures*, 146.

76. Galton, *Memories*, 310.

77. Galton, "Eugenics," 37.

78. Galton, "Possible Improvement," 2–3.

79. Richards, *Darwin*, 153.

80. Browne, "Making Darwin," 361.

81. Galton, *Hereditary Genius*, viii.

82. For a discussion of how Galton's ideas emerged out of this context, see Renwick, "Political Economy to Sociology."

83. Peart and Levy, *"Vanity of the Philosopher,"* xii.

84. Peart and Levy, *"Vanity of the Philosopher,"* 130.

85. Quoted in Gökyiğit, "Reception," 220.

86. Galton, "Possible Improvement," 34.

87. Galton, "Eugenics," 38.

88. Quoted in Brookes, *Extreme Measures*, 165.

89. Galton, *Hereditary Genius*, x.

90. Gökyiğit, "Reception."

91. Barany, "Savage Numbers," 241; Kevles, *In the Name of Eugenics*; Barkan, *Retreat of Scientific Racism*; Proctor, *Racial Hygiene*.

92. Donnelly, "We Have Always Been Biased."

93. Bulmer, *Francis Galton*.

94. Waller, "Putting Method First," 35.

95. Porter, *Rise of Statistical Thinking*, 110–46.

96. Lundgren, "Politics of Participation."

97. Galton, "Anthropometric Laboratory," 206, 208.

98. Richardson, *Howard Andrew Knox*, 48–49.

99. Pearson, "Laws of Inheritance," 159.

100. Galton, "Probability," 85.

101. Galton, "Probability," 94.

102. Galton, "Eugenics," 42.

103. Lundgren, "Politics of Participation."

7. The Sacrifice and Rebirth of "Man"

1. Veblen, "Evolutionary Science," 389–90.

2. Veblen, "Evolutionary Science," 393.

3. Hodgson, "Essence."

4. Weintraub, *Mathematical Science*; Morgan, "Economics," 283; Fourcade, *Economics and Societies*, xi; MacKenzie, *Engine*, 6–7.

5. Ross, "Changing Contours," 212.

6. Comic Strip, "Ripley's Believe It or Not," box 1, folder 11, Earle Edward Eubank Papers (hereafter cited as EEE), University of Chicago Special Collections Research Center.

7. Schmaus, Pickering, and Bourdeau, introduction, 10–12.

8. Manuel, *Prophets of Paris*; Lepenies, *Between Literature and Science*; Pickering, *Comte*.

9. Käsler, *Sociological Adventures*.

10. On Parsons, see Boskoff, "Systematic Sociology." On immigration, see Coser, *Refugee Scholars*; Steinmetz, "Ideas in Exile."

11. Turner and Turner, *Impossible Science*, 72–73.

12. Sala, "Rise of Sociology," 574.

13. Eubank to Ross, October 25, 1935, box 11, folder 10, EEE.

14. Chapoulie, Kornblum, and Wazer, *Chicago Sociology*, 11–35.

15. Quoted in Käsler, *Sociological Adventures*, 8.

16. Eubank to Ross, October 26, 1935, box 11, folder 10, EEE.

17. For the same acronym-related reasons the American Sociological Society changed its name from "Society" to "Association"; the text uses ASA as the acronym. On the revolt, see Kuklick, "'Scientific Revolution'"; Lengermann, "Founding"; Bannister, *Sociology and Scientism*, 188–230.

18. "Editor's Remarks," 315.

19. Lengermann, "Founding," 189.

20. Lengermann, "Founding," 189.

21. Ogburn, "Folkways," 306.

22. Ogburn, "Folkways," 300–301.

23. Ogburn, "Folkways," 302.

24. Bannister, *Sociology and Scientism*, 163–74.

25. Kenneth Barnhart, "Interview," box 1, folder 1, University of Chicago Department of Sociology Interviews 1972, University of Chicago Special Research Collections.

26. Barnhart, "Interview."

27. For a discussion of Hankins, see chapter 8.

28. "Uncollected Notes," box 2, folder 18, Luther Lee Bernard Papers (hereafter cited as LLB), University of Chicago Special Research Collections.

29. Bernard to Ellwood, October 7, 1927, box 2, folder 19, LLB.

30. Ellwood to Bernard, November 1, 1928. Quoted in LoConto, "Charles A. Ellwood," 120.

31. LoConto, "Charles A. Ellwood," 121.

32. LoConto, "Charles A. Ellwood," 113; Turner, "Life," 115–16.

33. Käsler, *Sociological Adventures*, 10.

34. MacLean and Williams, "'Ghosts of Sociologies Past.'"

35. Lengermann and Niebrugge, *Women Founders*, 65–93.

36. MacLean and Williams, "'Ghosts of Sociologies Past,'" 244.

37. Lengermann and Niebrugge, "Thrice Told."

38. Oberschall, *Empirical Sociology*, 204.

39. Oberschall, *Empirical Sociology*, 198.

40. Sala, "Rise of Sociology," 558.

41. Chriss, "Review Essay," 493.

42. Eubank, "Memo to Bernard," May 20, 1938, box 1, folder 2, LLB.

43. Käsler, *Sociological Adventures*, 11.

44. Steinmetz, "American Sociology," 328.

45. Oberschall, "Institutionalization," 214.

46. Diner, "Department and Discipline"; Bulmer, *Chicago School*; Abbot, *Department and Discipline*.

47. Odum, *American Masters*; Ross, *Origins*; Breslau, "American Spencerians."

48. Brackett Lewis to Brown, January 17, 1940, box 4, folder 4, EEE.

49. Eubank to Obrdlík, January 19, 1940, box 4, folder 4, EEE.

50. Quoted in Käsler, *Sociological Adventures*, 9.

51. Fourcade, *Economics and Societies*, 64.

52. Fourcade, *Economics and Societies*, 64, 66.

53. Hodgson, *How Economics Forgot History*, 113–33; Miller, "Richard T. Ely"; Bradizza, *Critique of Capitalism*; Fourcade, *Economics and Societies*, 78; Kurz, *Economic Thought*, 75; Rutherford, *Institutionalist Movement*; Berman, *Thinking*, 26–31.

54. Fine, "Richard T. Ely," 599.

55. Rutherford, "Institutional Economics."

56. Fourcade, *Economics and Societies*, ix.

57. Veblen, "Place of Science," 597.

58. Veblen, "Place of Science," 597.

59. Veblen, "Place of Science," 601.

60. Jevons, *Principles of Science*, 197.

61. Jevons, *Principles of Science*, 199, 200.

62. Maas, *Jevons*, xvii.

63. Quoted in Rutherford, "Wesley Mitchell," 64.

64. Rutherford, "Wesley Mitchell," 65.

65. Quoted in Rutherford, "Wesley Mitchell," 65.

66. Mitchell, "Quantitative Analysis," 4.

67. Mitchell, "Quantitative Analysis," 4.

68. Mitchell, "Quantitative Analysis," 4.

69. Mitchell, "Quantitative Analysis," 5.

70. Mitchell, "Quantitative Analysis," 6–7.

71. Mitchell, "Quantitative Analysis," 12.

72. Rutherford, "Institutional Economics," 179–80; Berman, *Thinking*, 27–29; Hodgson, "John R. Commons."

73. Berman, *Thinking*, 98.

74. Rothschild, *Economic Sentiments*, 218.

75. Yonay, *Soul of Economics*.

76. Koopman, "Measurement without Theory," 161.

77. Koopman, "Measurement without Theory," 166.

78. Berman, *Thinking*, 26.

79. Gonce, "F. H. Knight," 813.

80. Knight, "Fact and Value," 225.

81. Gonce, "F. H. Knight," 832.

82. See the discussion of Frank Hankins in chapter 8.

83. Knight, "Political Trend," 20.

84. Knight, "Fact and Value," 225.

85. Knight, "Fact and Value," 226–27.

86. Knight, "Fact and Value," 228.

87. Knight, "Fact and Value," 228.

88. Knight, "Fact and Value," 239.

89. Appelbaum, *Economists Hour*, chs. 2 and 9; Edwards and Montes, "Friedman in Chile."

90. Mäki, "Preface," xvii. For the full scope of the essay's influence, see Mäki, *Methodology of Positive Economics*.

91. Friedman, "Methodology," 12.

92. Friedman, "Methodology," 11–12.

93. Friedman, "Methodology," 32.

94. Friedman, "Methodology," 33.

95. For extended discussion of the philosophical implications of Freidman's argument about the billiard player, see Vromen, "Ontological Commitments," 196–98.

96. Friedman, "Methodology," 21.

97. Robbins, "Teaching of Economics," 590. Robbins would go on to lead a massive reform of the British school system. See Barr, *Shaping Higher Education*.

98. Robbins, "Teaching of Economics," 589.

99. Blakely, *We Built Reality*.

100. Backhouse and Medema, "Definition of Economics."

101. Papandreou, "Economics," 721. For an impassioned update on this idea, see McCloskey, "Wrong Track."

102. Papandreou, "Economics," 718.

103. Fourcade, *Economics and Societies*, 237.

104. Veblen, "Place of Science," 602.

105. Hacking, "Looping Effects."

8. Social Science by Other Means

1. Hankins, "Social Science," 3.

2. Hankins, "Social Science," 5.

3. Hankins, "Social Science," 7.

4. Leslie, *Cold War*. The role of government and corporate foundation support for this process has been well noted. See especially Porter, "Positioning Social Science." See also Solovey, *Shaky Foundations*.

5. Edwards, *Closed World*; Engerman and Unger, "Towards a Global History of Modernization"; Heyck, *Age of System*; Mueller, "Rockefeller Foundation."

6. Franchi and Güzeldere, "Machinations of the Mind," 18.

7. This is certainly not true of all social scientists. In particular, many twentieth-century sociologists and anthropologists who collected data "concerned themselves with . . . the inner life of another person." Lemov, *Database of Dreams*, 143. For more hopeful possibilities, see also Cohen-Cole, "Creative American"; Cohen-Cole, *Open Mind*.

8. Jennings, *Behavior of the Lower Organisms*, 335.

9. Jennings, *Behavior of the Lower Organisms*, 349.

10. Quoted in Schloegel and Schmidgen, "General Physiology," 630.

11. Schloegel and Schmidgen, "General Physiology," 633.

12. Leary, "Conceptual and Linguistic Activity," 15.

13. Watson, "Psychology," 166.

14. Watson, "Psychology," 167.

15. Quoted in Buckley, *Mechanical Man*, 57.

16. Quoted in Buckley, *Mechanical Man*, 155.

17. Watson, "Psychology," 170.

18. Leary, "Conceptual and Linguistic Activity," 23; Smith, *Human Sciences*, 665; Boden, *Mind as Machine*, 1:261; Lemov, *World as Laboratory*, ch. 4.

19. Hull, "Mind, Mechanism."

20. Hull, "Mind, Mechanism," 29.

21. Hull, "Mind, Mechanism," 29fn18.

22. For a rare connection between Hull's "robot approach" and artificial intelligence, see

Cordeschi, "Discovery," 219. For a revealing account of Hull's commitment to materialism, see Smith, "Models, Mechanism, and Explanation."

23. Rosenblatt, "Perceptron."

24. Smith, "Metaphors of Knowledge," 252.

25. Hull, *Principles of Behavior*, 27.

26. Hull, *Principles of Behavior*, 15.

27. Hull, *Principles of Behavior*, 18.

28. Hull, *Principles of Behavior*, 69.

29. Hull, *Principles of Behavior*, 304.

30. Hull, *Principles of Behavior*, 313.

31. Hull, *Principles of Behavior*, 323.

32. Hull, *Principles of Behavior*, 403.

33. Although their return often emerges in neurobiological approaches to psychology. While many examples exist in popular social science books, for the most extreme vision, see Sapolsky, *Behave*.

34. Koch, "Hull," 130.

35. Hull, *Principles of Behavior*, 304.

36. Hull, *Principles of Behavior*, 400.

37. Hull, *Principles of Behavior*, 400-401.

38. Leonard, *Von Neumann*, 302-3.

39. McCulloch and Pitts, "Logical Calculus," 101.

40. Cobb, *Idea of the Brain*, 178.

41. McCulloch and Pitts, "Logical Calculus," 101.

42. McCulloch and Pitts, "Logical Calculus," 115.

43. McCulloch and Pitts, "Logical Calculus," 113.

44. Abraham, *Rebel Genius*, 52.

45. Cobb, *Idea of the Brain*, 178; Abraham, *Rebel Genius*, 58-67.

46. Abraham, *Rebel Genius*, 62.

47. Abraham, *Rebel Genius*, 68-72;

48. Cobb, *Idea of the Brain*, 178.

49. McCulloch and Pitts, "Logical Calculus," 113.

50. Quoted in Parsons, *Structure*, 10fn1.

51. Parsons, *Structure*, 61.

52. Brinton, "Socio-astrology."

53. Brinton, "Socio-astrology," 253.

54. Brinton, "Socio-astrology," 248.

55. Nichols, "Interstitial Ascent."

56. Heyl, "'Pareto Circle,'" 322-26.

57. Brinton, "Residue of Pareto," 641.

58. Heyl, "Pareto Circle," 334.

59. Parsons, *Structure*, 6.

60. Parsons, *Structure*, 28, 30.

61. Parsons, *Structure*, 21.

62. Camic, "*Structure* after 50 Years," 43.

63. Parsons, *Structure*, 31.

64. Parsons, *Structure*, 35, 38.

65. Cobb, *Idea of the Brain*, 378–79.

66. Parsons, *Structure*, 28.

67. Parsons, *Structure*, 698.

68. Parsons, *Structure*, 77.

69. Parsons, *Structure*, 86.

70. Parsons, *Structure*, 125 (italics in original).

71. Cravens, "Column Right, March!"; Heyck, "Producing Reason."

72. Parsons, "Science Legislation," 241.

73. Parsons, "Science Legislation," 241.

74. Parsons, "Science Legislation," 243.

75. Parsons, "Science Legislation," 243–34.

76. Gerhardt, *Talcott Parsons*, 107.

77. Isaac, *Working Knowledge*.

78. Nichols, "Interstitial Ascent," 564.

79. Parsons, "Science Legislation," 245.

80. Parsons, "Science Legislation," 245.

81. Parsons, "Science Legislation," 248.

82. Solovey, *Social Science for What?*, 10.

83. Engerman, "Social Science"; Gilman, *Mandarins*; Heyck, *Herbert A. Simon*.

84. Quoted in Steinmetz, "American Sociology," 361.

85. Steinmetz, "American Sociology," 350–51.

86. Chriss, *Confronting Gouldner*, 120.

87. Schrum, *Instrumental University*, 73.

88. Isaac, *Working Knowledge*, 237.

89. Scott and Keates, *Schools of Thought*.

90. Copeland, "Intelligent Machinery," 265.

91. Dyson, *Turing's Cathedral*. Even after the "rediscovery" of Turing, there has been little biographical work to rival Hodges, *Alan Turing*, and its many subsequent editions. For a summary of biographical works, see Sumner, "Turing Today."

92. Copeland et al., *Turing Guide*, 3.

93. For the best overview of the many philosophical, psychological, and technical issues, see Shieber, *Turing Test*.

94. Bernhardt, *Turing's Vision*, 47.

95. Bernhardt, *Turing's Vision*, 47.

96. Turing, "Computable Numbers," 231.

97. Turing, "Intelligent Machinery," 17.

98. Turing, "Intelligent Machinery," 21.

99. Turing, "Intelligent Machinery," 3.

100. Turing, "Computing Machinery," 433.

101. Turing, "Computing Machinery," 459.

102. Staddon, *New Behaviorism*, 128; Abramson, "Descartes' Influence," 545.

103. Turing, "Intelligent Machinery," 8.

104. Smith, *Hidden History*, 15.

105. Smith, "National Observatory Transformed."

106. A number of these examples can be found in the work of Simon Schaffer, in particular "Astronomers Mark Time." See also Canales, "Exit the Frog."

107. Cohen, *Howard Aiken*, 164.

108. Minsky, "Some Methods," 6.

109. Quoted in Cohen-Cole, *Open Mind*, 160.

110. Turing, "Intelligent Machinery," 7–8.

111. Hodges, *Alan Turing*, 416.

112. Turing, "Lecture," 120.

113. Turing, "Lecture," 121.

114. Asimov, *I, Robot*, 198–224.

115. Turing, "Lecture," 120.

116. Isaac, "Tool Shock."

117. Robin, *Cold War Enemy*, 19–38.

118. Skinner, *Beyond Freedom*, 5.

119. Castells, *End of Millenium*, 3:367.

120. Minsky, *Computation*, 106.

9. Second-Rate Mathematicians

1. Fontenelle, *Digression*, 313.

2. Laplace, *Essai*, 117.

3. Taylor, "Interpretation," 7–8.

4. Taylor, "Interpretation," 9.

5. Taylor, "Interpretation," 9.

6. Tukey, "Future of Data Analysis," 6, 12–13.

7. Biographical details drawn from McCullagh, "John Wilder Tukey."

8. Lindee, *Rational Fog*, 210.

9. Brillinger, "John W. Tukey."

10. Brillinger, "John W. Tukey."

11. A diligent and thorough biography of Tukey that includes significant information on Tukey's life and technical contributions was self-published by Mark Jones Lorenzo in 2018 under the title *Adventures of a Statistician*.

12. Tukey, Max Mathews, and Arun Netravali, slide transparencies, April 9, 1984, series I, box 4, "Artificial Intelligence Research," John W. Tukey Papers, American Philosophical Society (hereafter cited as JTP).

13. McIlroy to Tukey, April 6, 1984, series I, box 4, "Artificial Intelligence Research," JTP.

14. Markoff, *Machines of Loving Grace*, 125–54.

15. Tukey to McCulloch, August 23, 1947, series I, box 21, JTP.

16. Minsky to Tukey, February 17, 1954, series I, box 21, JTP. The recommendation is undated.

17. "Machines," series I, box 20, JTP.

18. Committee on Behavioral Science, minutes, July 8, 1959, series I, box 23, JTP.

19. Committee on Behavioral Science, minutes, November 13, 1959, series I, box 23, JTP.

20. Tukey to Members of the Committee on Behavioral Sciences, memo, June 2, 1960, series I, box 23, JTP.

21. Tukey to Members of the Committee, memo, June 2, 1960.

22. Committee on Behavioral Sciences, draft report, June 9, 1960, series I, box 23, JTP.

23. For the overall challenges facing the social sciences vis-à-vis the natural sciences, see Solovey, "National Social Science Foundation," 63–67.

24. Committee on Behavioral Sciences, draft report, June 9, 1960, series I, box 23, JTP.

25. Tukey to Shapiro, March 7, 1962, series I, box 23, JTP.

26. Finch to Tukey, September 22, 1964, series I, box 23, JTP.

27. Solovey, "Project Camelot"; Rohde, *Armed with Expertise*, ch. 6.

28. Tukey to Finch, April 16, 1965, series I, box 23, JTP. Tukey had corresponded with the socialist American Association of Science Workers and had championed liberal causes during his life, so there is a strong possibility he distrusted the project on a moral rather than theoretical basis. But there is also no evidence that Tukey experienced a crisis on the level of Wiener, who objected to the militaristic co-option of cybernetics, and it is possible that the overall direction of the group set in 1959 had caused him to decline Finch's offer. See "American Association of Science Workers," series I, box 1, JTP; on Wiener, see Rid, *Rise of the Machines*, ch. 2.

29. Tukey, "Future of Data Analysis," 2.

30. Tukey, "Future of Data Analysis," 3.

31. Tukey, "Future of Data Analysis," 6; For the description of Tukey as "heretical," see Wiggins, "Interview," 3. For a recent popular work that places Tukey in the context of other "heretical scientists," see Wiggins and Jones, *How Data Happened*, 201–8.

32. Tukey, "Future of Data Analysis," 6.

33. Loynes, *Model Building*, 7.

34. Tukey, "Future of Data Analysis," 9 (italics in original).

35. Tukey, "Future of Data Analysis," 13.

36. Tukey, *Exploratory Data Analysis*.

37. Tukey, *Exploratory Data Analysis*, v.

38. Tukey, *Exploratory Data Analysis*, v.

39. Tukey, "Analyzing Data."

40. Lenhard, "Models and Statistical Inference," 86–89.

41. Huber, "Speculations," 177.

42. Huber, "Speculations," 178.

43. Loynes, *Model Building*, 5.

44. Zuboff, *Surveillance Capitalism*.

45. Gitelman, *"Raw Data."*

46. Isaac, "Epistemic Design," 80; Heyck, *Age of System*, 3.

47. Surowiecki, *Wisdom of Crowds*, 6.

48. Lazer et al., "Computational Social Science."

49. The subtitle was changed in subsequent editions.

50. Pentland, *Social Physics*, 4.

51. Pentland, *Social Physics*, 8.

52. Eagle and Greene, *Reality Mining*, 3.

53. Barron, *Reading Our Minds*. Quoted in Halpern, "Bull's-Eye," 62.

54. Halpern, "Bull's-Eye," 62.

55. Schweitzer, "Sociophysics," 40.

56. Schweitzer, "Sociophysics," 46.

57. Lazer et al., "Google Flu."

58. Lazer et al., "Computational Social Science," 1062.

59. Marantz, "Trouble in Paradise."

60. Marantz, "Trouble in Paradise."

61. Sobkowicz, "Social Simulation Models."

62. Pelley, Chasan, Weisz, and Flickinger, "60 Minutes Overtime."

63. Metz, "'Godfather of A.I.' Leaves Google."

64. O'Neil, *Weapons of Math Destruction*.

65. Sweeney, "Discrimination in Online Ad Delivery."

66. Executive Office of the President, "Big Data," 5–6.

67. Noble, *Algorithms of Oppression*, 5, 49.

68. Singer, "Amazon." For an overview of the "stakes" of such research, see Wiggins and Jones, *How Data Happened*, 3–12.

69. Iskenderova and Domahidi, "Systematic Literature Review."

70. Wilson, "Auditing for Bias."

71. De Meo, Pozzana, Prifti, and Provetti, "Finding Gender Bias."

72. Sivak and Smirnov, "Gender Bias."

73. Hajian, "Algorithmic Bias."

74. Kosinski, Stillwell, and Graepel, "Private Traits," 5802.

75. Kiciman, "Where Does Data Bias Come From?"

76. Kusner et al., "Counterfactual Fairness," 9.

77. Roose, "How A.I. Works."

78. Chaintreau, "Sneak Peek."

79. Robertson et al., "Auditing Partisan Audience Bias."

80. Sandvig et al., "Auditing Algorithms."

81. Sandvig et al., "Auditing Algorithms," 15.

82. Gray and Suri, *Ghost Work*.

83. Diesner, "Biases"; Block and Riesewieck, *Cleaners*.

84. To see just such a constellation of "endless queues of mundane tasks," see Crawford, *Atlas of AI*, 66.

85. Caliskan, Bryson, and Narayanan, "Semantics," 183.

86. Fourcade and Johns, "Loops, Ladders, and Links," 807.

87. Fourcade and Johns, "Loops, Ladders, and Links," 809.

88. Zuboff, *Surveillance Capitalism*, 8.

89. Pasquale, "Algorithmic Self"; Bouk, "National Data Center."

90. De Grazia, *Irresistible Empire*.

91. See Feldman and Riskin, "Biology Is Not Destiny," 43.

92. Duster, *Backdoor to Eugenics*.

93. Wade, *Troublesome Inheritance*.

94. Harden, "Genetics of Education." For the most concerted attack on progressives, see Leonard, *Illiberal Reformers*. For a defense of American progressives, see Hovenkamp, "Progressives."

95. Gould, *Mismeasure of Man*; Barkan, *Retreat of Scientific Racism*.

96. Jones, "How We Became Instrumentalists."

97. Wooldridge, *Brief History*, 2.

98. Ford, *Architects of Intelligence*, 48.

99. Ford, *Architects of Intelligence*, 328.

100. Mitchell, *Artificial Intelligence*, 250.

101. Mitchell, "Crashing the Barrier."

102. Metz, "Google A.I. Researcher."

103. Bengio and Marcus, "AI Debate."

104. Sapolsky, *Determined*.

Conclusion. The Intelligent Artifact

1. Strevens, *Knowledge Machine*, 260. See also Cowles, *Scientific Method*.

2. Strevens, *Knowledge Machine*, 275.

3. Solovey, *Social Science for What?*; Milo, *Good Enough*.

4. Quoted in Tarnoff and Weigel, *Voices from the Valley*, 110.

5. Quoted in Tarnoff and Weigel, *Voices from the Valley*, 112–13.

6. Wiener, *Uncanny Valley*, 186.

7. Ullman, *Life in Code*, 39.

8. Marx and Engels, *German Ideology*, 53.

9. Quoted in Ford, *Architects of Intelligence*, 25.

10. Cobb, *Idea of the Brain*, 373.

11. Quoted in Ford, *Architects of Intelligence*, 56.

12. Quoted in Ford, *Architects of Intelligence*, 63.

13. Ford, *Architects of Intelligence*, 100–101.

14. Quoted in Ford, *Architects of Intelligence*, 101.

15. Memorably put in the full title of Shapin, *Never Pure: Historical Studies of Science as if It Was Produced by People with Bodies, Situated in Time, Space, Culture, and Society, and Struggling for Credibility and Authority*.

16. Foucault, *Order of Things*, xxiii.

BIBLIOGRAPHY

Abbas, Niran B. *Thinking Machines: Discourses of Artificial Intelligence.* Hamburg: Lit Verlag, 2006.

Abbott, Andrew. *Department and Discipline: Chicago Sociology at One Hundred.* Chicago: University of Chicago Press, 1999.

Abraham, Tara H. *Rebel Genius: Warren S. McCulloch's Transdisciplinary Life in Science.* Cambridge, MA: MIT Press, 2016.

Abramson, Darren. "Descartes' Influence on Turing." *Studies in History and Philosophy of Science Part A* 42, no. 4 (2011): 544–51.

Adams, Matthew S. "Formulating an Anarchist Sociology: Peter Kropotkin's Reading of Herbert Spencer." *Journal of the History of Ideas* 77, no. 1 (2016): 49–73.

Adas, Michael. *Machines as the Measure of Men: Science, Technology, and Ideologies of Western Dominance.* 1989; repr., Ithaca, NY: Cornell University Press, 2014.

Adkins, Gregory Matthew. "When Ideas Matter: The Moral Philosophy of Fontenelle." *Journal of the History of Ideas* 61, no. 3 (2000): 433–52.

Aiton, Eric. *The Vortex Theory of Planetary Motions.* London: Macdonald, 1972.

d'Alembert, Jean le Rond. *Mélanges de littérature, d'histoire et de philosophie.* Edited by Martin Groult. Paris: Classiques Garnier, 2018.

Alanen, Lilli. *Descartes's Concept of Mind.* Cambridge, MA: Harvard University Press, 2003.

Albury, W.R. "The Order of Ideas: Condillac's Method of Analysis as a Political Instrument in the French Revolution." In *The Politics and Rhetoric of Scientific Method: Historical Studies*, edited by John A. Schuster and Richard R. Yeo, 203–26. Dordrecht: Kluwer, 1986.

Anderson, Chris. "The End of Theory: The Data Deluge Makes the Scientific Method Obsolete." *Wired*, June 23, 2000. https://www.wired.com/2008/06/pb-theory/.

Anderson, Chris. *The Long Tail: Why the Future of Business Is Selling Less of More.* New York: Hyperion, 2006.

Anderson, R. D. *European Universities from the Enlightenment to 1914.* Oxford: Oxford University Press, 2004.

Antognazza, Maria Rosa. *Oxford Handbook of Leibniz.* Oxford: Oxford University Press, 2013.

Antoine-Mahut, Delphine, and Sophie Roux. *Physics and Metaphysics in Descartes and in His Reception.* New York: Routledge, 2019.

Appelbaum, Binyamin. *The Economists' Hour: False Prophets, Free Markets, and the Fracture of Society.* New York: Little, Brown, 2019.

Archer, John E. *Social Unrest and Popular Protest in England, 1780–1840.* Cambridge: Cambridge University Press, 2000.

Ariew, Roger. "The Mathematization of Nature and the First Cartesians." In *The Language of Nature: Reassessing the Mathematization of Natural Philosophy in the Seventeenth Century*, edited by Geoffrey Gorham, Benjamin Hill, Edward Slowik, and C. Kenneth Walters, 112–33. Minneapolis: University of Minnesota Press, 2016.

Armenteros, Carolina. *The French Idea of History: Joseph de Maistre and His Heirs, 1794–1854.* Ithaca, NY: Cornell University Press, 2011.

Armenteros, Carolina, and Richard Lebrun. *Joseph de Maistre and His European Readers from Friedrich Von Gentz to Isaiah Berlin.* Boston: Brill, 2011.

Aronova, Elena, Christine von Oertzen, and David Sepkoski. "Historicizing Big Data." Introduction to *Data Histories*, edited by Elena Aronova, Christine von Oertzen, and David Sepkoski, vol. 32 of *Osiris* (2017): 1–17.

Ash, Michael G. "Psychology." In Porter and Ross, *Cambridge History of Science*, vol. 7, *The Modern Social Sciences*, 251–74.

Ashworth, William J. "The Calculating Eye: Baily, Herschel, Babbage and the Business of Astronomy." *British Journal for the History of Science* 27, no. 4 (1994): 409–41.

Asimov, Isaac. *I, Robot.* New York: Bantam Books, 1950.

Aspromourgos, Tony. "The Machine in Adam Smith's Economic and Wider Thought." *Journal of the History of Economic Thought* 34, no. 4 (2012): 475–90.

Aubin, David. "'Principles of Mechanics that Are Susceptible of Application to Society': An Unpublished Notebook of Adolphe Quetelet at the Root of His Social Physics." *Historia Mathematica* 41, no. 2 (2014): 204–23.

Avramov, Iordan. "An Apprenticeship in Scientific Communication: The Early Correspondence of Henry Oldenburg (1656–63)." *Notes and Records of the Royal Society of London* 53, no. 2 (1999): 187–201.

Backhouse, Roger E., and Steven G. Medema. "Retrospectives: On the Definition of Economics." *Journal of Economic Perspectives* 23, no. 1 (Winter 2009): 221–33.

Baker, Keith Michael. *Condorcet: From Natural Philosophy to Social Mathematics.* Chicago: University of Chicago Press, 1975.

Bannister, Robert C. *Sociology and Scientism: The American Quest for Objectivity, 1880–1940.* Chapel Hill: University of North Carolina Press, 1987.

Barany, Michael J. "Savage Numbers and the Evolution of Civilization in Victorian Prehistory." *British Journal for the History of Science* 47, no. 2 (2014): 239–55.

Barbey D'Aurevilly, Jules. *Les prophètes du passé.* 1851. Critical edition by David Cocksey. Paris: Sandre, 2006.

Barkan, Elazar. *The Retreat of Scientific Racism: Changing Concepts of Race in Britain and the United States between the World Wars.* Cambridge: Cambridge University Press, 1992.

Barnard, Howard Clive. *Education and the French Revolution.* Cambridge: Cambridge University Press, 1969.

Barnes, Harry E. "The Struggle of Races and Social Groups as a Factor in the Development of Political and Social Institutions: An Exposition and Critique of the Sociological System of Ludwig Gumplovicz." *Journal of Race Development* 9, no. 4 (1919): 394–419.

Barr, Nicholas, ed. *Shaping Higher Education.* London: London School of Economics, 2014. https://www.lse.ac.uk/economics/Assets/Documents/50YearsAfterRobbins.pdf.

Barrat, James. *Our Final Invention: Artificial Intelligence and the End of the Human Era.* New York: St. Martin's Press, 2013.

Barron, Daniel. *Reading Our Minds: The Rise of Big Data Psychiatry.* New York: Columbia Global Reports, 2021.

Barth, Brian. "The Defector." *New Yorker* 95, no. 38 (2019): 26–33.

Basulto, Dominic. "Artificial Intelligence Has an Amazing Future. Dystopian Movies Get It Wrong." *Washington Post*, May 16, 2014.

Beck, Naomi. "Social Darwinism." In Ruse, *Cambridge Encyclopedia of Darwin*, 195–201.

Beirne, Piers. "Adolphe Quetelet and the Origins of Positivist Criminology." *American Journal of Sociology* 92, no. 6 (1987): 1140–69.

Beiser, Frederick. *After Hegel: German Philosophy, 1840–1900*. Princeton, NJ: Princeton University Press, 2014.

Bénéton, Philippe. Foreword to *Critics of the Enlightenment: Readings in the French Counter-Revolutionary Tradition*, edited by Christopher Olaf Blum, vii–xv. Wilmington, DE: Crosscurrents, 2004.

Bengio, Yoshua, and Gary Marcus. "AI Debate." Hosted by Montreal.AI. Streamed live on December 23, 2019. YouTube video, 2:02:24. https://www.youtube.com/watch?v=Eeqw-FjqFvJA&ab_channel=Montreal.AI.

Bennett, James A. "The Mechanics' Philosophy and the Mechanical Philosophy." *History of Science* 24, no. 1 (1986): 1–28.

Bensaude-Vincent, Bernadette, and William Royall Newman, eds. *The Artificial and the Natural: An Evolving Polarity*. Cambridge, MA: MIT Press, 2007.

Berg, Maxine. *The Age of Manufactures, 1700–1820: Industry, Innovation, and Work in Britain*. London: Routledge, 1994.

Berg, Maxine. *The Machinery Question and the Making of Political Economy, 1815–1848*. Cambridge: Cambridge University Press, 1980.

Berlin, Isaiah. *The Crooked Timber of Humanity: Chapters in the History of Ideas*. 2nd ed. Princeton, NJ: Princeton University Press, 2013.

Berlin, Isaiah, and Henry Hardy. *The Roots of Romanticism*. 2nd ed. Princeton, NJ: Princeton University Press, 2013.

Berman, Elizabeth Popp. *Thinking Like an Economist: How Efficiency Replaced Equality in U.S. Public Policy*. Princeton, NJ: Princeton University Press, 2022.

Bernhardt, Chris. *Turing's Vision: The Birth of Computer Science*. Cambridge, MA: MIT Press, 2016.

Bertoloni Meli, Domenico. *Mechanism: A Visual, Lexical, and Conceptual History*. Pittsburgh: University of Pittsburgh Press, 2019.

Black, Jeff J. S. *Rousseau's Critique of Science: A Commentary on the Discourse on the Sciences and the Arts*. Lanham, MD: Lexington Books, 2009.

Blakely, Jason. *We Built Reality: How Social Science Infiltrated Culture, Politics, and Power*. Oxford: Oxford University Press, 2020.

Block, Hans, and Moritz Riesewieck, dirs. *The Cleaners*. Berlin, Germany: Gebrueder Beetz Filmproduktion, 2018. DVD.

Blum, Christopher Olaf, ed. *Critics of the Enlightenment: Readings in the French Counter-revolutionary Tradition*. Wilmington, DE: Crosscurrents, 2004.

Boden, Margaret A. *AI: Its Nature and Future*. Oxford: Oxford University Press, 2016.

Boden, Margaret A. *Mind as Machine: A History of Cognitive Science*. 2 vols. Oxford: Clarendon Press, 2006.

Bonald, Louis-Gabriel-Ambroise. "Préface à la Théorie du pouvoir politique." In Jules Barbey D'Aurevilly, *Les prophètes du passé*. 1851. Critical edition by David Cocksey. Paris: Sandre, 2006.

Bonald, Louis-Gabriel-Ambroise. *Lettres à Joseph de Maistre*. Edited by Michel Toda. Étampes, France: Clovis, 1997.

Bonald, Louis-Gabriel-Ambroise. *Œuvres completes de M. de Bonald*. 3 vols. Paris: J. P. Migne, 1864.

Boskoff, Alvin. "The Systematic Sociology of Talcott Parsons." *Social Forces* 28, no. 4 (1950): 393–400.

Bostrom, Nick. *Superintelligence: Paths, Dangers, Strategies.* Oxford: Oxford University Press, 2013.

Bouk, Dan. "The History and Political Economy of Personal Data over the Last Two Centuries in Three Acts." In *Data Histories*, edited by Elena Aronova, Christine von Oertzen, and David Sepkoski, vol. 32 of *Osiris* (2017): 85–106.

Bouk, Dan. *How Our Days Became Numbered: Risk and the Rise of the Statistical Individual.* Chicago: University of Chicago Press, 2015.

Bouk, Dan. "The National Data Center and the Rise of the Data Double." *Historical Studies in the Natural Sciences* 48, no. 5 (2018): 627–36.

Bourguet, Marie-Noëlle. *Déchiffrer la France: La statistique départementale à l'époque Napoléonienne.* Paris: Editions des Archives Contemporaines, 1988.

Bowen, Francis. *A Treatise on Logic.* Boston: Allyn and Bacon, 1880.

Bowler, Peter J. *Evolution: The History of an Idea.* Rev. ed. Berkeley: University of California Press, 1989.

Bowler, Peter J. "Herbert Spencer and Lamarckism." In *Herbert Spencer: Legacies*, edited by Mark Francis and Michael W. Taylor, 203–19. London: Routledge, 2015.

Bowler, Peter J., and Iwan Rhys Morus. *Making Modern Science: A Historical Survey.* Chicago: University of Chicago Press, 2005.

Bradizza, Luigi. *Richard T. Ely's Critique of Capitalism.* New York: Palgrave Macmillan, 2013.

Bradley, Owen. "Maistre's Theory of Sacrifice." In Lebrun, *Joseph de Maistre's Life*, 65–83.

Breckman, Warren, and Peter E. Gordon. *The Cambridge History of Modern European Thought.* Cambridge: Cambridge University Press, 2016.

Breslau, Daniel. "The American Spencerians: Theorizing a New Science." In Calhoun, *Sociology in America*, 39–62.

Brewster, David. *Memoirs of the Life, Writings, and Discoveries of Sir Isaac Newton.* Edinburgh: T. Constable, 1855.

Brillinger, David R. "John W. Tukey, a Biographical Memoir." National Academy of Sciences. 2018. https://www.nasonline.org/publications/biographical-memoirs/memoir-pdfs/tukey-john.pdf.

Brillinger, David R. "John W. Tukey: His Life and Professional Contributions." *Annals of Statistics* 30, no. 6 (2002): 1535–75.

Brinton, Crane. "The Residue of Pareto." *Foreign Affairs* 32, no. 4 (1954): 640–50.

Brinton, Crane. "Socio-astrology." *Southern Review* 3, no. 2 (1937): 243–66.

Brookes, Martin. *Extreme Measures: The Dark Visions and Bright Ideas of Francis Galton.* London: Bloomsbury, 2004.

Browne, Alice. "Descartes's Dreams." *Journal of the Warburg and Courtauld Institutes* 40 (1977): 256–73.

Browne, Janet. "Making Darwin: Biography and the Changing Representations of Charles Darwin." *Journal of Interdisciplinary History* 40, no. 3 (2010): 347–73.

Bruder, Jessica. *Nomadland: Surviving America in the Twenty-First Century.* New York: Norton, 2017.

Brynjolfsson, Erik, and Andrew McAfee. *The Second Machine Age: Work, Progress, and Prosperity in a Time of Brilliant Technologies.* New York: Norton, 2014.

Buckley, Kerry W. *Mechanical Man: John Broadus Watson and the Beginnings of Behaviorism.* New York: Guilford, 1989.

Bulmer, Martin. *The Chicago School of Sociology: Institutionalization, Diversity, and the Rise of Sociological Research.* Chicago: University of Chicago Press, 1984.

Bulmer, M. G. *Francis Galton: Pioneer of Heredity and Biometry.* Baltimore, MD: Johns Hopkins University Press, 2003.

Burrell, Jenna, and Marion Fourcade. "The Society of Algorithms." *Annual Review of Sociology* 47 (2021): 213–37.

Burrows, Simon. *French Exile Journalism and European Politics, 1792–1814.* Suffolk, UK: Royal Historical Society, 2000.

Buzon, Frédéric. "Beeckman, Descartes, and Physico-mathematics." In Garber and Roux, *Mechanization of Natural Philosophy*, 143–58.

Byrnes, Joseph F. *Catholic and French Forever: Religious and National Identity in Modern France.* University Park: Pennsylvania State University Press, 2005.

Byrnes, Joseph F. "Chateaubriand and Destutt de Tracy: Defining Religious and Secular Polarities in France at the Beginning of the Nineteenth Century." *Church History* 60, no. 3 (1991): 316–30.

Cabanis, André. *La presse sous le Consulat et l'Empire, 1799–1814.* 3 vols. Paris: Société des Études Robespierristes, 1975.

Calhoun, Craig J. *Sociology in America: A History.* Chicago: University of Chicago Press, 2007.

Caliskan, Aylin, Joanna J. Bryson, and Arvind Narayanan. "Semantics Derived Automatically from Language Corpora Contain Human-Like Biases." *Science* 356, no. 6334 (2017): 183–86.

Camic, Charles. "*Structure* after 50 Years: The Anatomy of a Charter." *American Journal of Sociology* 95, no. 1 (1989): 38–107.

Canales, Jimena. *Bedeviled: A Shadow History of Demons in Science.* Princeton, NJ: Princeton University Press, 2020.

Canales, Jimena. "Exit the Frog, Enter the Human: Physiology and Experimental Psychology in Nineteenth-Century Astronomy." *British Journal for the History of Science* 34, no. 2 (2001): 173–97.

Cannon, Susan Faye. *Science in Culture: The Early Victorian Period.* New York: Science History Publications, 1978.

Cantor, Geoffrey. "Science, Providence, and Progress at the Great Exhibition." *Isis* 103, no. 3 (2012): 439–59.

Carr, Nicholas G. *The Glass Cage: Automation and Us.* New York: Norton, 2014.

Carr, Nicholas. "Is Google Making Us Stupid?" *Atlantic Monthly*, July/August 2008. https://www.theatlantic.com/magazine/archive/2008/07/is-google-making-us-stupid/306868/.

Carson, John. "Culture of Intelligence." In Porter and Ross, *Cambridge History of Science*, vol. 7, *The Modern Social Sciences*, 635–48.

Carson, John. *The Measure of Merit: Talents, Intelligence, and Inequality in the French and American Republics, 1750–1940.* Princeton, NJ: Princeton University Press, 2007.

Cassirer, Ernst. *The Philosophy of the Enlightenment.* Translated by Fritz C. A. Keollen and James P. Petergrove. Princeton, NJ: Princeton University Press, 1951.

Castells, Manuel. *End of Millennium.* 3 vols. Malden, MA: Blackwell, 2000.

Chaintreau, Augustin. "A Sneak Peek at Big Data's Shadow Text." Keynote presentation at "Networks: The Creation and Circulation of Knowledge from Franklin to Facebook," American Philosophical Society, Philadelphia, June 6, 2019.

Chapoulie, Jean-Michel, William Kornblum, and Caroline Wazer. *Chicago Sociology.* New York: Columbia University Press, 2019.

Chriss, James J. *Confronting Gouldner: Sociology and Political Activism*. Boston: Brill, 2015.

Chriss, James J. "Review Essay: Addams, Ward et al.; American Sociology Past to Present." *Journal of Classical Sociology* 8, no. 4 (2008): 491–502.

Chu, Tim. *Human Purpose and Transhuman Potential: A Cosmic Vision for Our Future Evolution*. San Rafael, CA: Origins, 2014.

Claeys, Gregory. Introduction to *The Cambridge Companion to Nineteenth-Century Thought*, edited by Gregory Claeys, 1–7. Cambridge: Cambridge University Press, 2019.

Claeys, Gregory. "The 'Survival of the Fittest' and the Origins of Social Darwinism." *Journal of the History of Ideas* 61, no. 2 (2000): 223–40.

Clarke, Desmond M. *Descartes: A Biography*. New York: Cambridge University Press, 2006.

Cobb, Matthew. *The Idea of the Brain: A History*. London: Profile Books, 2020.

Cohen, I. Bernard. *Howard Aiken*. Cambridge, MA: MIT Press, 1999.

Cohen, I. Bernard. *The Triumph of Numbers: How Counting Shaped Modern Life*. New York: Norton, 2005.

Cohen-Cole, Jamie. "The Creative American: Cold War Salons, Social Science, and the Cure for Modern Society." *Isis* 100, no. 2 (2009): 219–62.

Cohen-Cole, Jamie. *The Open Mind: Cold War Politics and the Sciences of Human Nature*. Chicago: University of Chicago Press, 2013.

Cole, Joshua. *The Power of Large Numbers: Population, Politics, and Gender in Nineteenth-Century France*. Ithaca, NY: Cornell University Press, 2000.

Collard, Auguste. "Adolphe Quetelet et l'astronomie." *Ciel et Terre* 6 (1935): 209–23.

Compagnon, Antoine. *Les antimodernes: De Joseph de Maistre à Roland Barthes*. Paris: Gallimard, 2005.

"A Complete Guide to AI." *Wired*, February 2, 2018.

Condillac, Etienne Bonnot de. *Traité des systèmes*. Paris: Fayard, 1991.

Condorcet. *Essai sur l'application de l'analyse à la probabilité*. 1785; repr., New York: Chelsea Publishing, 1972.

Condorcet. *Œuvres de Condorcet*. 12 vols. Paris, 1849.

Condorcet. *Outlines of an Historical View of the Progress of the Human Mind*. Reprint of the first translated (anonymous) ed. London, 1795.

Cook, Sylvia Jenkins. *Working Women, Literary Ladies: The Industrial Revolution and Female Aspiration*. Oxford: Oxford University Press, 2008.

Copeland, Jack. "Intelligent Machinery." In Copeland et al., *Turing Guide*, 265–75.

Copeland, Jack, Jonathan P. Bowen, Mark Sprevak, and Robin Wilson. *The Turing Guide*. Oxford: Oxford University Press, 2017.

Cordeschi, Roberto. "The Discovery of the Artificial: Some Protocybernetic Developments, 1930–1940." *AI & Society* 5 (1991): 218–38.

Coser, Lewis A. *Refugee Scholars in America: Their Impact and Their Experiences*. New Haven, CT: Yale University Press, 1984.

Cowles, Henry M. *The Scientific Method: An Evolution of Thinking from Darwin to Dewey*. Cambridge, MA: Harvard University Press, 2020.

Crary, Jonathan. *Techniques of the Observer: On Vision and Modernity in the Nineteenth Century*. Cambridge, MA: MIT Press, 1990.

Cravens, Hamilton. "Column Right, March! Nationalism, Scientific Positivism, and the Conservative Turn of the American Social Sciences in the Cold War Era." In Solovey and Cravens, *Cold War Social Science*, 117–36.

Crawford, Kate. *Atlas of AI: Power, Politics, and the Planetary Costs of Artificial Intelligence.* New Haven, CT: Yale University Press, 2021.

Crépel, Pierre, and Jean-Nicolas Rieucau. "Condorcet's Social Mathematics, a Few Tables." *Social Choice and Welfare* 25, no. 23 (2005): 243–85.

Crosland, Maurice. "A Science Empire in Napoleonic France." *History of Science* 44, no. 1 (2006): 29–48.

Cumming, Ian. *Helvétius: His Life and Place in the History of Educational Thought.* London, Routledge, 1955.

Cunning, David. "Malebranche and Occasional Causes." *Philosophy Compass* 3, no. 3 (2008): 471–90.

Curran, Andrew S. *Diderot and the Art of Thinking Freely.* New York: Other Press, 2019.

Cushing, Max Pearson. "Baron d'Holbach: A Study of Eighteenth-Century Radicalism in France." PhD dissertation, Columbia University, 1914. Internet Archive.

Dahlbom, Bo, Svante Beckman, and Göran B. Nilsson. *Artifacts and Artificial Science.* Stockholm: Almqvist, 2002.

Darcel, Jean-Louis. "The Roads of Exile, 1792–1817." In Lebrun, *Joseph de Maistre's Life,* 15–31.

Darvall, Frank Ongley. *Popular Disturbances and Public Order in Regency England* [. . .]. London: Milford, 1934.

Darwin, Charles. *The Descent of Man, and Selection in Relation to Sex.* 2 vols. New York: New York University Press, 1989.

Dasgupta, Subrata. *It Began with Babbage: The Genesis of Computer Science.* New York: Oxford University Press, 2014.

Daston, Lorraine. *Classical Probability in the Enlightenment.* Princeton, NJ: Princeton University Press, 1988.

Daston, Lorraine. "D'Alembert's Critique of Probability Theory." *Historia Mathematica* 6, no. 3 (1979): 259–79.

Daston, Lorraine. "Enlightenment Calculations." *Critical Inquiry* 21, no. 1 (Autumn 1994): 182–202.

Daston, Lorraine, and Peter Galison. *Objectivity.* New York: Zone Books, 2007.

Davenport, Stewart. *Friends of the Unrighteous Mammon: Northern Christians and Market Capitalism, 1815–1860.* Chicago: University of Chicago Press, 2008.

de Chadarevian, Soraya, and Theodore M. Porter, eds. "Histories of Data and the Database." Special issue, *Historical Studies in the Natural Sciences* 48, no. 5 (2018).

de Chadarevian, Soraya, and Theodore M. Porter. "Scrutinizing the Data World." Introduction to "Histories of Data and the Database," special issue, *Historical Studies in the Natural Sciences* 48, no. 5 (2018): 549–56.

De Decker, Pierre. "De l'influence du libre arbitre de l'homme sur les faits sociaux." *Mémoires de l'Académie Royale des Sciences, des Lettres et des Beaux-Arts de Belgique* 21 (1848): 69–92.

De Grazia, Victoria. *Irresistible Empire: America's Advance through Twentieth-Century Europe.* Cambridge, MA: Belknap Press of Harvard University Press, 2005.

De Meo, Pasquale, Iacopo Pozzana, Ylli Prifti, and Alessandro Provetti. "Finding Gender Bias in Web-Based, High-Trust Interactions." Paper presented at the Second European Symposium Series on Societal Challenges in Computational Social Sciences, Cologne, Germany, December 2018.

Des Chene, Dennis. *Spirits and Clocks: Machine and Organism in Descartes.* Ithaca, NY: Cornell University Press, 2001.

Descartes, René. *Meditations on First Philosophy: With Selections from the Objections and Replies: A Latin-English Edition*. Translated and edited by John Cottingham. Cambridge: Cambridge University Press, 2013.

Descartes, René, and Ferdinand Alquié. *Œuvres philosophiques*. Paris: Bordas, 1996.

Desch, Michael C. *Cult of the Irrelevant: The Waning Influence of Social Science on National Security*. Princeton, NJ: Princeton University Press, 2019.

Desrosières, Alain. *The Politics of Large Numbers: A History of Statistical Reasoning*. Cambridge, MA: Harvard University Press, 1998.

Diaz-Bone, Rainer, and Emmanuel Didier. "Introduction: The Sociology of Quantification—Perspectives on an Emerging Field in the Social Sciences." *Historical Social Research* 41, no. 2 (2016): 7–26.

Dick, Stephanie. "Artificial Intelligence." *Harvard Data Science Review* 1, no. 1 (2019). https://doi.org/10.1162/99608f92.92fe150c.

Dick, Stephanie. "Of Models and Machines: Implementing Bounded Rationality." *Isis* 106, no. 3 (2015): 623–34.

Diderot, Denis. *Œuvres philosophiques*. Paris: Gallimard, 2010.

Diemer, Arnaud, and Hervé Guillemin. "L'économie politique au miroir de la physique: Adam Smith et Isaac Newton." *Revue d'histoire des sciences* 64, no. 1 (2011): 5–26.

Diesner, Jana. "Biases in Social Computing Data and Technology." Workshop at the Second European Symposium Series on Societal Challenges in Computational Social Sciences, Cologne, Germany, December 2018.

Diner, Steven J. "Department and Discipline: The Department of Sociology at the University of Chicago, 1892–1920." *Minerva* 13, no. 4 (1975): 514–53.

"Discours sur l'accord des sciences et des lettres." *Le Conservateur* 1 (1807): 63–68.

Dodd, William. *A Narrative of the Experiences and Sufferings of William Dodd, a Factory Cripple*. London, 1841.

Domingos, Pedro. *The Master Algorithm: How the Quest for the Ultimate Learning Machine Will Remake our World*. New York: Basic Books, 2015.

Donnelly, Kevin. *Adolphe Quetelet, Social Physics and the Average Men of Science, 1796–1874*. Pittsburgh: University of Pittsburgh Press, 2016.

Donnelly, Kevin. "On the Boredom of Science: Positional Astronomy in the Nineteenth Century." *British Journal for the History of Science* 47, no. 174 (2014): 479–503.

Donnelly, Kevin. "The Other Average Man: Science Workers in Quetelet's Belgium." *History of Science* 52, no. 4 (2014): 401–28.

Donnelly, Kevin. "Redeeming Belgian Science: Periodic Phenomena and Global Physics in Brussels, 1825–1870." *History of Meteorology* 8 (2017): 54–72.

Donnelly, Kevin. "We Have Always Been Biased: Measuring the Human Body from Anthropometry to the Computational Social Sciences." *PUBLIC* 60 (2020): 20–33.

Dormehl, Luke. *Thinking Machines: The Quest for Artificial Intelligence and Where It's Taking Us Next*. New York: TarcherPerigee, 2017.

Doyle, Bret J. Lalumia. "How (Not) to Study Descartes' *Regulae*." *British Journal for the History of Philosophy* 17, no. 1 (2009): 3–30.

Dupré, Louis K. *The Enlightenment and the Intellectual Foundations of Modern Culture*. New Haven, CT: Yale University Press, 2004.

Duster, Troy. *Backdoor to Eugenics*. New York: Routledge, 2003.

Dyson, George. *Turing's Cathedral: The Origins of the Digital Universe*. New York: Pantheon Books, 2012.

Dzelzainis, Ella. "Radicalism and Reform." In *The Oxford History of the Novel in English*, vol. 3, *The Nineteen Century Novel, 1820–1880*, edited by John Kucich and Jenny Bourne Taylor, 427–43. Oxford: Oxford University Press, 2012.

Eagle, Nathan, and Kate Greene. *Reality Mining: Using Big Data to Engineer a Better World*. Cambridge, MA: MIT Press, 2014.

Easley, Alexis. *Literary Celebrity, Gender, and Victorian Authorship, 1850–1914*. Newark: University of Delaware Press, 2011.

Edgerton, David. *The Shock of the Old: Technology and Global History since 1900*. Oxford: Oxford University Press, 2007.

"Editor's Remarks: When Sociology Reached the Masses." *Contemporary Sociology* 42, no. 3 (2013): 315–23.

Edwards, Paul N. *The Closed World: Computers and the Politics of Discourse in Cold War America*. Cambridge, MA: MIT Press, 1996.

Edwards, Sebastien, and Leonidas Montes. "Milton Friedman in Chile: Shock Therapy, Economic Freedom, and Exchange Rates." *Journal of the History of Economic Thought* 42 (March 2020): 105–32.

Ekbia, H. R. *Artificial Dreams: The Quest for Non-biological Intelligence*. Cambridge: Cambridge University Press, 2008.

Elliott, Paul. "Erasmus Darwin, Herbert Spencer, and the Origins of the Evolutionary Worldview in British Provincial Scientific Culture, 1770–1850." *Isis* 94, no. 1 (2003): 1–29.

Elmslie, Bruce. "Retrospectives: Adam Smith's Discovery of Trade Gravity." *Journal of Economic Perspectives* 32, no. 2 (2018): 209–22.

Emerson, Ralph Waldo. *The Method of Nature: An Oration, Delivered before the Society of the Adelphi, in Waterville College, in Maine, August 11, 1841*. Boston: S. G. Simpkins, 1841.

Engerman, David C. "Social Science in the Cold War." *Isis* 101, no. 2 (2010): 393–400.

Engerman, David C., and Corina R. Unger. "Towards a Global History of Modernization." *Diplomatic History* 33, no. 3 (2009): 375–85.

Epstein, Joshua M., and Robert Axtell. *Growing Artificial Societies: Social Science from the Bottom Up*. Washington, DC: Brookings Institution Press, 1996.

Estes, William Kaye, and A. T. Poffenberger. *Modern Learning Theory: A Critical Analysis of Five Examples*. New York: Appleton-Century Crofts, 1954.

Executive Office of the President. "Big Data: A Report on Algorithmic Systems, Opportunity, and Civil Rights Federal." 2016. https://obamawhitehouse.archives.gov/sites/default/files/microsites/ostp/2016_0504_data_discrimination.pdf.

Feldman, M. W., and Jessica Riskin. "Why Biology Is Not Destiny." *New York Review of Books*, April 21, 2022, 43–46.

Ferguson, Adam. *An Essay on the History of Civil Society*. 1767. New Brunswick, NJ: Transaction Publishers, 1995.

Fine, Sydney. "Richard T. Ely, Forerunner of Progressivism, 1880–1901." *Mississippi Valley Historical Review* 37, no. 4 (1951): 599–624.

Fjelland, Ragnar. "Why General Artificial Intelligence Will Not Be Realized." *Humanities and Social Sciences Communications* 7, no. 1 (2020): 1–9.

Flannery, Michael A. *Nature's Prophet: Alfred Russel Wallace and His Evolution from Natural Selection to Natural Theology*. Tuscaloosa: University of Alabama Press, 2018.

Foer, Franklin. *World without Mind: The Existential Threat of Big Tech*. New York: Penguin Press, 2017.

Foley, Vernard. *The Social Physics of Adam Smith*. West Lafayette, IN: Purdue University Press, 1976.

Ford, Martin. *Architects of Intelligence: The Truth about AI from the People Building It*. Birmingham, UK: Packt, 2018.

Fontenelle, Bernard Le Bovier de. *Digression sur les Anciens et les Modernes et autres textes philosophiques*. Édition sous la direction de Sophie Audidière. Paris: Classiques Garnier, 2015.

Fossati, William J. "Maximum Influence from Minimum Abilities: La Mettrie and Radical Materialism." In *The Failure of Modernism: The Cartesian Legacy and Contemporary Pluralism*, edited by Brendan Sweetman, 45–57. Washington, DC: Catholic University of America Press, 1999.

Foucault, Michel. "Nietzsche, Genealogy, History." In *Language, Counter-memory, Practice: Selected Essays and Interviews*, edited by Donald F. Bouchard, 139–64. Ithaca, NY: Cornell University Press.

Foucault, Michel. *The Order of Things: An Archaeology of the Human Sciences*. Paperback translation of *Les mots et les chose* (1966). New York: Vintage, 1994.

Fourcade, Marion. *Economies and Societies: Discipline and Profession in the Unites States, Britain, and France, 1890s to 1990s*. Princeton, NJ: Princeton University Press, 2010.

Fourcade, Marion, and Fleur Johns. "Loops, Ladders, and Links: The Recursivity of Social and Machine Learning." *Theory and Society* 49 (2020): 803–32.

Franchi, Stefano, and Güven Güzeldere. "Machinations of the Mind: Cybernetics and Artificial Intelligence from Automata to Cyborgs." In *Mechanical Bodies, Computational Minds: Artificial Intelligence from Automata to Cyborgs*, edited by Stefano Franchi and Güven Güzeldere, 15–152. Cambridge, MA: MIT Press, 2005.

Francis, Mark. *Herbert Spencer and the Invention of Modern Life*. Ithaca, NY: Cornell University Press, 2007.

Francis, Mark, and Michael Taylor. *Herbert Spencer: Legacies*. Durham, UK: Acumen, 2013.

Frawley, Maria. "Behind the Scenes of History: Harriet Martineau and 'The Lowell Offering.'" *Victorian Periodicals Review* 38, no. 2 (2005): 141–57.

Freedgood, Elaine. "Banishing Panic: Harriet Martineau and the Popularization of Political Economy." *Victorian Studies* 39, no. 1 (1995): 33–53.

Freedgood, Elaine. *Victorian Writing about Risk: Imagining a Safe England in a Dangerous World*. Cambridge: Cambridge University Press, 2000.

French, Robert M. "Dusting Off the Turing Test." *Science* 336, no. 6078 (2012): 164–65.

Frey, Carl Benedikt. *The Technology Trap: Capital, Labor, and Power in the Age of Automation*. Princeton, NJ: Princeton University Press, 2019.

Friedman, Milton. "The Methodology of Positive Economics." In *Essays in Positive Economics*, 3–43. Chicago: University of Chicago Press, 1966.

Friend, Tad. "Superior Intelligence." *New Yorker* 94, no. 13 (2018): 44.

Gagnier, Regenia. "Social Atoms: Working-Class Autobiography, Subjectivity, and Gender." *Victorian Studies* 30, no. 3 (1987): 335–63.

Galison, Peter. *Image and Logic: A Material Culture of Microphysics*. Chicago: University of Chicago Press, 1997.

Galton, Francis. *Essays in Eugenics*. London: Eugenics Education Society, 1909.

Galton, Francis. "Eugenics, Its Definition, Scope, and Aims." In *Essays in Eugenics*, 35–43.

Galton, Francis. *Hereditary Genius: An Inquiry into Its Laws and Consequences*. 2nd ed. London: Macmillan, 1892.

Galton, Francis. "Hereditary Talent and Character." *Macmillan's Magazine* 12, no. 68 (1865): 157–66.

Galton, Francis. *Memories of My Life*. London: Methuen, 1908.

Galton, Francis. "On the Anthropometric Laboratory at the Late International Health Exhibition." *Journal of the Anthropological Institute of Great Britain and Ireland* 14, no. 3 (1885): 205–20.

Galton, Francis. "The Possible Improvement of the Human Breed under Existing Conditions of Laws and Sentiment." In *Essays in Eugenics*, 1–34.

Galton, Francis. "Probability, the Foundation of Eugenics." In *Essays in Eugenics*, 73–99.

Garber, Daniel. "Descartes, Mechanics, and the Mechanical Philosophy." *Midwest Studies in Philosophy* 26, no. 1 (2002): 185–204.

Garber, Daniel. *Descartes' Metaphysical Physics*. Chicago: University of Chicago Press, 1992.

Garber, Daniel, and Sophie Roux, eds. *The Mechanization of Natural Philosophy*. Dordrecht: Springer, 2013.

Garrard, Graeme. *Counter-Enlightenments: From the Eighteenth Century to the Present*. London: Routledge, 2006.

Gascoigne, John. "The Study of Nature." In *The Cambridge History of Eighteenth-Century Philosophy*, vol. 2, edited by Knud Haakonssen, 854–72. Cambridge: Cambridge University Press, 2006.

Gaukroger, Stephen. *Descartes: An Intellectual Biography*. Oxford: Clarendon Press, 1995.

Gaukroger, Stephen. *Descartes' System of Natural Philosophy*. Cambridge: Cambridge University Press, 2002.

Gaukroger, Stephen. Introduction to *The Blackwell Guide to Descartes' Meditations*. Edited by Stephen Gaukroger, 1–5. Malden, MA: Blackwell, 2006.

Geissler, Heike. *Seasonal Associate*. South Pasadena, CA: Semiotext(e), 2018.

Gengembre, Gérard. *La contre-revolution, ou, l'histoire désespérante: Histoire des idées politiques*. Paris: Imago, 1989.

Gerhardt, Uta. *Talcott Parsons: An Intellectual Biography*. Cambridge: Cambridge University Press, 2002.

Gerrish, Sean, and Kevin Scott. *How Smart Machines Think*. Cambridge, MA: MIT Press, 2018.

Gilbert, Amy Margaret. "The Work of Lord Brougham for Education in England." PhD dissertation, University of Pennsylvania, 1922. Internet Archive.

Gildea, Robert. *Children of the Revolution: The French, 1799–1914*. Cambridge, MA: Harvard University Press, 2008.

Gillispie, Charles Coulston. "Probability and Politics: Laplace, Condorcet, and Turgot." *Proceedings of the American Philosophical Society* 116, no. 1 (1972): 1–20.

Gillispie, Charles Coulston. *Science and Polity in France: The Revolutionary and Napoleonic Years*. Princeton, NJ: Princeton University Press, 2004.

Gillispie, Charles Coulston. *Science and Polity in France at the End of the Old Regime*. Princeton, NJ: Princeton University Press, 1980.

Gillispie, Charles Coulston, Robert Fox, and Ivor Grattan-Guinness. *Pierre-Simon Laplace, 1749–1827: A Life in Exact Science*. Princeton, NJ: Princeton University Press, 1997.

Gilman, Nils. *Mandarins of the Future: Modernization Theory in Cold War America*. Baltimore, MD: Johns Hopkins University Press, 2003.

Gitelman, Lisa. *"Raw Data" Is an Oxymoron*. Cambridge, MA: MIT Press, 2013.

Gladwell, Malcolm. *The Tipping Point: How Little Things Can Make a Big Difference*. Boston: Little, Brown, 2000.

Gleick, James. *Isaac Newton*. New York: Vintage Books, 2003.

Glickstein, Jonathan A. *Concepts of Free Labor in Antebellum America*. New Haven, CT: Yale University Press, 1991.

Godfrey-Smith, Peter. "Spencer and Dewey on Life and Mind." In *The Philosophy of Artificial Life*, edited by Margaret Boden, 314–31. Oxford: Oxford University Press, 1996.

Gökyiğit, Emel Aileen. "The Reception of Francis Galton's *Hereditary Genius* in the Victorian Periodical Press." *Journal of the History of Biology* 27, no. 2 (1994): 215–40.

Goldman, Lawrence. "The Origins of British 'Social Science': Political Economy, Natural Science and Statistics, 1830–1835." *Historical Journal* 26, no. 3 (1983): 587–616.

Gonce, R.A. "F. H. Knight on Capitalism and Freedom." *Journal of Economic Ideas* 26, no. 3 (September 1992): 813–44.

Gorham, Geoffrey. *The Language of Nature*. Minneapolis: University of Minnesota Press, 2016.

Gottlieb, Anthony. *The Dream of Reason: A History of Western Philosophy from the Greeks to the Renaissance*. New York: Norton, 2000.

Gough, Hugh. *The Terror in the French Revolution*. New York: Palgrave Macmillan, 2010.

Gould, Stephen Jay. *The Mismeasure of Man*. New York: Norton, 1993.

Gould, Stephen Jay. "So Cleverly Kind an Animal." 1977. In *Ever Since Darwin: Essays in Natural History*, rep. ed., 260–67. New York: Norton, 2007.

Grattan-Guinness, Ivor. "The *Ecole Polytechnique*, 1794–1850: Differences over Educational Purpose and Teaching Practice." *American Mathematical Monthly* 112, no. 3 (March 2005): 233–50.

Gray, John. *Enlightenment's Wake: Politics and Culture at the Close of the Modern Age*. New York: Routledge, 1995.

Gray, Mary L., and Siddharth Suri. *Ghost Work: How to Stop Silicon Valley from Building a New Global Underclass*. Boston: Houghton Mifflin Harcourt, 2019.

Gray, Valerie. *Charles Knight: Educator, Publisher, Writer*. Burlington, VT: Ashgate, 2006.

Grendler, Paul F. "Universities Warring against Jesuit Schools." In *Jesuit Schools and Universities in Europe, 1548–1773: Brill's Research Perspectives in Jesuit Studies*, 100–106. Leiden: Brill, 2019. http://www.jstor.org/stable/10.1163/j.ctv2gjx0hq.33.

Grier, David Alan. *When Computers Were Human*. Princeton, NJ: Princeton University Press, 2005.

Griffith, Erin. "They Fled San Francisco. The A.I. Boom Pulled Them Back." *New York Times*, June 7, 2023. https://www.nytimes.com/2023/06/07/technology/ai-san-francisco-tech-industry.html.

Guairard, [M.]. Review of *De la manière d'étudier les mathématiques*, by P. H. Suzanne. *Mercure de France*, February 28, 1807: 404–10.

Guldi, Jo, and David Armitage. *The History Manifesto*. Cambridge: Cambridge University Press, 2014.

Gutin, Iliya. "In BMI We Trust: Reframing the Body Mass Index as a Measure of Health." *Social Theory & Health* 16, no. 3 (2018): 256–71.

Hacking, Ian. "The Looping Effects of Human Kinds." In *Causal Cognition: A Multidisciplinary Debate*, edited by Dan Sperber, David Premack, and Ann James Premack, 351–94. Oxford: Oxford University Press, 1995.

Hacking, Ian. "Making Up People." In *Reconstructing Individualism: Autonomy, Individuality, and the Self in Western Thought*, edited by Thomas C. Heller, Morton Sosna, and David E. Wellbery, 222–36. Stanford, CA: Stanford University Press, 1986.

Hacking, Ian. *The Taming of Chance*. Cambridge: Cambridge University Press, 1990.

Hahn, Roger. *The Anatomy of a Scientific Institution, the Paris Academy of Sciences, 1666–1803*. Berkeley: University of California Press, 1971.

Hahn, Roger. *Pierre Simon Laplace, 1749–1827: A Determined Scientist*. Cambridge, MA: Harvard University Press, 2005.

Haines, Valerie A. "Spencer, Darwin, and the Question of Reciprocal Influence." *Journal of the History of Biology* 24, no. 3 (1991): 409–31.

Hajian, Sara. "Algorithmic Bias: From Discovery to Mitigation." Paper presented at the Second European Symposium Series on Societal Challenges in Computational Social Sciences, Cologne, Germany, December 2018.

Halévy, Élie. *La formation du radicalisme philosophique*. 3 vols. Paris: Alcan, 1901.

Halpern, Sue. "The Bull's-Eye on Your Thoughts." *New York Review of Books* 70, no. 17 (November 2, 2023): 60–62.

Hammond, Nicholas. "Pascal's *Pensées* and the Art of Persuasion." In *The Cambridge Companion to Pascal*, edited by Nicholas Hammond, 235–52. Cambridge: Cambridge University Press, 2003.

Hankins, Frank Hamilton. *An Introduction to the Study of Society*. New York: Macmillan, 1927.

Hankins, Frank Hamilton. "Social Science and Social Action." *American Sociological Review* 4, no. 1 (1939): 1–16.

Hankins, Thomas L. *Jean D'Alembert: Science and the Enlightenment*. Oxford: Clarendon Press, 1970.

Hanley, Ryan Patrick. *Adam Smith: His Life, Thought, and Legacy*. Princeton, NJ: Princeton University Press, 2016.

Haraway, Donna Jeanne. *Simians, Cyborgs, and Women: The Reinvention of Nature*. New York: Routledge, 1991.

Harcourt, Bernard E. *Exposed: Desire and Disobedience in the Digital Age*. Cambridge, MA: Harvard University Press, 2015.

Harden, Kathryn Paige. "Why Progressives Should Embrace the Genetics of Education." *New York Times*, July 24, 2018. https://www.nytimes.com/2018/07/24/opinion/dna-nature-genetics-education.html.

Harrison, Carol E. *Romantic Catholics: France's Postrevolutionary Generation in Search of a Modern Faith*. Ithaca, NY: Cornell University Press, 2014.

Hayes, Julie Candler. *Reading the French Enlightenment: System and Subversion*. Cambridge: Cambridge University Press, 1999.

Hayles, N. Katherine. *How We Became Posthuman: Virtual Bodies in Cybernetics, Literature, and Informatics*. Chicago: University of Chicago Press, 1999.

Hayward, Jack. *Fragmented France: Two Centuries of Disputed Identity*. Oxford: Oxford University Press, 2007.

Heilbron, J. L. "Are Historians Fit to Rule?" *Isis* 107, no. 2 (2016): 350–52.

Heilbron, Johan. *The Rise of Social Theory*. Minneapolis: University of Minnesota Press, 1995.

Helmholtz, Hermann von. "On the Relation of Natural Science to General Science." In *Popular Lectures on Scientific Subjects*, translated by Edmund Atkinson, 1–32. New York: D. Appleton, 1885.

Helvétius, Claude-Adrien. *Œuvres complètes d'Helvétius*. 3 vols. Paris: V. Lepetit, 1818.

Henry, John. "Primary and Secondary Causation in Samuel Clarke's and Isaac Newton's Theories of Gravity." *Isis* 111, no. 3 (2020): 542–61.

Herschel, John. "Quetelet on Probabilities." *Edinburgh Review* 92 (1850): 1–57.

Hetherington, Norriss S. "Isaac Newton's Influence on Adam Smith's Natural Laws in Economics." *Journal of the History of Ideas* 44, no. 3 (1983): 497–505.

Heyck, Hunter. *Age of System: Understanding the Development of Modern Social Science*. Baltimore, MD: Johns Hopkins University Press, 2015.

Heyck, Hunter. *Herbert A. Simon: The Bounds of Reason in Modern America*. Baltimore, MD: Johns Hopkins University Press, 2005.

Heyck, Hunter. "Producing Reason." In Solovey and Cravens, *Cold War Social Science*, 99–116.

Heyl, Barbara S. "The Harvard 'Pareto Circle.'" *Journal of the History of the Behavioral Sciences* 4, no. 4 (1968): 316–34.

Himmelfarb, Gertrude. *Darwin and the Darwinian Revolution*. Garden City, NY: Doubleday, 1959.

Hine, Ellen McNiven. *A Critical Study of Condillac's Traité des systèmes*. The Hague: M. Nijhoff, 1979.

Hobsbawm, Eric. *The Age of Revolution, 1789–1848*. London: Weidenfeld and Nicolson, 1962.

Hobsbawm, Eric. "The Machine Breakers." *Past & Present* 1, no. 1 (1952): 57–70.

Hobsbawm, Eric, and George F. E. Rudé. *Captain Swing*. London: Lawrence and Wishart, 1969.

Hodges, Andrew. *Alan Turing: The Enigma*. New York: Simon and Schuster, 1983.

Hodgson, Geoffrey M. *How Economics Forgot History: The Problem of Historical Specificity in Social Science*. New York: Routledge, 2001.

Hodgson, Geoffrey M. "John R. Commons and the Foundations of Economics." *Journal of Economic Issues* 37, no. 3 (September 2003): 547–76.

Hodgson, Geoffrey M. "What Is the Essence of Institutional Economics?" *Journal of Economic Issues* 34, no. 2 (2000): 317–29.

Hofstadter, Richard. *Social Darwinism in American Thought: 1860–1915*. Philadelphia: University of Pennsylvania Press, 1944.

d'Holbach, Paul Henri Thiry. *Œuvres philosophiques*. Vol. 1. Paris: Alive, 1998.

Hollander, Samuel. "Retrospectives: Ricardo on Machinery." *Journal of Economic Perspectives* 33, no. 2 (2019): 229–42.

Holmes, Stephen. *The Anatomy of Antiliberalism*. Cambridge, MA: Harvard University Press, 1993.

Horkheimer, Max, and Theodor W. Adorno. *Dialectic of Enlightenment*. Edited by Gunzelin Schmid Noer. Translated by Edmund Jephcott. Stanford, CA: Stanford University Press, 2002.

Horn, Jeff. "Machine-Breaking in England and France during the Age of Revolution." *Labour/Le Travail* 55 (2005): 143–66.

Horn, Jeff. *The Path Not Taken: French Industrialization in the Age of Revolution, 1750–1830*. Cambridge, MA: MIT Press, 2006.

Horvitz, Eric. "Opening Statement." In *The Dawn of Artificial Intelligence*. Washington, DC: US Government Publishing Office, 2017. https://www.govinfo.gov/content/pkg/CHRG-114shrg24175/pdf/CHRG-114shrg24175.pdf.

Hovenkamp, Herbert. "The Progressives: Racism and Public Law." *Arizona Law Review* 59 (2017): 946–1003.

Howerth, W. "Basic Ideas of a Scientific Pedagogy." *Education* 23, no. 3 (November 1902): 129–41.

Huber, Peter J. "Speculations on the Path of Statistics." In *The Practice of Data Analysis: Essays in the Honor of John W. Tukey*, edited by David R. Brillinger, Luisa T. Fernholz, and Stephan Morgenthaler, 175–92. Princeton, NJ: Princeton University Press, 2014.

Huet, Marie-Hélène. *Mourning Glory: The Will of the French Revolution*. Philadelphia: University of Pennsylvania Press, 1997.

Hull, Clark L. "Mind, Mechanism, and Adaptive Behavior." *Psychological Review* 44, no. 1 (1937): 1–32.

Hull, Clark L. *Principles of Behavior: An Introduction to Behavior Theory*. New York: D. Appleton, 1943.

Hulliung, Mark. *The Autocritique of Enlightenment: Rousseau and the Philosophes*. Cambridge, MA: Harvard University Press, 1994.

Husain, Amir. *The Sentient Machine: The Coming Age of Artificial Intelligence*. New York: Scribner, 2017.

Ippolito, Christophe. "Avant-Propos." In *Résistances à la modernité dans la littérature Français de 1800 à nos jours*, edited by Christophe Ippolito, 13–22. Paris: L'Harmattan, 2010.

Irani, Lilly. "Justice for Data Janitors." *Public Books*, January 15, 2015. https://www.public-books.org/justice-for-data-janitors/.

Isaac, Joel. "Epistemic Design: Theory and Data in Harvard's Department of Social Relations." In Solovey and Cravens, *Cold War Social Science*, 79–95.

Isaac, Joel. "Tool Shock: Technique and Epistemology in the Postwar Social Sciences." *History of Political Economy* 42, no. 5 (2010): 133–64.

Isaac, Joel. *Working Knowledge: Making the Human Sciences from Parsons to Kuhn*. Cambridge, MA: Harvard University Press, 2012.

Iskenderova, Aliya, and Emese Domahidi. "Systematic Literature Review on Bias in Algorithmic Filtering." Poster presented at the Second European Symposium Series on Societal Challenges in Computational Social Sciences, Cologne, Germany, December 2018.

Israel, Jonathan. *The Enlightenment that Failed: Ideas, Revolution, and Democratic Defeat, 1748–1830*. Oxford: Oxford University Press, 2019.

Israel, Jonathan. *Radical Enlightenment: Philosophy and the Making of Modernity, 1650–1750*. Oxford: Oxford University Press, 2001.

Jacob, Margaret C. *The Radical Enlightenment: Pantheists, Freemasons, and Republicans*. London: Allen and Unwin, 1981.

Jacob, Margaret C. *Scientific Culture and the Making of the Industrial West*. Oxford: Oxford University Press, 1997.

Jacob, Margaret C., and Larry Stewart. *Practical Matter: Newton's Science in the Service of Industry and Empire, 1687–1851*. Cambridge, MA: Harvard University Press, 2004.

Jennings, H. S. *Behavior of the Lower Organisms*. New York: Columbia University Press, 1906.

Jevons, William Stanley. "Amusements of the People." In *Methods of Social Reform and Other Papers*, 1–27. London: Macmillan, 1883.

Jevons, William Stanley. "On the Mechanical Performance of Logical Inference." *Philosophical Transactions of the Royal Society of London* 160 (1870): 497–518.

Jevons, William Stanley. *The Principles of Economics: A Fragment of a Treatise on Industrial Mechanism and Other Papers*. 1905; repr., New York: Kelly, 1965.

Jevons, William Stanley. *The Principles of Science: A Treatise on Logic and Scientific Method*. 1874; repr., London: Macmillan, 1920.

Johannessen, Hélène Celdran. *Prophètes, sorciers, rumeurs: La violence dans trois romans de Jules Barbey D'Aurevilly (1808–1889)*. Amsterdam: Rodopi, 2008.

Jones, Matthew L. "How We Became Instrumentalists (Again): Data Positivism since World War II." *Historical Studies in the Natural Sciences* 48, no. 5 (2018): 673–84.

Jones, Matthew L. *Reckoning with Matter: Calculating Machines, Innovation, and Thinking about Thinking from Pascal to Babbage*. Chicago: University of Chicago Press, 2016.

Joy, Bill. "Why the Future Doesn't Need Us." *Wired*, April 1, 2000. https://www.wired.com/2000/04/joy-2/.

Jurt, Joseph. "Condorcet et les colonies." *Proceedings of the Meeting of the French Colonial Historical Society* 15 (1992): 9–21.

Kaczynski, Theodore. *Technological Slavery*. Edited by David Skribina. Port Townsend, WA: Feral House, 2010.

Kang, Minsoo. *Sublime Dreams of Living Machines: The Automaton in the European Imagination*. Cambridge, MA: Harvard University Press, 2011.

Kapor, Mitch, and Ray Kurzweil. "A Wager on the Turing Test: The Rules." *Kurzweil, Accelerating Intelligence*, April 9, 2002, https://www.kurzweilai.net/a-wager-on-the-turing-test-the-rules.

Käsler, Dirk. *Sociological Adventures: Earle Edward Eubank's Visits with European Sociologists*. New Brunswick, NJ: Transaction Books, 1991.

Katz, Yarden. *Artificial Whiteness: Politics and Ideology in Artificial Intelligence*. New York: Columbia University Press, 2020.

Kennedy, William F. "Lord Brougham, Charles Knight, and the Rights of Industry." *Economica* 29, no. 113 (1962): 58–71.

Kestner, Joseph A. *Protest and Reform: The British Social Narrative by Women, 1827–1867*. Madison: University of Wisconsin Press, 1985.

Kevles, Daniel J. *In the Name of Eugenics: Genetics and the Uses of Human Heredity*. Berkeley: University of California Press, 1985.

Kevles, Daniel J. "What's Manifest in the History of SciTech: Reflections on the History Manifesto." *Isis* 107, no. 2 (2016): 315–23.

Kiciman, Emre. "Where Does Data Bias Come From? With Notes on Fairness and the Uses and Limitations of Causal Reasoning." Keynote presented at the Second European Symposium Series on Societal Challenges in Computational Social Sciences, Cologne, Germany, December 2018.

Kim, Kwangsu. "Adam Smith's 'History of Astronomy' and View of Science." *Cambridge Journal of Economics* 36, no. 4 (2012): 799–820.

King, Helen Maxwell. *Les doctrines littéraires de la quotidienne, 1814–1830: Un chapitre de l'histoire du mouvement romantique en France*. Northampton, MA: Smith College, 1920.

King, J. E. *David Ricardo*. New York: Palgrave Macmillan, 2013.

Klinck, David. *The French Counterrevolutionary Theorist, Louis de Bonald (1754–1840)*. New York: Peter Lang, 1996.

[Knight, Charles]. *The Results of Machinery*. London: C. Knight, 1831.

Knight, Charles. *The Rights of Industry: Addressed to the Working-Men of the United Kingdom*. 1st American ed. Philadelphia: Carey and Hart, 1832.

Knight, Charles, and Harriet Martineau, eds. *Mind amongst the Spindles: A Miscellany, Wholly Composed by the Factory Girls*. Boston: Jordan, Swift, and Wiley, 1845.

Knight, Frank H. "Fact and Value in Social Science." In Knight, *Freedom and Reform*, 225–45.

Knight, Frank H. *Freedom and Reform: Essays in Economics and Social Philosophy*. Port Washington, NY: Kennikat, 1947.

Knight, Frank H. "Social Science and the Political Trend." In Knight, *Freedom and Reform*, 19–34.

Knight, Isabel F. *The Geometric Spirit: The Abbé de Condillac and the French Enlightenment*. New Haven, CT: Yale University Press, 1968.

Koch, Sigmund. "Clark C. Hull." In *Modern Learning Theory: A Critical Analysis of Five Examples*, 1–176. New York: Appleton, 1954.

Koopmans, Tjalling C. "Measurement without Theory." *Review of Economics and Statistics* 29, no. 3 (August 1947): 161–72.

Kors, Alan Charles. *D'Holbach's Coterie: An Enlightenment in Paris*. Princeton, NJ: Princeton University Press, 1976.

Kors, Alan Charles. *Epicureans and Atheists in France, 1650–1729*. Cambridge: Cambridge University Press, 2016.

Kosinski, Michal, David Stillwell, and Thore Graepel. "Private Traits and Attributes Are Predictable from Digital Records of Human Behavior." *Proceedings of the National Academy of Sciences* 110, no. 15 (2013): 5802–5.

Koyré, Alexandre. "Newton and Descartes." In *Newtonian Studies*, 55–114. Cambridge, MA: Harvard University Press, 1965.

Koyré, Alexandre. "The Significance of the Newtonian Hypothesis." In *Newtonian Studies*, 3–24. Cambridge, MA: Harvard University Press, 1965.

Kraus, Pamela A. "From Universal Mathematics to Universal Method: Descartes's 'Turn' in Rule IV of the Regulae." *Journal of the History of Philosophy* 21, no. 2 (1983): 159–74.

Krüger, Lorenz, Lorraine Daston, and Michael Heidelberger, eds. *The Probabilistic Revolution*. Cambridge, MA: MIT Press, 1987.

Kuhn, Thomas S. *The Structure of Scientific Revolutions*. 2nd ed. Chicago: University of Chicago Press, 1970.

Kuklick, Henrika. "A 'Scientific Revolution': Sociological Theory in the United States, 1930–1945." *Sociological Inquiry* 43, no. 1 (1973): 3–22.

Kurz, Heinz D. *Economic Thought: A Brief History*. Translated by Jeremiah Reimer. New York: Columbia University Press, 2016.

Kurzweil, Ray. *The Singularity Is Near: When Humans Transcend Biology*. New York: Viking, 2005.

Kusner, Matt J., Joshua R. Loftus, Chris Russell, and Ricardo Silva. "Counterfactual Fairness." Paper presented at the 31st Conference on Neural Information Processing Systems (NIPS 2017), Long Beach, CA, 2017. https://arxiv.org/pdf/1703.06856.pdf.

LaGrandeur, Kevin. *Androids and Intelligent Networks in Early Modern Literature and Culture*. New York: Routledge, 2013.

La Mettrie, Julien Offray de. *L'homme machine*. In *La Mettrie's "L'homme machine"*, edited by Adam Vartanian, 139–250. 1960; repr., Princeton, NJ: Princeton University Press, 2015.

Lanfrey, Pierre. *The History of Napoleon the First*. Vol. 2. New York: Macmillan, 1894.

Lange, Friedrich Albert. *The History of Materialism and Criticism of Its Present Importance*. Translated by Ernest Chester Thomas. 1888; repr., New York: Harcourt, Brace, 1925.

Lanier, Jaron. *You Are Not a Gadget: A Manifesto*. New York: Alfred A. Knopf, 2010.

Laplace, Pierre Simon. *Essai philosophique sur les probabilités*. 1825. 5th ed. Paris: C. Bourgois, 1986.

Larson, Erik J. *The Myth of Artificial Intelligence: Why Computers Can't Think the Way We Do*. Cambridge, MA: Belknap Press of Harvard University Press, 2021.

Latham, Michael E. *Modernization as Ideology: American Social Science and "Nation Building" in the Kennedy Era*. Chapel Hill: University of North Carolina Press, 2000.

Lauer, Rosemary. Review of *The Geometric Spirit: The Abbé de Condillac and the French Enlightenment*, by Isabel F. Knight. *Thomist* 33, no. 4 (1969): 780–83.

Lazer, David M. J., Alex Pentland, Duncan J. Watts, Sinan Aral, Susan Athey, Noshir Contractor, and Deen Freelon. "Computational Social Science: Obstacles and Opportunities." *Science* 369, no. 6507 (2020): 1060–62.

Lazer, David, Ryan Kennedy, Gary King, and Alessandro Vespignani. "The Parable of Google Flu: Traps in Big Data Analysis." *Science*, no. 6176 (2014): 1203–5.

Lazer, David, Alex Pentland, Lada Adamic, Sinan Aral, Albert-Laszlo Barabasi, Devon Brewer, and Nicholas Christakis. "Social Science: Computational Social Science." *Science* 323, no. 5915 (2009): 721–23.

Leonard, Thomas C. *Illiberal Reformers: Race, Eugenics, and American Economics in the Progressive Era.* Princeton, NJ: Princeton University Press, 2016.

Leary, David E. *Metaphors in the History of Psychology.* Cambridge: Cambridge University Press, 1990.

Leary, David E. "On the Conceptual and Linguistic Activity of Psychologists: The Study of Behavior from the 1890s to the 1990s and Beyond." *Behavior and Philosophy* 32, no. 1 (2004): 13–35.

Lebrun, Richard A. "Joseph de Maistre, Cassandra of Science." *French Historical Studies* 6, no. 2 (1969): 214–31.

Lebrun, Richard A., ed. *Joseph de Maistre's Life, Thought, and Influence: Selected Studies.* Montreal: McGill-Queen's University Press, 2001.

Legrand, Michel. "Jean-Paul Weber, la constitution du texte des Regulae." *Revue philosophique de Louvain,* 69 1971: 144–45.

Lemov, Rebecca M. *Database of Dreams: The Lost Quest to Catalog Humanity.* New Haven, CT: Yale University Press, 2015.

Lemov, Rebecca M. *World as Laboratory: Experiments with Mice, Mazes, Men.* New York: Hill and Wang, 2005.

Lengermann, Patricia M. "The Founding of the *American Sociological Review*: The Anatomy of a Rebellion." *American Sociological Review* 44, no. 4 (1979): 185–98.

Lengermann, Patricia M., and Gillian Niebrugge. "Thrice Told: Narratives of Sociology's Relation to Social Work." In Calhoun, *Sociology in America,* 63–114.

Lengermann, Patricia M., and Gillian Niebrugge. *The Women Founders: Sociology and Social Theory, 1830–1930; A Text/Reader.* Boston: McGraw-Hill, 1998.

Lenhard, Johannes. "Models and Statistical Inference: The Controversy between Fisher and Neyman-Pearson." *British Journal for the Philosophy of Science* 57, no. 1 (2006): 69–91.

Lennon, Thomas M. *The Battle of the Gods and Giants: The Legacies of Descartes and Gassendi, 1655–1715.* Princeton, NJ: Princeton University Press, 1993.

Leonard, Robert. *Von Neumann, Morgenstern, and the Creation of Game Theory: From Chess to Social Science, 1900–1960.* Cambridge: Cambridge University Press, 2010.

Lepenies, Wolf. *Between Literature and Science: The Rise of Sociology.* Cambridge: Cambridge University Press, 1988.

Lepore, Jill. *If, Then: How the Simulmatics Corporation Invented the Future.* New York: Liveright, 2020.

Leslie, Stuart W. *The Cold War and American Science: The Military-Industrial-Academic Complex at MIT and Stanford.* New York: Columbia University Press, 1993.

Lighthill, James. "Artificial Intelligence: A General Survey." In *Artificial Intelligence: A Paper Symposium,* 1–21. London: Science Research Council, 1973.

Lilti, Antoine. *The World of the Salons: Sociability and Worldliness in Eighteenth-Century Paris.* Oxford: Oxford University Press, 2015.

Limber, John. "Language in Child and Chimp?" In *Speaking of Apes: A Critical Anthology of*

Two-Way Communication with Man, edited by Tomas A. Sebeok and Jean Umiker-Sebeok, 197–220. New York: Plenum, 1980.

Lindee, M. Susan. *Rational Fog: Science and Technology in Modern War.* Cambridge, MA: Harvard University Press, 2020.

LoConto, David G. "Charles A. Ellwood and the End of Sociology." *American Sociologist* 42, no. 1 (2011): 112–28.

Lodge, Paul. "Leibniz's Close Encounter with Cartesianism in the Correspondence with De Volder." In *Leibniz and His Correspondents*, edited by Paul Lodge, 162–92. Cambridge: Cambridge University Press, 2004.

Lohr, Steve. "For Data Scientists, 'Janitor Work' Is Hurdle to Insights." *New York Times*, August 7, 2014. https://www.nytimes.com/2014/08/18/technology/for-big-data-scientists-hurdle-to-insights-is-janitor-work.html.

Lohr, Steve. "Researchers Seek Smarter Paths to A.I." *New York Times*, June 20, 2018. https://www.nytimes.com/2018/06/20/technology/deep-learning-artificial-intelligence.html.

Lottin, Joseph. *Quetelet, statisticien et sociologue.* Louvain: Institut supérieur de philosophie, 1912.

Loynes, R. M. *Model Building, a Current Statistical Preoccupation.* Sheffield, UK: University of Sheffield Press, 1969.

Lu, Yiewn. "Generative A.I. Can Add $4.4 Trillion in Value to Global Economy, Study Says." *New York Times*, June 14, 2023. https://www.nytimes.com/2023/06/14/technology/generative-ai-global-economy.html.

Lundgren, Frans. "The Politics of Participation: Francis Galton's Anthropometric Laboratory and the Making of Civic Selves." *British Journal for the History of Science* 46, no. 3 (2013): 445–66.

Lupton, Deborah. *The Quantified Self: A Sociology of Self-Tracking.* Cambridge: Polity, 2016.

Maas, Harro. *William Stanley Jevons and the Making of Modern Economics.* Cambridge: Cambridge University Press, 2005.

Machery, Pierre. "Le positivisme entre la revolution et la contre-revolution: Comte et Maistre." *Revue de Synthèse* 112, no. 1 (1991): 41–47.

MacKenzie, Donald. *An Engine, Not a Camera: How Financial Models Shape Markets.* Cambridge, MA: MIT Press, 2006.

MacLean, Vicky M., and Joyce E. Williams. "'Ghosts of Sociologies Past': Settlement Sociology in the Progressive Era at the Chicago School of Civics and Philanthropy." *American Sociologist* 43, no. 3 (2012): 235–63.

Mahon, Peter. *Posthumanism: A Guide for the Perplexed.* London: Bloomsbury Academic, 2017.

Mailly, Edouard. *Essai sur la vie et les ouvrages de L.-A.-J. Quételet.* Brussels: Hayez, 1875.

Maistre, Joseph Marie de. *Examen de la philosophie de Bacon ou l'on traite différentes de philosophe rationnelle.* 2 vols. Lyon: Pélagaud, 1845.

Maistre, Joseph Marie de. *Œuvres completes.* 14 vols. Geneva: Slatkine Reprints, 1979.

Mäki, Uskali, ed. *The Methodology of Positive Economics: Reflections on Milton Friedman's Legacy.* Cambridge: Cambridge University Press, 2009.

Mäki, Uskali. "Preface." In Mäki, *Methodology of Positive Economics*, xvii–xviii.

Mann, Adam. "Core Concept: Computational Social Science." *Proceedings of the National Academy of Sciences* 113, no. 3 (2016): 468–70.

Manuel, Frank E. *The Prophets of Paris.* Cambridge, MA: Harvard University Press, 1962.

Marantz, Andrew. "Trouble in Paradise." *New Yorker* 94, no. 24 (2019): 60–67.

Markoff, John. *Machines of Loving Grace: The Quest for Common Ground between Humans and Robots.* New York: HarperCollins, 2015.

Martineau, Harriet. *Autobiography*. 3 vols. London: Smith, Elder, 1877.

Martineau, Harriet. *The History of England during the Thirty Years' Peace: 1816–1846*. 2 vols. London: C. Knight, 1850.

Martineau, Harriet. *Illustrations of Popular Economy*. 9 vols. London: Fox, 1834.

Martineau, Harriet. *The Rioters, or, A Tale of Bad Times*. Wellington, UK: Houlston, 1827.

Marty, Olivier, and Ray J. Amirault. *Nicolas de Condorcet: The Revolution of French Higher Education*. Cham: Springer, 2020. https://link.springer.com/book/10.1007/978-3-030-43566-0.

Marx, Karl, Friedrich Engels, and C. J. Arthur. *The German Ideology*. 1862. London: Electric Book Co., 2001. Marx/Engels Archive. http://hiaw.org/defcon6/works/1862/letters/62_06_18.html.

Marx, Leo. *The Machine in the Garden: Technology and the Pastoral Ideal in America*. Oxford: Oxford University Press, 1964.

Mayor, Adrienne. *Gods and Robots: Myths, Machines, and Ancient Dreams of Technology*. Princeton, NJ: Princeton University Press, 2018.

Mayr, Ernst. *The Growth of Biological Thought: Diversity, Evolution, and Inheritance*. Cambridge, MA: Belknap Press of Harvard University Press, 1982.

McAleer, John, and John M. MacKenzie. *Exhibiting the Empire: Cultures of Display and the British Empire*. Manchester, UK: Manchester University Press, 2015.

McCarthy, John. "Artificial Intelligence: A Paper Symposium: Professor Sir James Lighthill." *Artificial Intelligence* 5, no. 3 (1974): 317–22.

McCarthy, John, Marvin Minsky, Nathaniel Rochester, and Claude E. Shannon. "A Proposal for the Dartmouth Summer Research Project on Artificial Intelligence." August 31, 1955. http://www-formal.stanford.edu/jmc/history/dartmouth/dartmouth.html.

McCloskey, Deirdre N. *Bourgeois Dignity: Why Economics Can't Explain the Modern World*. Chicago: University of Chicago Press, 2010.

McCloskey, Deirdre N. "Why Economics Is on the Wrong Track." In *The Economics of Economists: Institutional Settings, Individual Incentives, and Future Prospects*, edited by Alessandro Lanteri and Jack Vromen, 211–42. Cambridge: University of Cambridge Press, 2014.

McCorduck, Pamela. *Machines Who Think: A Personal Inquiry into the History and Prospects of Artificial Intelligence*. San Francisco: W. H. Freeman, 1979.

McCullagh, Peter. "John Wilder Tukey: 16 June 1915–26 July 2000." *Biographical Memoirs of Fellows of the Royal Society* 49 (2003): 537–55.

McCulloch, J. R. *A Treatise on the Principles and Practical Influence of Taxation and the Funding System*. 3rd ed. Edinburgh: A. and C. Black, 1863.

McCulloch, Warren S., and Walter Pitts. "A Logical Calculus of the Ideas Immanent in Nervous Activity, 1943." *Bulletin of Mathematical Biology* 52, no. 1–2 (1990): 99–115.

McDonough, Jeffrey K. "Freedom and Contingency." In *The Oxford Handbook of Leibniz*, edited by Maria Rosa Antognazza, 86–99. Oxford: Oxford University Press, 2018.

McMahon, Darrin M. *Enemies of the Enlightenment: The French Counter-Enlightenment and the Making of Modernity*. Oxford: Oxford University Press, 2001.

McMullin, Ernan. "The Impact of Newton's *Principia* on the Philosophy of Science." *Philosophy of Science* 68, no. 3 (2001): 279–310.

McNally, David. *Against the Market: Political Economy, Market Socialism and the Marxist Critique*. London: Verso, 1993.

McVickar, John. *Outlines of Political Economy*. New York: Wilder and Campbell, 1825.

Meeker, Natania. *Voluptuous Philosophy: Literary Materialism in the French Enlightenment*. New York: Fordham University Press, 2006.

Metz, Cade. "'The Godfather of A.I.' Leaves Google and Warns of Danger Ahead." *New York Times*, May 1, 2023. https://www.nytimes.com/2023/05/01/technology/ai-google-chat-bot-engineer-quits-hinton.html.

Metz, Cade. "A Second Google A.I. Researcher Says the Company Fired Her." *New York Times*, February 19, 2021. https://www.nytimes.com/2021/02/19/technology/google-eth-ical-artificial-intelligence-team.html.

Michotte, Paul. *Études sur les théories économiques qui dominèrent en Belgique de 1830 à 1886*. Louvain: C. Peters, 1904.

Mill, John Stuart. *On Liberty, and Other Essays*. Edited by John Gray. Oxford: Oxford University Press, 1991.

Mill, John Stuart. *A System of Logic*. 8th ed. New York: Harper, 1890. Internet Archive.

Miller, Tiffany Jones. "Richard T. Ely, the German Historical School of Economics, and the 'Socio-Teleological' Aspiration of the New Deal Planners." *Social Philosophy and Policy* 38, no. 1 (2021): 52–84.

Mills, John A. *Control: A History of Behavioral Psychology*. New York: New York University Press, 1998.

Milo, Daniel S. *Good Enough: The Tolerance for Mediocrity in Nature and Society*. Cambridge, MA: Harvard University Press, 2019.

Minsky, Marvin Lee. *Computation: Finite and Infinite Machines*. Englewood Cliffs, NJ: Prentice-Hall, 1967.

Minsky, Marvin. "Some Methods of Artificial Intelligence and Heuristic Programming." In *Mechanization of Thought Processes: Proceedings of Symposium held at the National Physical Laboratory*, vol. 1, 3–26. London: Her Majesty's Stationery Office, 1959.

Minsky, Marvin Lee, and Seymour Papert. *Perceptrons: An Introduction to Computational Geometry*. Cambridge, MA: MIT Press, 1969.

Mirowski, Phillip. *More Heat than Light: Economics as Social Physics, Physics as Nature's Economics*. Cambridge: Cambridge University Press, 1989.

Mishra, Pankaj. *Age of Anger: A History of the Present*. New York: Farrar, Straus and Giroux, 2017.

Mitchell, Melanie. *Artificial Intelligence: A Guide for Thinking Humans*. New York: Farrar, Straus and Giroux, 2019.

Mitchell, Melanie. "On Crashing the Barrier of Meaning in Artificial Intelligence." *AI Magazine* 41, no. 2 (2020): 86–92.

Mitchell, Wesley. "Quantitative Analysis in Economic Theory." *American Economic Review* 15, no. 1 (1925): 1–12.

Mokyr, Joel. *The Enlightened Economy: An Economic History of Britain, 1700–1850*. New Haven, CT: Yale University Press, 2009.

Moldoveanu, Mihnea C. *Inside Man: The Discipline of Modeling Human Ways of Being*. Stanford, CA: Stanford Business Books, 2011.

Montes, Leonidas. "Das Adam Smith Problem: Its Origins, the Stages of the Current Debate, and One Implication for Our Understanding of Sympathy." *Journal of the History of Economic Thought* 25, no. 1 (2003): 63–90.

Montes, Leonidas. "Newton's Real Influence on Adam Smith and Its Context." *Cambridge Journal of Economics* 32, no. 4 (2008): 555–76.

Morgan, Mary S. "Economics." In Porter and Ross, *Cambridge History of Science*, vol. 7, *The Modern Social Sciences*, 275–305.

Moscati, Ivan. *Measuring Utility: From the Marginal Revolution to Behavioral Economics.* Oxford: Oxford University Press, 2019.

Mosselmans, Burt. "Adolphe Quetelet, the Average Man, and the Development of Economic Methodology." *European Journal of the History of Economic Thought* 12, no. 4 (2005): 565–82.

Mosselmans, Burt. *William Stanley Jevons and the Cutting Edge of Economics.* New York: Routledge, 2007.

Mossner, Ernest Campbell, and Ian Simpson Ross, eds. *The Glasgow Edition of the Works and Correspondence of Adam Smith.* Oxford: Oxford University Press, 2014.

Mueller, Tim B. "The Rockefeller Foundation, the Social Sciences, and the Humanities in the Cold War." *Journal of Cold War Studies* 15, no. 3 (2013): 108–35.

Muller, Jerry Z. "Conservatism: The Utility of History and the Case against Rationalist Radicalism." In Breckman and Gordon, *Cambridge History of Modern European Thought*, 232–54.

Müller-Wille, Staffan. "Making and Unmaking Populations." *Historical Studies in the Natural Sciences* 48, no. 5 (2018): 604–15.

Murphy, Heather. "Coming Soon to a Police Station Near You: The DNA 'Magic Box.'" *New York Times*, January 21, 2019. https://www.nytimes.com/2019/01/21/science/dna-crime-gene-technology.html.

Mykhailova, Iryna. "D'Holbach's Legacy in the Russian Empire and the Soviet Union." In *The Great Protector of Wits: Baron d'Holbach and His Time*, edited by Laura Nicolì, 300–330.

Neff, Gina, and Dawn Nafus. *Self-Tracking.* Cambridge, MA: MIT Press, 2016.

Newcomb, Alyssa. "Meet the Computer That Passed the Turing Test." *ABC News,* June 9, 2014. https://abcnews.go.com/Technology/meet-computer-passed-turing-artificial-intelligence-test-pretending/story?id=24054640.

Nichols, Lawrence T. "The Interstitial Ascent of Talcott Parsons: Cross-Disciplinary Collaboration and Careerism at Harvard, 1927–1951." *American Sociologist* 50, no. 4 (2019): 563–88.

Nietzsche, Friedrich. *On the Genealogy of Morals.* Translated by Walter Kaufmann and R. J. Hollingdale. 1887; repr., New York: Vintage, 1969.

Nilsson, Nils J. *The Quest for Artificial Intelligence: A History of Ideas and Achievements.* Cambridge: Cambridge University Press, 2010.

Nisbet, Robert A. "De Bonald and the Concept of the Social Group." *Journal of the History of Ideas* 5, no. 3 (1944): 315–31.

Noble, Safiya Umoja. *Algorithms of Oppression: How Search Engines Reinforce Racism.* New York: New York University Press, 2018.

Noë, Alva. *The Entanglement: How Art and Philosophy Make Us What We Are.* Princeton, NJ: Princeton University Press, 2023.

Norman, Jesse. *Adam Smith: Father of Economics.* New York: Basic, 2018.

Nourbakhsh, Illah Reza. *Robot Futures.* Cambridge, MA: MIT Press, 2013.

Nourbakhsh, Illah Reza, and Jennifer Keating. *AI & Humanity.* Cambridge, MA: MIT Press, 2020.

Oberschall, Anthony. *The Establishment of Empirical Sociology: Studies in Continuity, Discontinuity, and Institutionalization.* New York: Harper and Row, 1972.

O'Brien, D. P. *J. R. McCulloch: A Study in Classical Economics.* New York: Barnes and Noble, 1970.

O'Connell, Mark. *To Be a Machine: Adventures among Cyborgs, Utopians, Hackers, and the Futurists Solving the Modest Problem of Death.* New York: Doubleday, 2017.

Odum, Howard Washington. *American Masters of Social Science: An Approach to the Study of the Social Sciences through a Neglected Field of Biography.* New York: Henry Holt, 1927.

Offer, John. *Herbert Spencer and Social Theory.* New York: Palgrave Macmillan, 2010.

Ogburn, William F. "Presidential Address: The Folkways of a Scientific Sociology." *Scientific Monthly* 30, no. 4 (April 1930): 300–306.

O'Gieblyn, Meghan. *God, Human, Animal, Machine: Technology, Metaphor, and the Search for Meaning.* New York: Doubleday, 2021.

Oliveira, Arlindo L. *The Digital Mind: How Science Is Redefining Humanity.* Cambridge, MA: MIT Press, 2017.

"An Open Letter to the United Nations Convention on Certain Conventional Weapons." Future of Life Institute, accessed July 9, 2018. https://futureoflife.org/autonomous-weapons-open-letter-2017/.

O'Neal, John C. *The Authority of Experience: Sensationist Theory in the French Enlightenment.* University Park: Pennsylvania State University Press, 1996.

O'Neil, Cathy. "The Ivory Tower Can't Keep Ignoring Tech." *New York Times,* November 14, 2017. https://www.nytimes.com/2017/11/14/opinion/academia-tech-algorithms.html.

O'Neil, Cathy. *Weapons of Math Destruction.* New York: Crown, 2016.

Oražem, Claudia. *Political Economy and Fiction in the Early Works of Harriet Martineau.* Frankfurt am Main: Peter Lang, 1999.

Orwicz, Michael R. *Art Criticism and Its Institutions in Nineteenth-Century France.* Manchester, UK: Manchester University Press, 1994.

Osler, Margaret J. *Divine Will and the Mechanical Philosophy: Gassendi and Descartes on Contingency and Necessity in the Created World.* Cambridge: Cambridge University Press, 1994.

Ozouf, Mona. *Festivals and the French Revolution.* Cambridge, MA: Harvard University Press, 1988.

Pannabecker, John R. "Technocracy and the École Polytechnique: Bruno Belhoste, *La formation d'une technocratie.*" *Technology and Culture* 46, no. 3 (2005): 618–22.

Parsons, Talcott. "Science Legislation and the Social Sciences." *Political Science Quarterly* 62, no. 2 (June 1947): 241–49.

Parsons, Talcott. *The Structure of Social Action: A Study in Social Theory with Special Reference to a Group of Recent European Writers.* 2nd ed. Glencoe, IL: Free Press, 1949.

Pasquale, Frank. "The Algorithmic Self." *Hedgehog Review* 17, no. 1 (2015): 30.

Paul, Charles B. *Science and Immortality: The Éloges of the Paris Academy of Sciences (1699–1791).* Berkeley: University of California Press, 1980.

Paul, Diane B. "The Selection of the 'Survival of the Fittest.'" *Journal of the History of Biology* 21, no. 3 (1988): 411–24.

Peacock, Thomas Love. *Crotchet Castle.* Edited by Raymond Wright. New York: Penguin Classics, 1986.

Pearson, Karl. "On the Laws of Inheritance in Man." *Biometrika* 3, no. 2/3 (1904): 131–90.

Pearson, Roger. *Unacknowledged Legislators: The Poet as Lawgiver in Post-Revolutionary France.* Oxford: Oxford University Press, 2016.

Peart, Sandra. *The Economics of W. S. Jevons.* London: Routledge, 1996.

Peart, Sandra. "On Making and Remaking Ourselves and Others: Mill to Jevons and Beyond on Rationality, Learning, and Paternalism." *Review of Behavioral Economics* 8 (2021): 221–37.

Peart, Sandra. "W. S. Jevons's Methodology of Economics: Some Implications of the Procedures for 'Inductive Quantification.'" *History of Political Economy* 25, no. 3 (Fall 1993): 435–60.

Peart, Sandra, and David M. Levy. *The "Vanity of the Philosopher": From Equality to Hierarchy in Postclassical Economics*. Ann Arbor: University of Michigan Press, 2005.

Pelley, Scott, Aliza Chasan, Aaron Weisz, and Ian Flickinger. "60 Minutes Overtime: The Risks and Promise of Artificial Intelligence, According to the 'Godfather of AI' Geoffrey Hinton." *CBS News*, October 8, 2023. https://www.cbsnews.com/news/artificial-intelligence-risks-dangers-geoffrey-hinton-60-minutes/.

Pentland, Alex. *Social Physics: How Good Ideas Spread—The Lessons from a New Science*. New York: Penguin Press, 2014.

Peterson, Linda H. "From French Revolution to English Reform: Hannah More, Harriet Martineau, and the 'Little Book.'" *Nineteenth-Century Literature* 60, no. 4 (2006): 409–50.

Phillips, Henry. "Pascal's Reading and the Inheritance of Montaigne and Descartes." In *The Cambridge Companion to Pascal*, edited by Nicholas Hammond, 20–39. Cambridge: Cambridge University Press, 2003.

Pickering, Mary. *Auguste Comte: An Intellectual Biography*. Cambridge: Cambridge University Press, 1993.

Pius IX (Pope). "The Syllabus of Errors." Papal Encyclicals. 1864. https://www.papalencyclicals.net/pius09/p9syll.htm.

Poniewozik, James. "TV's War with the Robots Is Already Here." *New York Times*, May 10, 2023. https://www.nytimes.com/2023/05/10/arts/television/writers-strike-artificial-intelligence.html.

Poovey, Mary. *A History of the Modern Fact: Problems of Knowledge in the Sciences of Wealth and Society*. Chicago: University of Chicago Press, 1998.

Papandreou, A. G. "Economics and the Social Sciences." *Economic Journal* 60, no. 240 (December 1950): 715–23.

Popkin, Jeremy D. *A New World Begins: The History of the French Revolution*. New York: Basic Books, 2019.

Popkin, Richard H. "Leonora Cohen Rosenfield." *Eighteenth-Century Studies* 18, no. 4 (1985): 465–71.

Porter, Roy, ed. *The Cambridge History of Science*. Vol. 4, *Eighteenth Century Science*. Cambridge: Cambridge University Press, 2003.

Porter, Theodore M. "Positioning Social Science in Cold War America." Foreword to Solovey and Cravens, *Cold War Social Science*, ix–xvi.

Porter, Theodore M. "Objectivity and Authority: How French Engineers Reduced Public Utility to Numbers." *Poetics Today* 12, no. 2 (1991): 245–65.

Porter, Theodore M. *The Rise of Statistical Thinking, 1820–1900*. Princeton, NJ: Princeton University Press, 1986.

Porter, Theodore M. *Trust in Numbers: The Pursuit of Objectivity in Science and Public Life*. Princeton, NJ: Princeton University Press, 1995.

Porter, Theodore M., and Dorothy Ross. *The Cambridge History of Science*. Vol. 7, *The Modern Social Sciences*. Cambridge: Cambridge University Press, 2003.

Porter, Theodore M., and Dorothy Ross. "Writing the History of Social Science." Introduction to Porter and Ross, *Cambridge History of Science*, vol. 7, *The Modern Social Sciences*, 1–10.

Pranchère, Jean-Yves. "The Persistence of Maistrian Thought." In Lebrun, *Joseph de Maistre's Life*, 290–326.

Proctor, Robert. *Racial Hygiene: Medicine under the Nazis*. Cambridge, MA: Harvard University Press, 1988.

"Prospectus." *Mercure de France,* January 1807, p. 8.

Proudfoot, Dianna. "The Turing Test—From Every Angle." In Copeland et al., *Turing Guide,* 287–300.

"Quelques réflexions sur les sciences et les lettres." *Mercure de France,* February 28, 1807, 391–98.

Quetelet, Adolphe. "Dominique-François-Jean Arago." *Annuaire de l'Académie Royale de Belgique,* 1855: 157–97.

Quetelet, Adolphe. *Histoire des sciences mathématique et physique chez les Belges.* Brussels: Hayez, 1871.

Quetelet, Adolphe. "Mémoire sur les lois des naissances et de la mortalité à Bruxelles." *Nouveaux memoires de l'académie* 3 (1826): 493–512.

Quetelet, Adolphe. *Notice sur Alexis Bouvard.* Brussels: Hayez, 1844.

Quetelet, Adolphe. "Rapport décennal des travaux de l'Académie Royale." *Bulletin de l'Académie Royale des Sciences, des Lettres et des Beaux-Arts de Belgique* 7 (1840): 271–342.

Quetelet, Adolphe. *Sciences mathématiques et physiques chez les Belges, au commencement du XIXe siècle.* Brussels: Buggenhout, 1866.

Quetelet, Adolphe. *Sur l'homme et les développement de ses facultés.* Brussels: Bachelier, 1835.

Quetelet, Adolphe. "Sur la statistique morale et les principes qui doivent en former la base." *Nouveaux memoires de l'Académie Royale des Sciences et Belles-Lettres de Bruxelles* 21 (1848): 3–68.

Rabouin, David. "Mathesis universalis et algebre generale dans les *Regulae ad Directionem Ingenii* de Descartes." *Revue d'histoire des sciences* 69, no. 2 (2017): 259–309.

Raby, Peter. *Alfred Russel Wallace: A Life.* Princeton, NJ: Princeton University Press, 2001.

Radin, Joanna. "'Digital Natives': How Medical and Indigenous Histories Matter for Big Data." In *Data Histories,* edited by Elena Aronova, Christine von Oertzen, and David Sepkoski, vol. 32 of *Osiris* (2017): 43–64.

Redman, Deborah A. *The Rise of Political Economy as a Science: Methodology and the Classical Economists.* Cambridge, MA: MIT Press, 1997.

Reese, Byron. *The Fourth Age: Smart Robots, Conscious Computers, and the Future of Humanity.* New York: Atria Books, 2018.

Renwick, Chris. "From Political Economy to Sociology: Francis Galton and the Social-Scientific Origins of Eugenics." *British Journal for the History of Science* 44, no. 3 (2011): 343–69.

Richards, Robert John. *Darwin and the Emergence of Evolutionary Theories of Mind and Behavior.* Chicago: University of Chicago Press, 1987.

Richardson, John T. E. *Howard Andrew Knox: Pioneer of Intelligence Testing at Ellis Island.* New York: Columbia University Press, 2011.

Rid, Thomas. *Rise of the Machines: A Cybernetic History.* New York: Norton, 2016.

Rioux-Beauline, Mitia. "What Is Cartesianism? Fontenelle and the Subsequent Construction of Cartesian Philosophy." In *The Oxford Handbook of Descartes and Cartesianism,* edited by Steven Nadler, Tad M. Schmaltz, and Delphine Antoine-Mahut, 481–95. Oxford: Oxford University Press, 2019.

Riskin, Jessica. "The Defecating Duck, or, the Ambiguous Origins of Artificial Life." *Critical Inquiry* 29, no. 4 (Summer 2003): 599–633.

Riskin, Jessica. "Eighteenth Century Wetware." In *The Artificial and the Natural: An Evolving Polarity,* edited by Bernadette Bensaude-Vincent and William R. Newman, 239–75. Cambridge, MA: MIT Press, 2007.

Riskin, Jessica. *The Restless Clock: A History of the Centuries-Long Argument over What Makes Living Things Tick.* Chicago: University of Chicago Press, 2016.

Riskin, Jessica. *Science in the Age of Sensibility: The Sentimental Empiricists of the French Enlightenment*. Chicago: University of Chicago Press, 2002.

Robbins, Lionel. "The Teaching of Economics in Schools and Universities." *Economic Journal* 65, no. 260 (December 1955): 579–93.

Robertson, Ronald, Shan Jiang, Kenneth Joseph, Lisa Friedland, David Lazer, and Christo Wilson. "Auditing Partisan Audience Bias within Google Search." *Proceedings of the ACM on Human-Computer Interaction* 2 (2018): 1–22.

Robespierre, Maximilien. *Œuvres de Robespierre*. Paris: F. Cournol, 1866.

Robin, Ron Theodore. *The Making of the Cold War Enemy: Culture and Politics in the Military-Industrial Complex*. Princeton, NJ: Princeton University Press, 2001.

Rohde, Joy. *Armed with Expertise: The Militarization of American Social Research during the Cold War*. Ithaca, NY: Cornell University Press, 2013.

Roland, Alex, and Philip Shiman. *Strategic Computing: DARPA and the Quest for Machine Intelligence, 1983–1993*. Cambridge, MA: MIT Press, 2002.

Roncaglia, Alessandro. *The Age of Fragmentation: A History of Contemporary Economic Thought*. Cambridge: Cambridge University Press, 2019.

Roose, Kevin. "Maybe We Will Finally Learn More about How A.I. Works." *New York Times*, October 18, 2023. https://www.newyorktimes.com/2023/10/18/technology/how-ai -works-stanford.html.

Rose, Frank. *Into the Heart of the Mind: An American Quest for Artificial Intelligence*. New York: Vintage Books, 1984.

Rosenblatt, Frank. "The Perceptron: A Probabilistic Model for Information Storage and Organization in the Brain." *Psychological Review* 65, no. 6 (1958): 386–408.

Ross, Dorothy. "Changing Contours of the Social Science Disciplines." In Porter and Ross, *Cambridge History of Science*, vol. 7, *The Modern Social Sciences*, 205–37.

Ross, Dorothy. *The Origins of American Social Science*. Cambridge: Cambridge University Press, 1991.

Rothschild, Emma. *Economic Sentiments: Adam Smith, Condorcet, and the Enlightenment*. Cambridge, MA: Harvard University Press, 2001.

Rouse, William B. *Computing Possible Futures*. Oxford Scholarship Online. Oxford: Oxford University Press, 2019.

Rousseau, Jean-Jacques, and Michel Launay. *Œuvres complètes de Rousseau*. 3 vols. Paris: Éditions du Seuil, 1971.

Rousset, Alphonse. *Dictionnaire géographique, historique et statistique des communes de la Franche-Comté*. Vol. 4. Paris: A. Robert, 1856.

Roux, Sophie. "An Empire Divided: French Natural Philosophy." In Garber and Roux, *Mechanization of Natural Philosophy*, 55–95.

Ruestow, Edward Grant. *Physics at Seventeenth and Eighteenth-Century Leiden: Philosophy and the New Science in the University*. International Archives of the History of Ideas. Series Minor, 11. The Hague: Nijhoff, 1973.

Ruse, Michael, ed. *The Cambridge Encyclopedia of Darwin*. Cambridge: Cambridge University Press, 2013.

Ruse, Michael. "Evolution before Darwin." In Ruse, *Cambridge Encyclopedia of Darwin*, 39–45.

Russell, Stuart J., and Peter Norvig. *Artificial Intelligence: A Modern Approach*. Upper Saddle River, NJ: Prentice-Hall, 2010.

Rutherford, Malcolm. "Institutional Economics: Then and Now." *Journal of Economic Perspectives* 15, no. 3 (Summer 2001): 173–94.

Rutherford, Malcolm. *The Institutionalist Movement in American Economics, 1918–1947.* Cambridge: Cambridge University Press, 2011.

Rutherford, Malcolm. "Wesley Mitchell: Institutions and Quantitative Methods." *Eastern Economic Journal* 13, no. 1 (1987): 63–73.

Sailor, Danton B. "Cudworth and Descartes." *Journal of the History of Ideas* 23, no. 1 (1962): 133–40.

Sala, Roberto. "The Rise of Sociology: Paths of Institutionalization in Germany and the United States around 1900." *Geschichte und Gesellschaft* 43, no. 4 (2017): 557–84.

Sanders, Valerie, and Gaby Weiner. *Harriet Martineau and the Birth of Disciplines: Nineteenth-Century Intellectual Powerhouse.* New York: Routledge, 2017.

Sandvig, Christian, Kevin Hamilton, Karrie Karahalios, and Cedric Langbort. "Auditing Algorithms: Research Methods for Detecting Discrimination on Internet Platforms." Paper presented at a preconference at the 64th Annual Meeting of the International Communication Association, Seattle, WA, May 22, 2014.

Sapolsky, Robert. *Behave: The Biology of Humans at Our Best and Worst.* New York: Penguin, 2018.

Sapolsky, Robert. *Determined: A Science of Life without Free Will.* New York: Penguin, 2023.

Sarasohn, Lisa T. *Gassendi's Ethics: Freedom in a Mechanistic Universe.* Ithaca, NY: Cornell University Press, 1996.

Sarton, George. "Preface to Volume XXIII of *Isis* (*Quetelet*)." *Isis* 23, no. 1 (1935): 6–24.

Sasaki, Chikara. *Descartes's Mathematical Thought.* Boston: Kluwer, 2003.

Schabas, Margaret. "Adam Smith's Debts to Nature." *History of Political Economy* 35 (2003): 262–81.

Schabas, Margaret. *The Natural Origins of Economics.* Chicago: University of Chicago Press, 2005.

Schabas, Margaret. *A World Ruled by Number: William Stanley Jevons and the Rise of Mathematical Economics.* Princeton, NJ: Princeton University Press, 1990.

Schaffer, Simon. "Astronomers Mark Time: Discipline and the Personal Equation." *Science in Context* 2, no. 1 (1988): 115–45.

Schaffer, Simon. "Babbage's Intelligence: Calculating Engines and the Factory System." *Critical Inquiry* 21, no. 1 (1994): 203–27.

Schaffer, Simon. "Scientific Discoveries and the End of Natural Philosophy." *Social Studies of Science* 16, no. 3 (1986): 387–420.

Schloegel, Judy Johns, and Henning Schmidgen. "General Physiology, Experimental Psychology, and Evolutionism." *Isis* 93, vol. 4 (2002): 614–45.

Schmaus, Warren, Mary Pickering, and Michel Bourdeau. "Introduction: The Significance of Auguste Comte." In *Love, Order, and Progress: The Science and Philosophy of Auguste Comte,* edited by Michel Bourdeau, Mary Pickering, and Warren Schmaus, 3–24. Pittsburgh: University of Pittsburgh Press, 2018.

Schøsler, Jørn. "Rousseau et Diderot, critiques de la philosophie égalitaire d'Helvétius." *Revue Romane* 15, no. 1 (1980): 68–93.

Schrecker, Cherry. *Transatlantic Voyages and Sociology: The Migration and Development of Ideas.* Burlington, VT: Ashgate, 2010.

Schrum, Ethan D. *The Instrumental University: Education in Service of the National Agenda after World War II.* Ithaca, NY: Cornell University Press, 2019.

Schubring, Gert. "The Impact of Napoleonic Structural Reforms of the Educational System in Europe." *Annali dell'instito storico italo-germanico in Trento* 21 (1995): 435–43.

Schuster, John A., and Richard R. Yeo, eds. *The Politics and Rhetoric of Scientific Method: Historical Studies*. Dordrecht: D. Reidel, 1986.

Schweber, Libby. *Disciplining Statistics: Demography and Vital Statistics in France and England, 1830–1885*. Durham, NC: Duke University Press, 2006.

Schweitzer, Frank. "Sociophysics." *Physics Today* 71, no. 2 (2018): 40–46.

Scott, A. O. "When the Movies Pictured A.I., They Imagined the Wrong Disaster." *New York Times*, February 22, 2023. https://www.nytimes.com/2023/02/22/movies/ai-movies-microsoft-bing-robots.html.

Scott, James C. *Seeing Like a State: How Certain Schemes to Improve the Human Condition Have Failed*. New Haven, CT: Yale University Press, 2008.

Scott, Joan Wallach, and Debra Keates. *Schools of Thought: Twenty-Five Years of Interpretive Social Science*. Princeton, NJ: Princeton University Press, 2001.

Segala, Marco. "Jean-Baptiste Biot's Collaboration with the *Mercure de France*: Popularization and the Philosophical Analysis of Science." *Revue d'histoire des sciences* 66, no. 1 (2013): 107–36.

Seguin, Maria Susana. "Histoire de l'Académie Royale des Sciences." In Fontenelle, *Digression sur les anciens et les modernes et autres textes philosophiques*, 319–25.

Seidman, Steven. *Liberalism and the Origins of European Social Theory*. London: Blackwell, 1983.

Shapin, Steven. *Never Pure: Historical Studies of Science as if It Was Produced by People with Bodies, Situated in Time, Space, Culture, and Society, and Struggling for Credibility and Authority*. Baltimore, MD: Johns Hopkins University Press, 2010.

Shapin, Steven. *The Scientific Life: A Moral History of a Late Modern Vocation*. Chicago: University of Chicago Press, 2008.

Shapin, Steven. *The Scientific Revolution*. Chicago: University of Chicago Press, 2018.

Shermer, Michael. *In Darwin's Shadow: The Life and Science of Alfred Russel Wallace*. Oxford: Oxford University Press, 2002.

Shieber, Stuart M. *The Turing Test: Verbal Behavior as the Hallmark of Intelligence*. Cambridge, MA: MIT Press, 2004.

Shu, Xiaoling. *Knowledge Discovery in the Social Sciences: A Data Mining Approach*. Berkeley: University of California Press, 2020.

Sica, Alan. "Defining Disciplinary Identity: The Historiography of U.S. Sociology." In Calhoun, *Sociology in America*, 713–32.

Singer, Natasha. "Amazon Is Pushing Facial Technology That a Study Says Could Be Biased." *New York Times*, January 24, 2019. https://www.nytimes.com/2019/01/24/technology/amazon-facial-technology-study.html.

Sivak, Elizabeth, and Ivan Smirnov. "Gender Bias in Sharenting." Paper presented at the Second European Symposium Series on Societal Challenges in Computational Social Sciences, Cologne, Germany, December 2018.

Siegfried, Susan L. "The Politicization of Art Criticism in the Post-revolutionary Press." In Orwicz, *Art Criticisms and Its Institutions in Nineteenth-Century France*, 9–28.

Simon, Herbert A. *The Sciences of the Artificial*. 2nd ed. Cambridge, MA: MIT Press, 1981.

Skinner, B. F. *Beyond Freedom and Dignity*. New York: Bantam/Vintage Books, 1972.

Smith, Adam. *Essays, Philosophical and Literary*. London, 1880. Internet Archive.

Smith, Adam. *An Inquiry into the Nature and Causes of the Wealth of Nations*. Oxford: Oxford University Press, 1993.

Smith, C. U. M. "Evolution and the Problem of Mind: Part I. Herbert Spencer." *Journal of the History of Biology* 15, no. 1 (1982): 55–88.

Smith, Christopher. *The Hidden History of Bletchley Park: A Social and Organisational History, 1939–1945*. New York: Palgrave Macmillan, 2015.

Smith, Craig. "The Essays on Philosophical Subjects." In *Adam Smith: His Life, Thought, and Legacy*, edited by Ryan Patrick Hanley, 89–104. Princeton, NJ: Princeton University Pres, 2016.

Smith, David Warner. *Helvétius: A Study in Persecution*. Oxford: Clarendon Press, 1965.

Smith, Justin E. H. *Irrationality: A History of the Dark Side of Reason*. Princeton, NJ: Princton University Press, 2019.

Smith, Laurence D. "Metaphors of Knowledge and Behavior in the Behaviorist Tradition." In *Metaphors in the History of Psychology*, edited by David E. Leary, 239–66. Cambridge: Cambridge University Press, 1990.

Smith, Laurence D. "Models, Mechanism, and Explanation in Behavior Theory: The Case of Hull versus Spence." *Behavior and Philosophy* 18, no. 1 (Spring/Summer 1990): 1–18.

Smith, Robert W. "A National Observatory Transformed: Greenwich in the Nineteenth Century." *Journal for the History of Astronomy* 22, no. 1 (1991): 5–20.

Smith, Roger. *The Norton History of the Human Sciences*. New York: Norton, 1997.

Snyder, Laura J. *The Philosophical Breakfast Club: Four Remarkable Friends Who Transformed Science and Changed the World*. New York: Broadway Books, 2011.

Snyder, Laura J. *Reforming Philosophy: A Victorian Debate on Science and Society*. Chicago: University of Chicago Press, 2006.

Sobkowicz, Pawel. "Social Simulation Models at the Ethical Crossroads." *Science and Engineering Ethics* 25, no. 1 (2019): 143–57.

Solovey, Mark. "Project Camelot and the 1960s Epistemological Revolution: Rethinking the Politics-Patronage-Social Science Nexus." *Social Studies of Science* 31, no. 2 (2001): 171–206.

Solovey, Mark. *Social Science for What? Battles over Public Funding for the "Other Sciences" at the National Science Foundation*. Cambridge, MA: MIT University Press, 2020. https://doi.org/10.7551/mitpress/12211.001.0001.

Solovey, Mark. "Senator Fred Harris's National Social Science Foundation Proposal: Reconsidering Federal Science Relations, and American Liberalism during the 1960s." *Isis* 102, no. 1 (2011): 54–82.

Solovey, Mark. *Shaky Foundations: The Politics-Patronage-Social Science Nexus in Cold War America*. New Brunswick, NJ: Rutgers University Press, 2013.

Solovey, Mark, and Hamilton Cravens, eds. *Cold War Social Science: Knowledge Production, Liberal Democracy, and Human Nature*. New York: Palgrave Macmillan, 2012.

Somers, James. "Is AI Riding a One-Trick Pony?" *Technology Review* 120, no. 6 (2017): 36.

Spallanzani, Mariafranca. "First Philosophy, Metaphysics, and Physics." In *Physics and Metaphysics in Descartes and in His Reception*, edited by Delphine Antoine-Mahut and Sophie Roux, 13–32. New York: Routledge, 2019.

Spencer, Herbert. *An Autobiography*. New York: D. Appleton, 1904.

Spencer, Herbert. "The Nebular Hypothesis." In *Essays Scientific, Political, and Speculative*, vol. 1, 239–99. London: Williams and Northgate, 1868.

Spencer, Herbert. *The Principles of Psychology*. 2 vols. New York: D. Appleton, 1872.

Spencer, Herbert. *Social Statics: Or, the Conditions Essential to Human Happiness Specified, and the First of Them Developed*. London: John Chapman, 1851.

Spencer, Herbert. *A Theory of Population, Deduced from the General Law of Animal Fertility*. London: Longman, 1852. Originally published in *Westminster Review* 47 (April 1852): 468–501. Citations refer to the Longman edition. Internet Archive.

Spinoza, Baruch. *Complete Works.* Indianapolis, IN: Hackett, 2002.

Spufford, Francis, and Jenny Unglow, eds. *Cultural Babbage: Technology, Time, and Invention.* London: Faber, 1996.

Staddon, J. E. R. *The New Behaviorism: Mind, Mechanism, and Society.* Philadelphia, PA: Psychology Press, 2001.

Stallo, J. B. *The Concepts and Theories of Modern Physics.* New York: Appleton, 1884.

Steinmetz, George. "American Sociology before and after World War II: The (Temporary) Settling of a Disciplinary Field." In Calhoun, *Sociology in America,* 314–66.

Steinmetz, George. "Ideas in Exile: Refugees from Nazi Germany and the Failure to Transplant Historical Sociology into the United States." *International Journal of Politics, Culture, and Society* 23, no. 1 (2010): 1–27.

Steinmetz, George. *Sociology and Empire: The Imperial Entanglements of a Discipline.* Durham, NC: Duke University Press, 2013.

Sternhell, Zeev. *Les anti-lumieres: Du XVIIIe siècle à la guerre froide.* Paris: Fayard, 2006.

Stevens, Hallam. "A Feeling for the Algorithm: Working Knowledge and Big Data in Biology." In *Data Histories,* edited by Elena Aronova, Christine von Oertzen, and David Sepkoski, vol. 32 of *Osiris* (2017): 151–74.

Stigler, Stephen M. *The History of Statistics: The Measurement of Uncertainty before 1900.* Cambridge, MA: Harvard University Press, 1986.

Stigler, Stephen M. *Statistics on the Table: The History of Statistical Concepts and Methods.* Cambridge, MA: Harvard University Press, 1999.

Stone, Debora. *Counting: How We Use Numbers to Decide What Matters.* New York: Liveright, 2020.

Strevens, Michael. *The Knowledge Machine: How Irrationality Created Modern Science.* New York: Liveright, 2020.

Strugnell, Anthony. *Diderot's Politics: A Study of the Evolution of Diderot's Political Thought after the Encyclopédie.* The Hague: Nijhoff, 1973.

Stuckenberg, J. H. W. "The Relation of Science to Religion." *Quarterly Review of the Evangelical Lutheran Church* (July 1886): 327–45.

Sumner, James. "Turing Today." *Notes and Records of the Royal Society of London* 66, no. 3 (2012): 295–300.

Surowiecki, James. *The Wisdom of Crowds.* New York: Anchor Books, 2005.

Sutherland, Kathryn. "Explanatory Notes and Commentary." In Adam Smith, *An Inquiry into the Nature and Causes of the Wealth of Nations,* 465–602. Oxford: Oxford University Press, 1993.

Sweeney, Latanya. "Discrimination in Online Ad Delivery." *Communications of the ACM* 11, no. 3 (2013): 10–29. https://queue.acm.org/detail.cfm?id=2460278.

Sweet, James H. "Is History History? Identity Politics and Teleologies of the Present." *Perspectives,* August 17, 2022. https://www.historians.org/publications-and-directories/perspectives-on-history/september-2022/is-history-history-identity-politics-and-teleologies-of-the-present.

Sweetman, Brendan. *The Failure of Modernism: The Cartesian Legacy and Contemporary Pluralism.* Washington, DC: American Maritain, 1999.

Sytsma, David S. *Richard Baxter and the Mechanical Philosophers.* Oxford: Oxford University Press, 2017.

Tani, Jun. *Exploring Robotic Minds: Actions, Symbols, and Consciousness as Self-Organizing Dynamic Phenomena.* Oxford: Oxford University Press, 2017.

Tarnoff, Ben, and Moira Weigel, eds. *Voices from the Valley: Tech Workers Talk about What They Do—and How They Do It.* New York: FSG Originals, 2020.

Taylor, Charles. "Interpretation and the Sciences of Man." *Review of Metaphysics* 25, no. 1 (1971): 3–51.

Thompson, E. P. *The Making of the English Working Class.* New York: Vintage Books, 1963.

Thomson, Herbert F. "Adam Smith's Philosophy of Science." *Quarterly Journal of Economics* 79, no. 2 (1965): 212–33.

Tischler, Henry L. *Introduction to Sociology.* 11th ed. Belmont, CA: Wadsworth Cengage, 2013.

Toda, Michel. "Introduction." In Bonald, *Lettres à Joseph de Maistre*, 5–37.

Townsend, Peter. *The Dark Side of Technology.* Oxford: Oxford University Press, 2016.

Tukey, John W. "Analyzing Data: Sanctification or Detective Work?" *American Psychologist* 24, no. 2 (1969): 83–91.

Tukey, John W. *Exploratory Data Analysis.* Reading, MA: Addison-Wesley, 1977.

Tukey, John W. "The Future of Data Analysis." *Annals of Mathematical Statistics* 33, no. 1 (1962): 1–67.

Turing, Alan. "Computing Machinery and Intelligence." *Mind* 49, no. 236 (1950): 433–60.

Turing, Alan. "Intelligent Machinery." In *Machine Intelligence 5*, edited by Bernard Meltzer and Donald Michie, 3–23. Edinburgh: Edinburgh University Press, 1969.

Turing, Alan. "Lecture to the London Mathematical Society on 20 February 1947." In *A. M. Turing's ACE Report and Other Related Papers*, edited by B. E. Carpenter and R. W. Doran, 106–24. Cambridge, MA: MIT Press, 1986.

Turing, Alan. "On Computable Numbers, with an Application to the Entscheidungsproblem." *Proceedings of the London Mathematical Society* 42, no. 1 (1937): 230–65.

Turkle, Sherry. *The Second Self.* New York: Simon and Schuster, 1984.

Turner, Stephen. "A Life in the First Half-Century of Sociology: Charles Ellwood and the Division of Sociology." In Calhoun, *Sociology in America*, 115–54.

Turner, Stephen P. *The Search for a Methodology of Social Science.* Dordrecht: Springer, 1986.

Turner, Stephen P., and Jonathan H. Turner. *The Impossible Science: An Institutional Analysis of American Sociology.* Newbury Park, CA: Sage Publications, 1990.

Ullman, Ellen. *Life in Code: A Personal History of Technology.* New York: Farrar, Straus and Giroux, 2017.

Van Berkel, Klaas. *Isaac Beeckman on Matter and Motion: Mechanical Philosophy in the Making.* Baltimore, MD: Johns Hopkins University Press, 2013.

van Lunteren, Frans. "Clocks to Computers: A Machine-Based 'Big Picture' of the History of Modern Science." *Isis* 107, no. 4 (2016): 762–76.

Van Meenen, Pierre-François. "De l'influence du libre arbitre de l'homme sur les faits sociaux." *Mémoires de l'Académie Royale des sciences, des lettres et des beaux-arts de Belgique* 21 (1848): 93–112.

van Wyhe, John. "Alfred Russel Wallace." In Ruse, *Cambridge Encyclopedia of Darwin*, 165–72.

"Varieties." *Journal des débats et des décrets*, February 22, 1800: 3–4.

Vartanian, Aram. *Diderot and Descartes: A Study of Scientific Naturalism in the Enlightenment.* Princeton, NJ: Princeton University Press, 1953.

Vartanian, Aram. "Diderot and Maupertuis." *Revue internationale de philosophie* 38, no. 148/149 (1984): 46–66.

Vartanian, Aram. "La Mettrie and Diderot Revisited: An International Encounter." *Diderot Studies* 21 (1983): 155–97.

Vartanian, Aram. *Science and Humanism in the French Enlightenment*. Charlottesville, VA: Rookwood, 1999.

Veblen, Thorstein. "The Place of Science in Modern Civilization." *American Journal of Sociology* 11, no. 5 (March 1906): 585–609.

Veblen, Thorstein. "Why Is Economics Not an Evolutionary Science?" *Quarterly Journal of Economics* 12, no. 4 (July 1898): 373–97.

"Viewpoint: Clocks to Computers." *Isis* 107, no. 4 (2016): 762–804.

Von Oertzen, Christine. "Datafication and Spatial Visualization in Nineteenth-Century Census Statistics." *Historical Studies in the Natural Sciences* 48, no. 5 (2018): 568–80.

Von Oertzen, Christine. "The Historicity of Data: Concepts, Tools, and Practices during the Nineteenth Century." *Naturwissenschaften, Technik und Medizin* 25 (2017): 407–34.

Voskuhl, Adelheid. *Androids in the Enlightenment: Mechanics, Artisans, and Cultures of the Self*. Chicago: University of Chicago Press, 2013.

Vromen, Jack. "Ontological Commitments of Evolutionary Economics." In *The Economic Worldview: Studies in the Ontology of Economics*, edited by Uskali Mäki, 179–224. Cambridge: Cambridge University Press, 2001.

Wade, Ira O. *The Structure and Form of the French Enlightenment*. Princeton, NJ: Princeton University Press, 1977.

Wade, Nicholas J. *A Troublesome Inheritance: Genes, Race and Human History*. New York: Penguin Press, 2014.

Wallace, Alfred R. "The Origin of Human Races and the Antiquity of Man Deduced from the Theory of 'Natural Selection.'" *Journal of the Anthropological Society of London* 2 (1864): clviii–clxxxvii.

Waller, John C. "Putting Method First: Re-appraising the Extreme Determinism and Hard Hereditarianism of Sir Francis Galton." *History of Science* 40, no. 1 (2002): 35–62.

Watson, John B. "Psychology as the Behaviorist Views It." *Psychological Review* 20, no. 2 (1913): 158–77.

Weber, Jean-Paul. *La constitution du texte des regulae*. Paris: Société d'édition d'enseignement supérieur, 1964.

Weikart, Richard. *From Darwin to Hitler: Evolutionary Ethics, Eugenics, and Racism in Germany*. New York: Palgrave Macmillan, 2004.

Weintraub, E. Roy. *How Economics Became a Mathematical Science*. Durham, NC: Duke University Press, 2002.

Wellman, Kathleen Anne. *La Mettrie: Medicine, Philosophy, and Enlightenment*. Durham, NC: Duke University Press, 1992.

Wernimont, Jacqueline. *Numbered Lives: Life and Death in Quantum Media*. Cambridge, MA: MIT Press, 2018.

West, E.G. "Adam Smith's Two Views on the Division of Labour." *Economica* 31, no. 121 (1964): 23–32.

Westfall, Richard S. *The Construction of Modern Science: Mechanisms and Mechanics*. Cambridge: Cambridge University Press, 1971.

"When Sociology Reached the Masses." *Contemporary Sociology* 42, no. 3 (2013): 315–23.

White, Michael V. "The Moment of Richard Jennings: The Production of Jevons's Marginalist Economic Agent." In *Natural Images in Economic Thought: Markets Read in Tooth and Claw*, edited by Phillip Mirowski, 197–230. Cambridge: Cambridge University Press, 1994.

Wiener, Anna. *Uncanny Valley: A Memoir*. New York: Farrar, Straus and Giroux, 2020.

Wiggins, Chris. "Interview." In *Data Scientists at Work*, edited by Sebastien Guitierrez, 1–18. Berkeley, CA: Apress, 2014.

Wiggins, Chris, and Matthew L. Jones. *How Data Happened: A History from the Age of Reasons to the Age of Algorithms*. New York: Norton, 2023.

Williams, Bernard. *Truth & Truthfullness: An Essay in Genealogy*. Princeton, NJ: Princeton University Press, 2003.

Williams, David. *Condorcet and Modernity*. Cambridge: Cambridge University Press, 2004.

Wilson, Christo. "Auditing for Bias in Search Engines." Paper presented at the Second European Symposium Series on Societal Challenges in Computational Social Science, Cologne, Germany, December 5–7, 2018.

Wilson, Margaret Dauler. *Ideas and Mechanism: Essays on Early Modern Philosophy*. Princeton, NJ: Princeton University Press, 1999.

Wittrock, Björn, Johan Heilbron, and Lars Magnusson. "The Rise of the Social Sciences and the Formation of Modernity." In *The Rise of the Social Sciences and the Formation of Modernity*, edited by Johan Heilbron, Lars Magnusson, and Björn Wittrock, 1–33. Dordrecht, Netherlands: Kluwer Academic, 1998.

Wokler, Robert. *Rousseau, the Age of Enlightenment, and Their Legacies*. Edited by Bryan Garsten. Princeton, NJ: Princeton University Press, 2012.

Woloch, Isser. *The New Regime: Transformations of the French Civic Order, 1789–1820s*. New York: Norton, 1994.

Wooldridge, Michael J. *A Brief History of Artificial Intelligence: What It Is, Where We Are, and Where We Are Going*. New York: Flatiron Books, 2021.

Woolf, Stuart. "Statistics and the Modern State." *Comparative Studies in Society and History* 31, no. 3 (1989): 588–604.

Woudenberg, René van, Rik Peels, and Jeroen de Ridder. "Putting Science on the Philosophical Agenda." Introduction to *Scientism: Prospects and Problems*, edited by Rik Peels, Jeroen de Ridder, and René van Woudenberg, 1–27. Oxford: Oxford University Press, 2018.

Yeo, Richard. *Defining Science: William Whewell, Natural Knowledge, and Public Debate in Early Victorian Britain*. Cambridge: Cambridge University Press, 1993.

Yonay, Yuval P. *The Struggle over the Soul of Economics: Institutionalists and Neoclassical Economists in America between the Wars*. Princeton, NJ: Princeton University Press, 1998.

Zékian, Stéphanie. "Siècle des lettres contre siècle des sciences: Décisions mémorielles et choix épistémologiques au début du xixe siècle." *Fabulat-LhT* 8 (2011). https://doi.org/10.58282/lht.234.

Zuboff, Shoshana. *The Age of Surveillance Capitalism: The Fight for a Human Future at the New Frontier of Power*. New York: PublicAffairs, 2019.

INDEX

Note: Page references in *italics* refer to figures.